Handbook of Inland Aquatic Ecosystem Management

Applied Ecology and Environmental Management

A SERIES

Series Editor
Sven E. Jørgensen
Copenhagen University, Denmark

Handbook of Inland Aquatic Ecosystem Management
Sven E. Jørgensen, Jose Galizia Tundisi, and Takako Matsumura Tundisi

Eco-Cities: A Planning Guide
Zhifeng Yang

Introduction to Systems Ecology
Sven E. Jørgensen

Handbook of Ecological Indicators for Assessment of Ecosystem Health, Second Edition
Sven E. Jørgensen, Fu-Liu Xu, and Robert Costanza

Surface Modeling: High Accuracy and High Speed Methods
Tian-Xiang Yue

Handbook of Ecological Models Used in Ecosystem and Environmental Management
Sven E. Jørgensen

ADDITIONAL VOLUMES IN PREPARATION

Handbook of Inland Aquatic Ecosystem Management

Sven Jørgensen
Jose Galizia Tundisi
Takako Matsumura Tundisi

CRC Press
Taylor & Francis Group
Boca Raton London New York

CRC Press is an imprint of the
Taylor & Francis Group, an **informa** business

CRC Press
Taylor & Francis Group
6000 Broken Sound Parkway NW, Suite 300
Boca Raton, FL 33487-2742

First issued in paperback 2019

© 2013 by Taylor & Francis Group, LLC
CRC Press is an imprint of Taylor & Francis Group, an Informa business

No claim to original U.S. Government works

ISBN-13: 978-1-4398-4525-7 (hbk)
ISBN-13: 978-0-367-86566-5 (pbk)

Visit the Taylor & Francis Web site at
http://www.taylorandfrancis.com

and the CRC Press Web site at
http://www.crcpress.com

Contents

Introduction

This handbook has two parts. Part I covers the basic scientific knowledge that is needed in environmental and ecological management of aquatic ecosystems, while Part II presents the toolboxes that are applied to achieve integrated and holistic environmental and ecological management.

This handbook focuses on the following types of aquatic inland ecosystems: lakes, reservoirs, ponds, rivers, wetlands, lagoons, and estuaries. Chapter 1 gives an overview of these seven types of ecosystems. Chapters 2 and 3 present the biological, ecological, chemical, and physical characteristics of lakes and reservoirs with an emphasis on ecosystem properties and ecological processes. The origins of lakes and reservoirs are also covered. Chapter 4 deals with rivers, with an emphasis on ecosystem properties and ecological, chemical, and physical processes. Chapters 5 and 6 cover lagoons, estuaries, and wetlands, aiming to understand the characteristics, properties, and responses of these complex ecosystems and to describe the many interacting processes that characterize all aquatic inland ecosystems. Chapters 7 and 8 describe the climatic perspectives of the aquatic inland water ecosystems—Chapter 7 covers tropical ecosystems while Chapter 8 deals with temperate ecosystems. The differences between aquatic ecosystems in the tropical and temperate zones are important to understand, because the very comprehensive knowledge we have about temperate ecosystems can be transferred to tropical ecosystems, if we understand how the differences in climate influence the ecological processes. Chapters 1 through 8 all focus on the ecological aspects of aquatic inland ecosystems.

In environmental and ecological management, however, we need to also develop mass balances in accordance with conservation principles to be able to calculate important concentrations, as well as important processes such as biomagnification and bioaccumulation. The basic scientific considerations on which the conservation principles and their applications are based are covered in Chapter 9. Furthermore, it is necessary to understand all the possible aquatic chemical processes and how to use basic aquatic chemistry to calculate the concentrations of the relevant chemical species. Chapters 10 and 11 focus on aquatic chemistry. Chapters 9 through 11 present the basic science that we use as tools to quantify in environmental management. The calculation methods that are covered in these three chapters are of utmost importance for integrated and holistic environmental and ecological management. Therefore, in addition to the scientific background material, these methods have been described in detail by citing relevant examples to aid in proper application.

The brief Chapter 12 provides an overview of the needs and importance of limnology for the management of inland waters. The first chapter of Part II,

Chapter 13, provides an overview of the environmental problems of aquatic inland water ecosystems and the sources of these problems. Clearly, the first step in environmental management is to define the problems and quantify their sources. Chapter 14 presents the steps for up-to-date environmental management, including the toolboxes used to identify the problem and set up a diagnostic interpretation of the problem and its sources. Three different toolboxes can be used to identify the problem: mass balances (covered in Chapter 9), use of ecological indicators (reviewed in Chapter 18), and ecological models (discussed in Chapter 19). Once a diagnosis has been developed, the next step is to find a solution to the problem. Chapter 14 identifies four toolboxes that can be used to find an environmental management solution to the problem: environmental technology, cleaner technology, ecotechnology, and environmental legislation. Chapter 15 presents methods based on environmental technology, while cleaner technology is discussed in Chapter 16. Chapter 17 provides an overview of ecotechnology. Environmental legislation is not covered in this handbook because it is varies from country to country. It would therefore not be possible to write a general chapter on this toolbox. To sum up, the chapters in Part II present how to solve an environmental problem in a holistic manner. A proper application of all the steps as they are presented in Chapter 14 requires, however, basic knowledge of ecological, physical, chemical, and biological aspects, which is covered in the chapters of Part I.

This handbook is entitled *Handbook of Inland Aquatic Ecosystem Management* to underline the fact that saline lakes, the brackish estuaries, and saline and brackish coastal lagoons are included. As freshwater ecosystems are dominant among aquatic inland water ecosystems, they are, of course, treated in more detail than saline and brackish ecosystems.

Authors

Sven Erik Jørgensen is a professor of environmental chemistry at University of Copenhagen, Copenhagen, Denmark. He is an honorable doctor of science at Coimbra University, Portugal, and at Dar es Salaam University, Tanzania. He was editor in chief of *Ecological Modelling* from the journal's inception in 1975 to 2009. In 2004 and 2005, Dr. Jørgensen was awarded the prestigious Stockholm Water Prize, the Prigogine Prize, and the Einstein Professorship by the Chinese Academy of Sciences. In 2007, he received the Pascal medal and was elected a member of the European Academy of Sciences. He has written close to 350 papers, has edited or written 64 books, and has given lectures and courses in ecological modeling, ecosystem theory, and ecological engineering worldwide.

Jose Galizia Tundisi is a retired professor of environmental sciences at the University of São Paulo, São Carlos, Brazil. He currently serves as president of the International Institute of Ecology and Environmental Management, a nonprofit organization at São Carlos. He also serves as full professor of environmental quality at University Feevale, Novo Hamburgo, Rio Grande do Sul, Brazil.

He has served as a consultant in 40 countries in water and resource management, and limnology, aquatic biology, and watershed management. He was a member of the scientific board of the International Lake Environment Committee (ILEC) in Japan for 20 years. He was president of the National Research Council of Brazil. He is a full member of the Brazilian Academy of Sciences and a member of staff of the Ecology Institute in Germany. He has also published 320 papers, 31 books, and has organized many international conferences in Brazil and in other counties.

Takako Matsumura Tundisi is a retired professor of ecology at the Federal University of São Carlos, Brazil. She currently serves as scientific director of the International Institute of Ecology and Environmental Management, a nonprofit organization at São Carlos.

She developed several projects in limnology of tropical lakes and reservoirs and coordinated a large-scale study of 220 reservoirs in São Paulo State for the Biota-FAPESP program. She has also supervised several MSc and PhD students. Since 1999, she has been chief editor of the *Brazilian Journal of Biology*, an international journal dedicated to neotropical biology and ecology. She has also published 125 scientific papers and 5 books.

Part I

Limnology and Ecology of Inland Waters

1

Overview of Inland Aquatic Ecosystems and Their General Characteristics

1.1 Introduction: A Short Overview

The freshwaters of the continents are a collection of standing (lentic) or flowing waters (lotic) that vary in origin, shape, volume, and areas. Limnology is the science that studies these inland water ecosystems, their characteristics and mechanisms of function, and the interaction between their physical, chemical, and biological components.

Lakes—These are shallow (<5 m average depth) or deep (>5 m average depth) aquatic systems with areas that vary from 0.1 to thousands of square kilometers with fresh, brackish, or saline water accumulated in depressions that have different origins. Lakes can circulate all time or can stratify periodically or during short times. The chemical composition of the lake water is related to the hydrogeological characteristics of their watersheds.

From the water stored in lakes, in the world, 50% is freshwater and 50% is saline water. The largest saline lake is the Caspian Sea (see Table 1.4). Saline lakes are important aquatic ecosystems from the limnological and ecological points of view because they have specialized flora and fauna and specific biodiversity. Their mechanisms of functioning are important as models for aquatic ecosystem studies. Saline lakes are present on every continent and are frequently the only surface waters in dry climatic regions (Hammer 1986).

Ponds—These are small natural depressions filled with water and are shallow (<3 m) with permanent vertical circulation and areas that range from 0.1 to 100 m^2. The water contained in the ponds can be fresh, saline, or brackish, and they can be permanent or temporary bodies of water.

Wetlands—These are shallow areas with soils saturated with water depth of less than 1 m to 2 or 3 m with extensive areas of floating or emerged vegetation of aquatic plants. Some wetlands have extensive areas covered by trees standing in water. These wetlands can be *lentic* associated with a lake or *lotic* associated with a river. Wetlands can be permanently inundated or can be temporarily dry (see Chapter 6).

Bogs—These are shallow inundated areas with brown acidic waters and accumulated organic matter covered by vegetation (shrubs and grasses).

Artificial reservoirs—These are shallow (average depth <5 m) or deep (average depth >5 m) artificial lentic ecosystems, with small (0.1–100 m²) or large (1–3000 km²) areas constructed by man in main rivers or their tributaries. Morphometric characteristics, hydrological functioning, and ecological features differ from natural lakes or ponds (see Chapter 2). Figures 1.1 and 1.2 show the number of reservoirs built in the twentieth century and the number of reservoirs in different continents. Figure 1.3 shows the El Cajon reservoir in Honduras as an example of the many reservoirs built during the last 50 years.

Rivers, streams, and creeks—These are flowing water ecosystems, or lotic, that differ in size, shape, volume, and discharge and connect the water cycle in

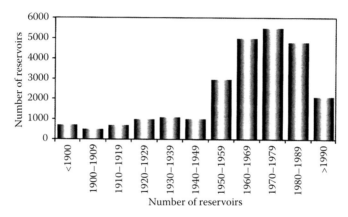

FIGURE 1.1
Number of reservoirs built in the world during the twentieth century. (From Agostinho, A.A. et al., *Ecologia e manejo de recursos pesqueiros em reservatórios do Brasil*, Eduem, Maringá, Brazil, 2007, 501pp.)

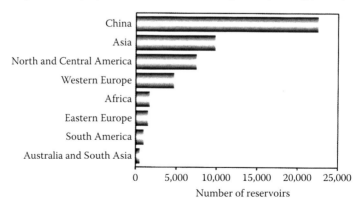

FIGURE 1.2
Number of large reservoirs in different continents. (From Agostinho, A.A. et al., *Ecologia e manejo de recursos pesqueiros em reservatórios do Brasil*, Eduem, Maringá, Brazil, 2007, 501pp.)

FIGURE 1.3
El Cajon Reservoir in Honduras.

the watersheds. Rivers are the largest flowing aquatic system, in all continents for thousands of kilometers. Streams and creeks have smaller sections, less discharge, but are important components of the hydrographic network of watersheds. Streams, creeks, and even some rivers can be temporary (intermittent streams and rivers) with flowing water only during parts of the year, especially in arid or semiarid regions. *Interrupted* streams flow alternately on and below the surface (Chapter 4).

Springs—These are small flowing waters (a few meters wide, with reduced flow), resulting from the flow of ground water.

Estuaries—The region where the freshwater from inland meets the coastal water from the sea is an ecotone, an estuary. This is a complex region with a mixing zone of freshwater and saline water with several intermediate zones of varying salinities and a gradient from the river to the coastal zone. Estuaries vary in size, area, morphometry, and extent. This variation occurs with modifications subsequent to the original formation. Physical, chemical, and biological features vary along with the gradient of salinity (Chapter 5).

1.1.1 General Features of the Aquatic Inland Ecosystems

These aquatic ecosystems are characterized by three interfaces: *the air–water interface, the sediment–water interface*, and the *organism–water interface*. Throughout these interfaces exchanges of energy, elements, and substances occur. The properties that influence aquatic ecosystems are shown in Table 1.1.

Tables 1.2 through 1.4 show, respectively, the classification of lakes and ponds by surface area, the 10 largest rivers of the world and the 12 largest lakes in the world. Table 1.5 shows the 10 largest reservoirs in the world.

TABLE 1.1

Properties and Characteristics of Aquatic Ecosystems

Natural properties

Regional properties—Watershed characteristics
    Climate
    Geology
    Soil
    Regional geomorphology
    Topography
    Hydrology

Lake, reservoir, or river characteristics
    Morphometry
    Volume
    Flushing rate/discharge—(for rivers)
    Retention time (for lakes or reservoirs)
    Stratification or circulation patterns (for lakes and reservoirs)
    Water quality (nutrients, turbidity, salinity, biogeochemistry)
    Biological/ecological properties (biodiversity, biomass, productivity,
      trophic structure)

Human impacts
    Hydrological alterations
    Sedimentation rate from the watershed
    Toxin discharge in the aquatic systems
    Habitat disruption
    Harvesting of biomass
    Nutrient input (eutrophication)
    Deforestation of the watershed
    Climate change

Source: Modified from Kalff, J., *Limnology*, Prentice Hall, Upper Saddle River, NJ, 2003, 592pp.

TABLE 1.2

Lakes and Ponds Classified by Surface Area

Lake/Pond	Surface Area (km²)	Total Surface Area (km²)
Great lakes	>10,000	997,000
Large lakes	100–10,000	686,000
Medium lakes	1–100	642,000
Small lakes	0.1–1	288,000
Large ponds	0.01–0.1	190,000
Other small ponds	>0.01	An unknown area

Source: Modified from Meybeck, M., Global distribution of lakes, in Lerman, A. and Gat, J. (Eds.), *Physics and Chemistry of Lakes*, Springer-Verlag, Berlin, Germany, 1995, pp. 1–32.

TABLE 1.3

Ten Largest Rivers of the World

River	Country/Continent	Mean Discharge (10^3 m³ s)	Drainage Area (10^3 km²)	Sediment Yield (ton km²/year)
Amazon	Brazil, Colombia, Peru (South America)	212.5	6.062	67
Congo	Zaire, Congo (Africa)	39.7	3.968	18
Yangtze	China (Southeast Asia)	21.8	1013	553
Ganges	Bangladesh, India (Indian subcontinent)	19.8	553	1.469
Mississippi	United States (North America)	17.3	3.185	109
Orinoco	Brazil, Colombia, Venezuela (South America)	17.0	939	103
Paraná	Argentina, Brazil, Bolivia (South America)	14.9	2.278	40
Zambezi	Several African countries		1.280	78
Danube	Europe	6.2	806	27
Nile	Egypt, Ethiopia, Sudan, Uganda	2.8	2944	42

Sources: Modified from Welcomme, R.L., *Fisheries Ecology of Floodplain Rivers.* Longman, London, and New York, pp. 317, 1985; Kalff, J., *Limnology*, Prentice Hall, Upper Saddle River, NJ, 2003, 592pp.

TABLE 1.4

Twelve Largest Lakes in the World

Lake/Origin	Surface Area (km²)	Volume (km³)	Maximum Depth (m)
Caspian (T)	374.000	78.200	1.025
Superior (G + T)	82.100	12.230	406
Michigan (T)	57.750	4.920	281
Baikal (T)	31.500	22.995	1.741
Chad (T)	25.900	Variable	5
Tanganyika (T)	32.000	17.827	1.471
Victoria (T)	62.940	2.518	80
Titicaca (T)	8.562	827	284
Ontario (G)	19.000	1.637	244
Erie (G)	25.657	483	13
Malawi (T)	22.490	6.140	706
Aral (T)[a]	43.000	1.451	—

Source: Modified from Kalff, J., *Limnology*, Prentice Hall, Upper Saddle River, NJ, 2003, 592pp.

T, Tectonic; G, Glacial; Volcanic, V.

[a] Area and volume of Aral Sea great greatly reduced by multiple uses (mainly irrigation).

TABLE 1.5

Ten Largest Reservoirs in the World

Reservoir/Country	Surface Area (km²)	Maximum Volume (km³)
Volta reservoir	8.480	148
Bratsk (Russia)	5.500	58
Kariba (Zimbabwe)	5.120	160
Cabora Bassa (Mozambique)	4.450	66
Ilha Solteira (Brazil)	4.214	21
Nasser (Egypt, Sudan)	2.700	157
Guri (Venezuela)	1.784	135
Sobradinho (Brazil)	1.230	34
Serra da Mesa (Brazil)	1.084	55
Three Gorges (China)	2.600	39

Sources: Modified from Straskraba, M. and Tundisi, J.G., *Reservoir Water Quality Management*, Guidelines for Lake Management, Vol. 9, ILEC, UNEP, Kusatsu, Japan, 1999, 229pp.; Kalff, J., *Limnology*, Prentice Hall, Upper Saddle River, NJ, 2003, 592pp.

1.2 Reservoirs in Brazil: An Example of Large-Scale Construction of Artificial Aquatic Systems

The construction and operation of large reservoirs in Brazil are illustrative examples of the impacts (positive and negative) of large artificial ecosystems on watersheds and rivers. Initially built with the purpose of producing hydroelectricity, these reservoirs are used for multiple activities such as energy production, navigation, irrigation, fisheries, recreation, tourism, and water for public supply. The multiple uses of the reservoirs' waters require complex management operation, because it involves technology, optimization, and regulation of activities and control systems based on operational rules that should maintain water volumes and water levels and water quality in the reservoir and downstream of the dam. It includes furthermore protection and recovery of biodiversity. But the large dams built in Brazil had several positive impacts on the regional economy, supporting development by the generation of energy as well as providing opportunities for employment, diversification of economic activities, and better sanitation infrastructure. The main impacts of these reservoirs are (1) changes in the ecological services of the rivers, (2) reduction of biodiversity, (3) water quality degradation, and (4) changes in the hydrosocial cycle for the human population of the watershed. Table 1.6 shows the main, large Brazilian reservoirs and their characteristics.

TABLE 1.6

Largest Brazilian Reservoirs

Reservoir	Hydrographic Basin	Area (km²)	Height (m)	Volume (10⁶ m³)	Discharge (m³/s)	Energy Production (MW)
Sobradinho	São Francisco	4.214	43	34.100	22.850	1.050
Tucurui	Tocantins	2.875	93	45.500	100.000	4.000
Balbina	Amazonas	2.360	39	17.500	6.450	250
Porto Primavera	Paraná	2.250	38	18.500	52.000	4.540
Serra da Mesa	Tocantins	1.784	144	55.200	15.000	1.275
Furnas	Paraná	1.440	127	22.950	13.000	1.216
Itaipu	Paraná	1.350	196	29.000	61.400	12.600
Ilha Solteira	Paraná	1.195	74	21.166	40.000	3.444
Três Marias	São Francisco	1.142	75	21.000	8.700	396

1.2.1 Inland Waters as Water Resources

Lakes, reservoirs, rivers, wetlands, and estuaries are intensively used for the development of human activities. The main uses are described in Table 1.7.

Therefore, all the aquatic inland ecosystems have important ecological, economical, cultural, and social functions all over the world (Jørgensen et al. 2005). Their protection, recovery, and careful management are fundamental to the well being of the human population. The economic value of water is considered today as the fundamental criteria for development of countries

TABLE 1.7

Main Uses of Inland Aquatic Ecosystems

Drinking water supply

Irrigation

Fish production and other biomass
  production (aquaculture)

Industrial water

Water supply in urban regions

Recreation

Tourism

Hydropower production

Flood control

Navigation

Sites for maintenance of aquatic biodiversity

Water purifiers of low cost

Training and education

Aesthetic values

and regions. This economic value is direct and indirect. The production of hydroelectricity is one of the most important indirect economic values of water.

The maintenance of ecosystem services of the inland aquatic ecosystems is of prime importance and their management should consider this sustainable approach as the basis for their conservation and recovery.

Human activities have affected global lake distribution and produced changes in morphometry, mean depth, and water quality. Examples of such drastic effects are the loss of 84% of the volume of Aral Sea by excessive use in irrigation, increase in the total area of reservoirs that changed the stored water, and changes in hydrological cycles in all continents due to excessive land use and exploitation of the aquatic biota (Meybeck 1995, Jørgensen et al. 2005).

Another impact that is widespread and is causing several changes in the biodiversity and the food chains of the inland aquatic ecosystems is the accidental or on purpose introduction of exotic species.

References

Agostinho, A.A.; Stones, L.C.; and Peliace, F.M. 2007. *Ecologia e manejo de recursos pesqueiros em reservatórios do Brasil*. Eduem, Maringá, Brazil, 501pp.

Hammer, T. 1986. *Saline Lakes Ecosystems of the World*. Dr. W. Junk Publishers, Dordrecht, the Netherlands, 616pp.

Jørgensen, S.E.; Lofler, H.; Rast, R.; and Straskraba, M. 2005. *Lake and Reservoir Management*. Developments in Water Sciences, Vol. 54. Elsevier, Amsterdam, the Netherlands, 502pp.

Kalff, J. 2003. *Limnology*. Prentice Hall, Upper Saddle River, NJ, 592pp.

Meybeck, M. 1995. Global distribution of lakes. In: Lerman, A. and Gat, J. (Eds.), *Physics and Chemistry of Lakes*. Springer-Verlag, Berlin, Germany, pp. 1–32.

Straskraba, M. and Tundisi, J.G. 1999. *Reservoir Water Quality Management*. Guidelines for Lake Management, Vol. 9. ILEC, UNEP, Kusatsu, Japan, 229pp.

2

Lakes and Reservoirs as Ecosystems

2.1 Lakes and Reservoirs Have Many Interactive Factors

A lake or a reservoir is intimately related with a watershed—a biogeophysical boundary—that is a drainage system connecting these natural or artificial bodies of water. A river basin or watershed is "defined as the entire area drained by a major river system and by its main tributaries" (Revenga et al. 1998). The watershed, thus, controls several mechanisms of functioning of lakes and reservoirs. Pristine watersheds not subjected to extensive modification by human activities contribute with low input of nutrients, suspended material, or toxic substances to the lakes and reservoirs. Degraded watersheds have a great impact on the functioning of the natural or artificial ecosystems that are connected to them. Figure 2.1 shows the main 106 watersheds of the world.

Therefore, any lake or reservoir in a watershed depends upon a matrix of *geological, hydrogeochemical, climatologically* natural characteristics (Figure 2.2) and of the human activities imposed upon these natural features.

Watersheds can be evaluated by their values and the ecosystem services that they can promote such as the freshwater supply or erosion control (Revenga et al. 1998, IIEGA/SVMA 2010). The biological value of the watershed is its biodiversity uniqueness, the genetic diversity, its spatial heterogeneity, and the ecological complexity it harbors. Fish species and degree of endemism, endemic bird areas, degree of urbanization, population density, and water availability are ecological values of a watershed that are considered fundamental for comparative purposes. Watershed conditions are related to areas of original or natural landscape, forest cover, and extension of riparian forest, soil erosion, irrigated cropland, water demands, and water uses. The number of area and extension of natural lakes, wetlands, and artificial reservoirs is also a measure of watershed condition.

Therefore, when describing the lakes and reservoirs as ecosystems their connection with the watershed, its status, value, and condition should be taken into account.

The vulnerability of the watersheds reflects in the pristine or degraded condition of a lake or reservoir within this watershed. In general, the vulnerability is related to the degree of urbanization; volume of nontreated effluents

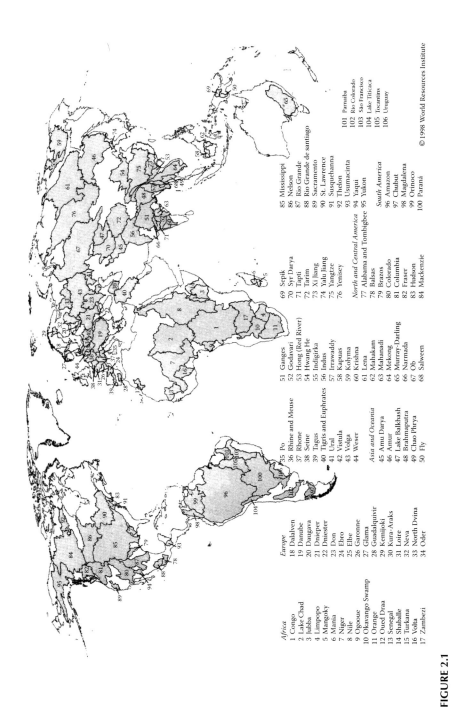

FIGURE 2.1
The main 106 watersheds of the world. (From Revenga, C. et al., *Watersheds of the World*, World Resource Institute, Washington, DC, 164pp., 1998.)

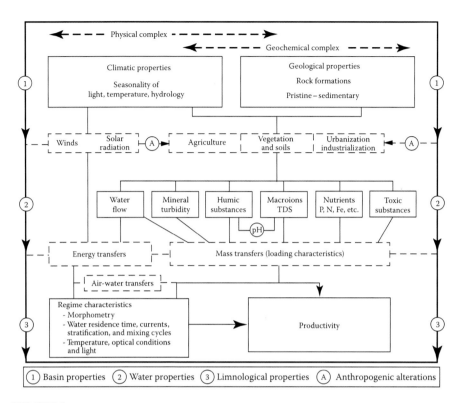

FIGURE 2.2
Matrix of interactions between natural characteristics that determine the productivity of a lake or reservoir. (From Vollenweider, R.A., *Schweiz. Z. Hydrol.*, 37, 53, 1987.)

(domestic or industrial); degree of deforestation; endemic species at risk; and degree of pollution and contamination of soil, water, and air. The forest cover of the watersheds is a very sensitive component related to its status and vulnerability, and it is a measure of the intensity of the impact or the resilience capacity to impacts. The area of natural forests, mosaics of vegetation, and wetlands is a measure of the resilience capacity of a watershed. Another important measure of the vulnerability of a watershed and as a consequence to the lakes and reservoirs is the status of riparian forest along the rivers, the lakes, and the reservoirs. Riparian forests have specific and very important functions in the watersheds especially related to maintaining the water quantity and regulating the water quality (Likens 1992, Paula Lima and Zakia 2001, Rodrigues and Leitão Filho 2001). Tundisi and Matsumura Tundisi (2010) have made an extensive series of measurements at two watersheds in a subtropical region of Brazil and showed how the maintenance of a good natural forest cover was fundamental for the good water quality of rivers and reservoirs in these watersheds. Figure 2.3 shows the role of riparian forest as a regulation and control factor of water quantity and quality in a tributary of a lake or a reservoir. The flux of nutrients, the water quality, and the atmospheric water quantity

Conceptual scheme of a riparian forest and its dynamic interactions with aquatic systems in a watershed

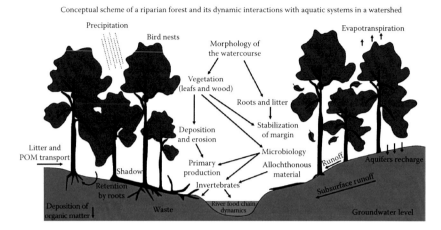

FIGURE 2.3
(See color insert.) Role of riparian forest as a control and regulating factor in a watershed. (Modified from Likens, G.E., The ecosystem approach: Its use and abuse, in: Kinne, O. (Series Ed.), *Excellence in Ecology*, Ecology Institute, Oldendorf/Luke, Germany, 166pp., 1992; Paula Lima, W. and Zakia, M.J.B., Hidrobiologia de Matas Ciliares, in: Rodrigues, R.R. and Leitão Filho, H. (Eds.), *Matas Ciliares: Conservação e recuperação*, EDUSP, FAPESP, São Paulo, Brazil, pp. 33–44, 320pp., 2001.)

(by evapotranspiration), as well as the rivers', lakes', and reservoirs' water volume and the groundwater quality and quantity are all maintained by the presence of this riparian vegetation in the tributaries and near the limnic systems of the watersheds (Tundisi and Matsumura-Tundisi 2010).

The chemical composition of the water quality is regulated by the riparian vegetation and wetlands, and this has an enormous impact if the river reservoir or lakes are a source of water for public supply. By controlling and maintaining a good water quality (generally with low conductivity, low suspended matter, and low concentration of phosphorus and nitrogen), the cost of treatment of this water for human supply is generally very low (2 or 3 U.S.$/1.000 m^3). With increasing degradation due to the removal of wetlands and riparian forests, the cost of water treatment increases up to 100 or 200 U.S.$/1.000 m^3 (Tundisi et al. 2006).

A lake and a reservoir function under the impact of the external loading but they have their own internal mechanisms of vertical and horizontal dynamics that are related with the circulation of matter, biogeochemical cycles, and the relation to production/decomposition of organic matter. In a lake or reservoir, the following interfaces are fundamental: the air/surface–water interface, the sediment–water interface, and the organism–water interface. Main regions and compartments of a lake or reservoir are present in Figure 2.4.

The exchanges of substances and elements between the water and these interfaces characterize the daily, seasonal cycles of physical, chemical, and biological events in these aquatic ecosystems. Figure 2.5 shows the main fluxes of the interaction of the chemical and biological pools and the relationships

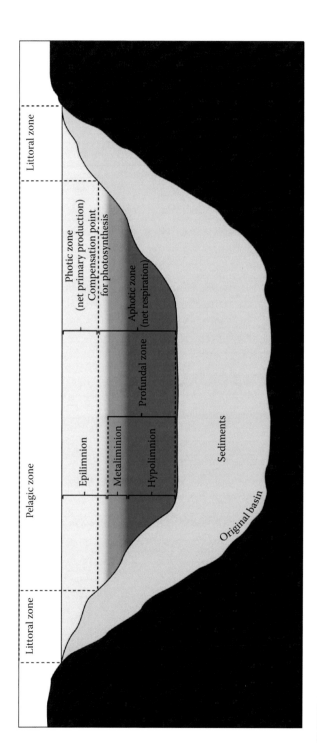

FIGURE 2.4
(See color insert.) Main regions and compartments of a lake or reservoir. (Modified from Kalff, J., *Limnology*, Prentice Hall, Upper Saddle River, NJ, 592pp., 2003.)

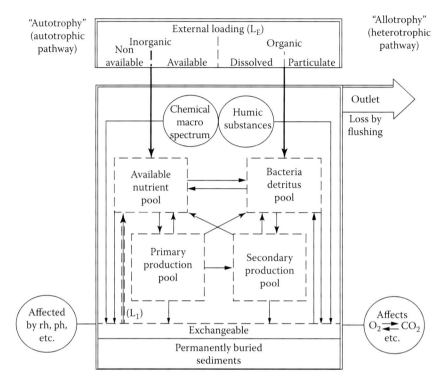

FIGURE 2.5
Internal fluxes of interactions in lakes and reservoirs. (From Vollenweider, R.A., *Schweiz. Z. Hydrol.*, 37, 53, 1987.)

such as autotrophy and allotrophy in a lake or reservoir. The intensity of the flux between the various pools depends upon a series of factors: geographical location, altitude, morphometry and origin, wind stress over the surface, and the impact of human activities.

The horizontal and vertical distribution of communities is depending upon the mechanisms of horizontal and vertical circulation, density currents, and degree of stratification (if permanent or temporary).

2.2 Pulse Effects in Lakes and Reservoirs

Lakes and reservoirs are subjected to several types of sudden changes of natural or man-induced origin, which affects many physical, chemical, and biological variables. Pulses of natural origin come from rapid changes in meteorological conditions such as rainfall or wind stress. These pulses can be seasonal, frequent, or infrequent. Heavy rainfall can cause a very rapid change in the physical, chemical, and biological conditions of a reservoir or a lake, especially in watersheds where deforestation was extensive and several tons/hectare of suspended matter are discharged in a few hours.

Tundisi and Matsumura-Tundisi (2008) described how sudden rainfall in Barra Bonita reservoir was the cause of a mass mortality of fishes due to oxygen depletion; phytoplanktonic primary production was reduced to less than 20% of the carbon produced per square meter due to strong reduction of light penetration as a consequence of the amount of suspended material discharged into the reservoir (largely clay).

Pulses of *artificial* origin may be caused by manipulation of water levels during periods of regulation of water flows for several uses or withdrawal of water from different levels for supply of drinking water.

The magnitude of the pulses differs geographically in the case of natural forcing functions, because they depend on the seasonal variation of these forcing functions such as rainfall or wind stress. Tundisi et al. (2004, 2006, 2008 and Tundisi and Matsumura-Tundisi (2010) demonstrated the effect of cold fronts in the vertical structure of reservoirs in the southeast of Brazil. Cold fronts coming from the Antarctic are an important source of wind stress in the southeast of Brazil. Pulses of river water in floodplain lakes change the vertical structure and modify the patterns of vertical distribution of dissolved oxygen, phytoplankton, and zooplankton (Tundisi et al. 1984), Junk (2006).

Cavalcanti and Kousky (2009) showed that there is seasonality in the passage of the cold fronts in the South American continent and in the southeastern region of South America: between 40 and 60 events occurred yearly during the last 22 years. Cold fronts impact the functioning of reservoirs or lakes by changing thermal structure, vertical distribution of chemical variables such as dissolved oxygen, nutrients and biological variables such as: chlorophyll phytoplankton composition changed from Chlorophyceae dominance to diatom dominance during periods of strong winds (6–10 km/h) as a consequence of cold fronts, at Lobo/Broa reservoir.

This shift in species composition and of phytoplankton group compositions has a strong effect in eutrophic lakes and reservoirs; for example, Tundisi et al. (2006) showed the changes in phytoplankton composition in reservoirs of the metropolitan region of São Paulo: during the period of stratification, cyanobacteria blooms predominate, and during the periods of vertical mixing promoted by the cold front effect, the blooms disappeared and the phytoplankton community was dominated by Chlorophyceae and Diatomaceae.

2.3 Vertical and Longitudinal (or Horizontal) Processes in Reservoirs and Their Complexity

Patterns of vertical and horizontal organization of reservoirs depend on their structural characteristics and mode of operation. A reservoir constructed on the main stream of a river is an artificial structure that affects the upstream and downstream ecosystem of the river. Thus, the technical features of the reservoir, such as level of the outlet, volume of the reservoir, average depth, maximum depth, and retention time, are all characteristics of the reservoir that determine the water quality; the distribution of the phytoplankton, zooplankton, and benthic community; and the organization and distribution of fish assemblages and macrophytes. The organization of a reservoir and its spatial features are described in Figure 2.6. It is a classical conceptualization that encompasses the riverine zone, the transition zone, and the lacustrine zone.

As artificial ecosystems, reservoirs are subjected to natural and manmade forcing functions that determine their dynamic characteristics: horizontal gradients in physical, chemical, and biological variables and vertical oscillations due to the inputs of mechanical energy from the tributaries. The operational characteristics of the reservoirs such as the retention time and the level of the outlet of the water withdrawal are other fundamental dynamic features of the reservoirs. The cycles of thermal stratification and

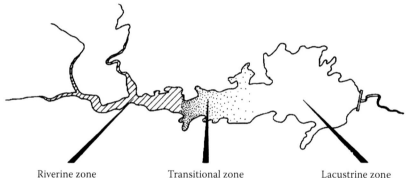

Riverine zone	Transitional zone	Lacustrine zone
• Narrow, channelized basin	• Broader, deeper basin	• Broad, deep, lake-like basin
• Relatively high flow	• Reduced flow	• Little flow
• High susp. solids, turbid, low light avail., $Zp < zm$	• Reduced susp. solids, less turbid, light avail. increased	• Rel. clear, light more avail. at depth, $Zp > Zm$
• Nutrient supply by advection, rel. high nutrients	• Advective nutrient supply reduced	• Nutrient supply by internal recycling, rel. low nutrients
• Light-limited PPR	• PPR/m^3 Rel. high	• Nutrient-limited PPR
• Cell losses primarily by sedimentation	• Cell lossed by sedimentation and grazing	• Cell losses primarily by grazing
• Organic matter supply pri-allochthonous, $P < R$	• Intermediate	• Organic matter supply primarily autochthonous, $P > R$
• More "eutrophic"	• Intermediate	• More "oligotrophic"

FIGURE 2.6
Longitudinal zonation in environmental factors controlling light and nutrient availability for phytoplankton production, algal productivity and standing crop, organic matter supply, and trophic status in an idealized reservoir. Represents the riverine zone, the transition zone, and the lacustrine zone. (Modified from Kimmel, B.L. and Groeger, A.W., Factors controlling phytoplankton production in lakes and reservoirs: A perspective, in: *Lake and Reservoirs Management*, EPA 440/5/84 001, U.S. EPA, Washington, DC, pp. 277–281, 1984.)

destratification that occur in a reservoir are a result of changing meteorological conditions such as solar and atmospheric radiation and wind effects. The vertical temperature structure and its oscillation are reflected in the vertical distribution of other variables such as dissolved oxygen, nutrients and the biological components, phytoplankton, and zooplankton.

The plunge point of the tributaries in the reservoir sets up horizontal gradients that depending on the density of the water can be described as an *overflow* (surface), interflow (mid depths), or underflow (bottom flow); these density inflows result mainly in processes of transport by advection, convection, turbulence, or diffusion (Ford 1990). These transport systems impact reservoir water quality by modifying the concentrations of dissolved oxygen, conductivity concentration, and nutrients, as well as phosphorus and nitrogen (particulate or dissolved) or pollutants such as toxic metals or organic substances.

Horizontal variations in temperature may occur due to different in the tributaries of the reservoir. For example, Tundisi and Matsumura-Tundisi (1990) described tributaries' inflows with lower temperatures due to the dense riparian forest cover in Barra Bonita reservoir resulting in extensive horizontal differences up to 2°C less in this reservoir as compared with open regions with no forest cover.

Horizontal gradients of concentrations of several chemical compounds occur in reservoirs as demonstrated by Armengol et al. (1999). These horizontal gradients are a result of a hydrodynamic process and have, as consequences, the horizontal distribution of communities that are spatially organized.

Figure 2.7 shows a longitudinal profile from the river to the dam in the SAU reservoir as presented by Armengol et al. (1999). This figure represents gradients of environment changes along the main axis of this reservoir in Spain. The longitudinal changes in the plankton community composition are coupled with biological activity such as the bacterial activity. Phytoplankton abundance measured by chlorophyll and cell number shows a pattern of longitudinal variability as demonstrated in Figure 2.8. Generally, in the transition zone between the riverine conditions and the lacustrine conditions of the reservoir, the best combination of nutrient concentration and light is achieved (Kimmel et al. 1990).

Longitudinal organizations in reservoirs considering physical, chemical, and biological variables were described for several of these artificial ecosystems (Kimmel et al. 1990). Tundisi and Matsumura-Tundisi (1995) described how, in a relatively small and shallow (7 km², 3 m average depth—Lobo/Broa in Brazil) reservoir, a gradient of physical, chemical, and biological variables occurred. In the upstream reservoir, macrophytes predominated with fast decomposition rates of leaves and roots after death. This is a structure of a detritus food chain. The upstream reservoir is a nursery ground for fishes, zooplanktons, and macroinvertebrates. In the lower reservoir with no macrophytes, a food chain based on phytoplankton production predominates.

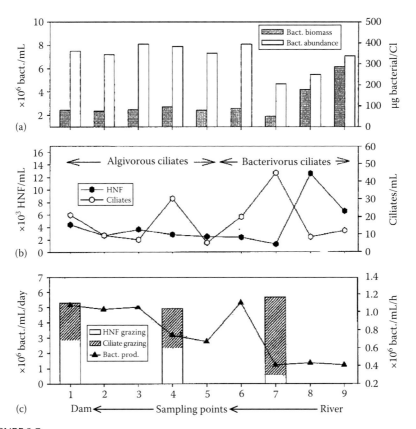

FIGURE 2.7
Longitudinal profile from the river to the dam. (a) Abundance and biomass of bacteria,
(b) abundance of HNF and ciliates, (c) comparison of bacterial production and protistan grazing divided into HNF and ciliate bacterivory. Shows a longitudinal profile from the river to the
dam in SAU reservoir as presented by Armengol et al. (1999).

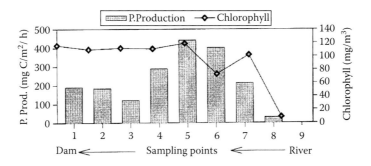

FIGURE 2.8
Longitudinal profile from the river to the dam of chlorophyll a concentration and primary pro-
duction. (From Armengol, J. et al., Longitudinal process in Canyon type reservoirs: The case of
SAU (N.E. Spain), in: Tundisi, J.G. and Straskraba, M. (Eds.), *Theoretical Reservoir Ecology and Its
Application*, IIE, BAS, Backhuys Publishers, Leiden, the Netherlands, pp. 313–345, 583pp., 1999.)

2.4 Differences between Lakes and Reservoirs

Even considering the limnological similarities between lakes and reservoirs, there are some differences that have to be described, because these are fundamental for management purposes. Shallow reservoirs (average depth <5 m) are excluded from this comparison because they have similar characteristics as shallow lakes. Therefore, this comparison is made only between deep lakes and deep reservoirs (average depth >5 m).

As artificial ecosystems constructed by man, reservoirs are built in a river with the objective of one main use or multiple uses: hydroelectricity, irrigation, water supply, recreation, or navigation. In general, reservoirs are built in watersheds or regions where natural lakes are scarce; for example, in Spain, where natural lakes are very few, more than 700 small and large reservoirs were constructed with the purpose to supply drinking water or for irrigation (Margalef et al. 1976).

Drainage basin—The drainage basin of reservoirs is generally much larger that the drainage basin of natural lakes. The area of the drainage basin is much larger than the inundated area of the reservoir. When the dam is located in the main stream of a river, in general the reservoir is elongated especially if there are many tributaries in the watershed; for example, Matsumura-Tundisi and Tundisi (2005) described how 114 tributaries at Barra Bonita reservoir (São Paulo State, Brazil) resulted in an oblong shape of the reservoir.

Lakes have a more circular shape, are less irregular, and their area is much smaller when compared with the watershed area.

The *water level* fluctuation in reservoirs is much higher than in natural lakes. These can be irregular fluctuations due to the extension of dry periods or the flooding periods as a result of heavy rainfall. In natural lakes, the water level fluctuations are much smaller and more stable.

Horizontal gradients—Reservoirs have extensive horizontal gradients due to inputs of tributaries, water withdrawal from the outlets, and zones of river influence or lacustrine characteristics.

In general, natural lakes have much less reduced horizontal gradients or this is geographically limited to a few tributaries that develop density currents.

Suspended matter loads—Since reservoirs in many regions are used as artificial ecosystems to stimulate economic development, the watershed uses are very intense and the contribution of suspended matter is thus much higher in reservoirs than in lakes.

Underwater currents—The tributaries of the reservoirs generally develop underwater, at surface water, or in midwater currents that promote horizontal gradients. This is not the case for natural lakes, where the tributaries do not have such a great impact.

Outflow—Reservoirs have many outlets for water withdrawal and this is regulated by their use. The withdrawal can be made throughout the surface or the hypolimnion. Natural lakes have generally a stable surface outflow.

Retention time (relationship of volume and flushing rate)—The retention time of a reservoir varies very much from a few days to months or years. This retention time is also extremely variable between reservoirs. In lakes, the retention time is much more stable. The retention time in reservoirs varies with the multiple uses and the water needs.

Deposition and distribution of sediments—In a reservoir, the deposition of sediments occurs in the riverine zone or in the tributaries' inlets to the reservoirs. It is low in the lacustrine zone. Variable rates of sediment deposition are dependent on the seasonal events such as increase of drainage during high rainfall or dry periods. In lakes, the deposition of sediments is more regular and stable and has a limited dispersal. The sediment deposition in the tributaries' inlets to the reservoir creates areas of high concentration of organic matter and sites for invertebrate development (mollusks, insect larvae, and oligochaeta). Also extensive periphytic growth of algae occurs in these regions.

2.5 External Nutrient Loading and Nutrient Dynamics

Due to the intensive use of watersheds, the nutrient input to the reservoirs is generally much higher than the nutrient input to natural lakes. Horizontal gradients in nutrient distribution occur in reservoirs. Biogeochemical cycles are accelerated in reservoirs due to tributary contribution and higher suspended material loads. The stocking of fishes in reservoirs is also from the lake ecosystems. Heavy stock of fishes and intensive cultivation of fishes in the reservoirs can accelerate the biogeochemical cycles, especially for nitrogen (due to the excretion of ammonia by fishes) and for phosphorus (due to accumulation of fish feed on the bottom sediments). In periods of low water level, several meters or kilometers of sediments become exposed to air, accelerating the decomposition of organic matter. In the next rainfall period, with the increase of drainage there is a pulse of nutrients to the reservoirs, enhancing primary production of phytoplanktons. These processes are rare for natural lakes. Bacterial activity is more intense in reservoirs than in lakes due to this extensive sediment decomposition at the mouth of the tributaries. These areas become exposed during periods of low water level in reservoirs by an accelerating organic matter decomposition and biogeochemical cycling of elements and substances.

Dissolved oxygen—In general, horizontal variability in dissolved oxygen is higher in reservoirs than in lakes. Inflows of tributaries with low oxygen concentration due to high particulate organic matter concentration are common in reservoirs. In lakes, there are smaller horizontal gradients of dissolved oxygen. Also, in lakes, concentration of dissolved oxygen is less dependent of inflows and outflows as it occurs in reservoirs.

Light penetration—It is generally higher in lakes than in reservoirs. In the riverine zones of reservoirs, there is very low light penetration. Horizontal gradients in light penetration in reservoirs are more common than in lakes. Euphotic depth increases in the lacustrine zone of reservoirs.

The littoral zone—It is very irregular in reservoirs due to higher water level fluctuations than in lakes. Each tributary influences littoral zones in the reservoirs. In lakes, the littoral zone is more constant with generally higher levels of primary production of phytoplanktons and macrophytes.

Biodiversity—Reservoirs are young ecosystems with the biodiversity depending on the duration of the filling phase, the capacity of the local flora and fauna to colonize this artificial ecosystem, and the retention time. Depending on the substrate remaining after the beginning of the inundation, biodiversity of periphytic algae and benthic organisms can be high in reservoirs. Production of organic matter by phytoplanktons is high during the filling phase and then decreases. In lakes, diversity is high. Primary production of phytoplanktons is relatively constant in lakes and is subjected to seasonal variation of forcing functions, such as solar radiation and precipitation.

2.6 Succession in Lakes and Reservoirs

Reservoirs in many regions are built with the purpose to stimulate the regional development. The succession in a reservoir is greatly influenced by human activities in the watershed and is generally accelerated by multiple uses of water and the degree of agricultural activities, industrial development, and urbanization. In lakes, the influence of the watershed uses is smaller, less intensive, and the hydrological cycle is much more in equilibrium because of several years (hundreds of years or ever millennia) of interaction with the human population.

The main characteristics of reservoirs that are key factors for their management are as follows:

- *The filling phase and the situation of the inundation area*—The duration of the filling phase is a fundamental factor for the development of the reservoir as an ecosystem; this is coupled with the status of the inundation area of the future reservoir. For example, in the Amazonian reservoirs the presence of the forest in the inundated area resulted in anoxic water that lasted for several years (Tundisi and Matsumura-Tundisi 2010). If the organic matter in the future area of inundation is not removed, the water quality of the reservoir will be degraded from the very beginning of the filling phase.

- *The retention time*—The retention time of the reservoir regulates the biogeochemical cycles, the primary production, and the biodiversity. High retention times (greater than 1 year) retain nutrients, degrade water quality, and influence biodiversity (Jørgensen et al. 2005). Low retention time is fundamental for a reservoir that supplies drinking water, as it may imply a better water quality.

- *The watershed inputs to the reservoir*—Input of nutrients and organic matter (dissolved and particulate) to the reservoirs depend upon the watershed uses. The control of these inputs is of importance for the management of the reservoirs.

2.7 Lake and Reservoir Sedimentation

The intensive use of watershed due to deforestation and rate of urbanization has significant impacts in the contribution of sediment inflow. The sediment yield produced by the watershed can cause long-term morphological changes in the tributaries to the lake or reservoirs, modify the rates of inflow, and reduce the volume of water. In the case of reservoirs, erosion and sediment deposition toward the dam can reduce its stability, affecting the operation of outflows and the safety of the dam. Several factors that determine the sediment yield of a watershed are given below:

- Rainfall amount, intensity, and seasonal cycle
- Soil type and geological formation
- Ground cover
- Land use
- Topography
- Erosion rate
- Drainage network density
- Slope, size, shape, and alignment of channels
- Runoff
- Sediment characteristics, such as grain size and mineralogy

After removal from the watershed surface, sediment particles are transported through the river system into the lake or reservoir. The volume of sediment and the deposition depends upon the hydraulic characteristics of the river, the velocity of the currents, the hydrodynamics of flows, and the morphometry of the river channel and its shape at the inflow point to the lake or reservoir. The surface erosion and soil loss can be calculated by the *universal soil loss equation*.

$$A = RKLSCP$$

where
    A is the computed soil loss in (ton/ha)/year
    R is the rainfall factor
    K is the soil-erodibility factor
    L is the slope-length factor
    S is the slope-steepness factor
    C is the cropping-management factor
    P is the erosion-control practice factor

The rainfall factor **R** accounts for differences in rainfall intensity–duration–frequency for different locations, that is, the average number of erosion-index units in a year of rain.

Source: From Yang, C.T., Reservoir sedimentation, in: Yang, C.T. (Eds.), *Sediment Transport: Theory and Practice*, Krieger Publishing Company, Melbourne, FL, pp. 267–315, 2003.

The density of the deposited sediment and the trap efficiency of the natural or artificial ecosystem are the main factors that control the volume of the sediment deposited. Trap efficiency depends on the density of sediment particles, the retention time of a lake or reservoir, and the outflow mechanism. The texture and size of the deposited sediment particles and the compaction of the sediment at the bottom are other important factors.

Reservoirs in a cascade act as a sediment trap. Reservoirs trap generally as much as 80% of the sediment influx.

The accumulation of sediment in lakes and reservoirs reduces their volume impairing multiple uses and changing water quality. The amount of sediment input to these artificial or natural inland ecosystems is a key process for their management. Structural measures of the watershed are reforestation, construction of small impoundments in the tributaries of the reservoirs or lake, construction of sedimentation basins to store sediments, and stream bank stabilization to reduce bank erosion. The maintenance of a mosaic of wetlands in the watershed combined with a dense riparian forest cover in the tributaries is a nonstructural effective measure to reduce sedimentation.

The lost storage capacity of a lake or reservoir can be recovered by using techniques of flushing, diversion, or siphoning sediment.

2.8 Shallow Lakes

Shallow lakes can be found in all continents and watersheds. These shallow (<5 m average depth) bodies of water that are often polymictic with only occasional stratifications at surface waters vary widely in area (from less than 0.1 km² to more than 1.000 km²) and are often subject to strong wind effects that cause resuspension of sediments, inorganic particles, and algal cells and colonies. The balance of resuspension and sedimentation is important in shallow lakes as it is decisive for the nutrient concentrations of the lake water. Turbidity caused by the resuspension becomes an important lake property in shallow lakes (Scheffer 1998).

Nutrient dynamic processes are generally important and complex quantitative processes. The biogeochemical cycles of phosphorus, nitrogen, and carbon, including the exchanges of these nutrients between the water and the sediment determine the primary production and therefore the possibilities for eutrophication. The decomposition of detritus in the sediments of shallow lakes can be very fast due to high rates of mineralization. Pulses of ammonium (NH_4^+) can easily occur in shallow lakes. Phosphorus and nitrogen releases from detritus decomposition in the water after phytoplankton blooms and macrophytes are other sources of nutrients producing rapid changes in the concentrations of carbon, nitrogen, and phosphorus.

The seasonal dynamics of shallow lakes is influenced by external forcing functions such as wind, solar radiation, air temperature, and also by pulses of biomass growth and decomposition. The growth of macrophytes in the shores of shallow lakes can be extensive with high rates of biomass growth and decomposition. In some less turbid regions of a shallow lake, submerged macrophytes can grow very fast. They are an important nursery ground for fish larvae and invertebrates.

The nutrient load is one of the main factors that influences the shallow lakes' productivity and ecological dynamics. The external nutrient load can be very high, accelerating the eutrophication process, changing the equilibrium of the lake, and moving the lake toward a eutrophic state with high nutrient concentrations in the sediment and water and an almost permanent bloom of cyanobacteria or macrophyte vegetation. Controlling the nutrient load is one of the main management procedures for shallow lakes. This has to be done at the watershed level by integrated holistic management (see also Chapter 14). Internal load can be controlled by removing sediment or the biomass of macrophytes, or other ecological engineering methods can be applied; see Chapter 16. A removal of the sediment is under all circumstances a useful measure for small ponds, small reservoirs, or fish ponds. Figure 2.9 shows the forcing functions that control these ecosystems (Talling 2001).

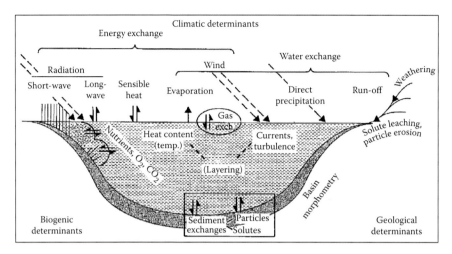

FIGURE 2.9
Diagrammatic vertical section of a shallow lake, showing components of environmental control by climatic–atmospheric, geological, and biological determinants. (From Talling, J.F. and Lemoalle, J.L., *Ecological Dynamics of Tropical Inland Waters*, Cambridge University Press, Cambridge, U.K., 1998.)

2.9 Future Research Needs for Lakes and Reservoirs as Tools for Advanced Management of These Ecosystems

The knowledge on the hydrodynamics of lakes and reservoirs coupled with studies on the water quality is still limited to a few groups of these ecosystems mainly in temperate regions. Studies on mixed layer energetic coupled with simulation of vertical processes and measurements of horizontal variations in physical, chemical, and biological variables are extremely necessary in order to optimize the planning of management and control operations. Better knowledge of internal wave propagation coupled with studies of horizontal transport of pollutants will help to clarify the motion patterns and transport phenomena of large tropical and small lakes and reservoirs. The spatial variability of wind stress affects the deepening of water masses producing several mechanisms of distribution of suspended material and pollutants at different depths. Selective withdrawal of water in reservoirs that affects upstream the reservoir and downstream the river or the next reservoir in the cascade need to be better known (Imberger and Patterson 1990).

The interaction of the reservoir and lakes with the watershed also needs to be known in more detail; especially the inputs and the quantitative contribution of these sources (point and nonpoint sources) need to be better known (IIEGA/SVMA 2010).

Climate changes affect watersheds, lakes, and reservoirs. Extreme hydrological pulses, long periods of dryness, and inputs of climatologically forcing functions such as strong winds affect circulation, distribution of communities, and biogeochemical cycles. Expected rise in temperature due to climatic changes can stimulate the growth and dominance of cyanobacteria in lakes and reservoirs as discussed by Paerl and Huisman (2008). New technologies for the management of cyanobacteria blooms are needed in articulation with the implementation of strategies aimed at controlling and reducing nutrient inputs into the aquatic ecosystem from the watersheds.

Comparative studies on the response of reservoirs and lakes to external inputs are fundamental to understanding their dynamics and the key factors that govern their processes. This is of importance for their management.

Reservoirs and lakes have the same responses to the inputs from the watersheds. But the responses of the reservoir can be more chaotic and more irregular due to their manipulation such as the variable retention time and levels of outlets. Biotic stability is much higher in lakes with external inputs (Wetzel 1990).

2.10 Eutrophication Problem

Many aquatic ecosystems suffer from eutrophication: lakes, reservoirs, estuaries, lagoons, fjords, and bays. It is a worldwide problem and is, together with the oxygen depletion problem, probably the most serious pollution problems of aquatic ecosystems. The word eutrophy is generally taken to mean "nutrient rich." Nauman introduced in 1919 the concepts of oligotrophy and eutrophy, distinguishing between oligotrophic lakes containing little planktonic algae and eutrophic lakes containing much phytoplankton. The eutrophication of aquatic ecosystems all over the world has increased rapidly during the last decade due to increased urbanization and consequently increased discharge of nutrient per capita. The production of fertilizers has grown exponentially in this century and the concentration of phosphorus in many lakes reflects this.

The word eutrophication is used increasingly in the sense of artificial addition of nutrients, mainly nitrogen and phosphorus, to water. Eutrophication is generally considered to be undesirable, but this is not always true. The green color of eutrophied lakes makes swimming and boating less safe due to the increased turbidity, and from an aesthetic point of view the chlorophyll concentration should not exceed $100\,mg/m^3$. However, the most critical effect from an ecological point of view is the reduced oxygen content of the hypolimnion caused by the decomposition of dead algae, particularly in the fall. Eutrophic aquatic ecosystems sometimes show a high oxygen concentration during the summer time at the surface but a low concentration of oxygen in the hypolimnion that may be lethal to fishes. The oxygen depletion in the hypolimnion will often imply that the eutrophication is more difficult to abate, because anaerobic sediment will more easily release its content of phosphorus. Iron(III) is reduced to iron(II) by anaerobic conditions. As iron(III) has a very insoluble phosphorus salt, while iron(II) phosphate is readily soluble, the phosphorus release by anaerobic conditions is dependent on the composition of the sediment, particularly of course the iron content. One of the most applied lake restoration methods is the pumping of air or oxygen to hypolimnion, which is used to reduce significantly the release of phosphorus from the sediment.

Water is 75%–90% of the total wet weight for plant tissue. It means that except for oxygen and hydrogen, the composition on dry weight basis would be 4–10 times higher. For phytoplanktons, the contents of carbon, nitrogen, and phosphorus on dry weight basis are, respectively, approximately 40%–60%, 6%–8%, and 0.75%–1.0%. Phosphorus is considered as the major cause of eutrophication in lakes, as it was formerly the growth-limiting factor for algae in the majority of lakes. Its use has increased tremendously during the last decade. Nitrogen is limiting in a number of East African lakes as a result of the nitrogen depletion of soils by intensive erosion in the past. Nitrogen is furthermore often limiting in the coastal ecosystems, at least a part of the year. However, today nitrogen may become limiting in lakes as a result of the tremendous increase in the phosphorus concentration caused by discharge of wastewater, which contains relatively more phosphorus than nitrogen. While algae uses 5–10 times more nitrogen than phosphorus (see the contents of these two elements in phytoplanktons mentioned previously), wastewater generally contains only three times as much nitrogen as phosphorus. In lakes, a considerable amount of nitrogen is furthermore lost by denitrification (nitrate $\rightarrow N_2$). Lakes that have received wastewater for a longer period may therefore be limited by nitrogen. In environmental management, the key question is however not which element is the limiting factor but which element can most easily be controlled as limiting factor. As phosphorus generally can be removed easier or by less cost, it is often the most effective abatement of eutrophication, particularly of lakes, to remove phosphorus very effectively in the wastewater discharged to the lake. Further details are given later.

It is a good management strategy for an abatement of eutrophication in aquatic ecosystems to find quantitatively all the sources of nitrogen and phosphorus. It is based on this information, in most cases, that it is easy to find possible solutions and the corresponding costs. In this context, it is beneficial to apply an ecological model. Eutrophication models have been developed and applied for all these ecosystems. Many eutrophication models have been applied for environmental management, particularly for lakes and reservoirs. Jørgensen (2011) gives a good overview of the available ecological models including eutrophication models.

2.11 Growth of Phytoplankton

The growth of phytoplanktons is the key process of eutrophication, and it is therefore important to understand the interacting processes that regulate growth. Primary production has been measured in great detail in a number of aquatic ecosystems and presents the synthesis of organic matter, and the overall process can be summarized as follows:

$$\text{Light} + 6CO_2 + 6H_2O \rightarrow C_6H_{12}O_6 + 6O_2 \tag{2.1}$$

The composition of phytoplanktons is not constant. The composition of phytoplanktons and plants, in general, reflects to a certain extent the concentration of the water. If, for example, the phosphorus concentration is high, the phytoplanktons will take up relatively more phosphorus—this is called luxury uptake. Phytoplankton consists mainly of carbon, oxygen, hydrogen, nitrogen, and phosphorus; without these elements no algal growth will take place. This leads to the concept of the limiting nutrient, which has been developed by Liebig as the law of the minimum. However, the concept has been considerably misused due to oversimplification. First of all, growth might be limited by more than one nutrient. The composition as mentioned previously is not constant but varies with the composition of the environment. Furthermore, growth is not at its maximum rate until the nutrients are used and is then stopped, but the growth rate slows down as the nutrients become depleted.

The sequences of events leading to *eutrophication* has often been described as follows: Oligotrophic waters will have a ratio of N:P greater than or equal to 10, which means that phosphorus is less abundant than nitrogen for the needs of phytoplanktons. If sewage is discharged into the aquatic ecosystem, the ratio will decrease, since the N:P ratio for municipal wastewater is 3:1, and consequently nitrogen will be less abundant than phosphorus relative to the needs of phytoplanktons, as indicated earlier. In this situation, however, the best remedy for the excessive growth of algae is not necessary the removal of nitrogen from the sewage, because the mass balance might then show that nitrogen-fixing algae will give an uncontrollable input of nitrogen into the system. It is particularly the case if the aquatic ecosystem is a lake or reservoir. It is necessary to set up mass balances for each of the nutrients as already pointed out, and these will often reveal that the input of nitrogen from nitrogen-fixing blue–green algae, precipitation, and tributaries is contributing too much to the mass balance for the removal of nitrogen from the sewage to have any effect. On the other hand, the mass balance may reveal that the phosphorus input (often more than 95%) comes mainly from sewage, which means that it is better management to remove phosphorus from the sewage than nitrogen. Thus, it is not important in environmental management which nutrient is most limiting but which nutrient can most easily be made to limit the algal growth.

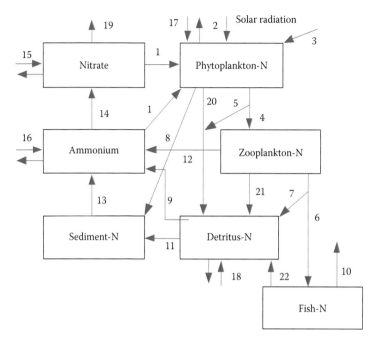

FIGURE 2.10
Conceptual diagram of a nitrogen cycle in an aquatic ecosystem. The processes that connect the state variables and forcing functions are (1) uptake of nitrate and ammonium by algae; (2) photosynthesis; (3) nitrogen fixation; (4) grazing with loss of undigested matter; (5–7) predation and loss of undigested matter; (8) settling of algae; (9) mineralization; (10) fishery; (11) settling of detritus; (12) excretion of ammonium from zooplankton; (13) release of nitrogen from the sediment; (14) nitrification; (15–18) inputs/outputs; (19) denitrification; and (20–22) mortality of phytoplankton, zooplankton, and fish.

The conceptual diagram Figure 2.10 shows the nitrogen cycle and Figure 2.11 gives the phosphorus cycle of aquatic ecosystems and illustrate the processes behind the cycling of these nutrients. As clearly pointed out earlier, it is always beneficial to use mass balances to choose the most important components, such as forcing functions and state variables, to be considered in the management context. Let us illustrate the needed mass balance considerations by use of an example. Let us anticipate that it is an open question, whether birds should be included in the selected management strategy (and in the eutrophication model). Birds may contribute considerably to the inputs of nutrients by their droppings. If the nutrients—nitrogen and phosphorus—coming from the birds' droppings are insignificant compared with the amounts of nutrients coming from drainage water, precipitation, and wastewater, inclusion of birds in the management strategy is an unnecessary complication that would only contribute to the uncertainty. There are, however, a few cases where birds may contribute as much as 25% or at least more than 5% of the total inputs of nutrients. In such cases, it is of course important to include birds in the management and as a model component or

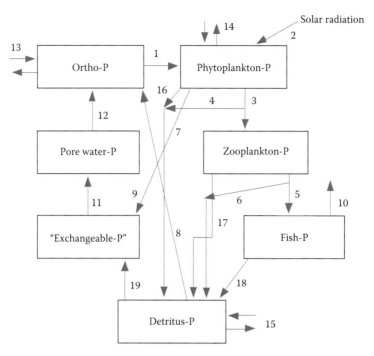

FIGURE 2.11
Phosphorus cycle. The processes are (1) uptake of phosphorus by algae, (2) photosynthesis, (3) grazing with loss of undigested matter, (4, 5) predation with loss of undigested material, (6), (7), and (9) settling of phytoplanktons, (8) mineralization, (10) fishery, (11) mineralization of phosphorous organic compounds in the sediment, (12) diffusion of pore water P, (13–15) inputs/outputs, (16–18) represent mortalities, and (19) settling of detritus.

at least as an important forcing function. A mass balance is always needed to uncover the main sources of a pollution problem—in the example of the inputs of nutrients. It is, however, rare (a good guess based on the author's experience is about 1% of aquatic ecosystems) that it is necessary to consider the dropping of birds as a significant source of eutrophication.

The so-called Michaelis–Menten's equation can be applied to describe the growth of phytoplanktons, $g_r = g_{rmax} NS/(k_n + NS)$, or $g_r = g_{rmax} PS/(k_p + PS)$, dependent on which nutrient is limiting N or P. g_{rmax}, k_n, and k_p are parameters. If both nutrients are limiting in different periods of the year, the formulation is

$$g_r = g_{rmaxmin}\left(\frac{NS}{(k_n+NS)}, \frac{PS}{(k_p+PS)}\right) \qquad (2.2)$$

Product or average of several limiting factors have also been proposed and applied.

For the influence of the temperature, there are two possible formulations:

$$K_t{}^\wedge(\text{TEMP} - 20)\text{(the so-called Arrhenius equation)} \qquad (2.3)$$

or

$$\exp\left(A * \frac{(\text{TEMP} - \text{OPT})}{(\text{TEMP} - \text{MAXTEMP})}\right) \qquad (2.4)$$

K_t is a parameter, which in most cases is between 1.04 and 1.06, in average 1.05. OPT is the optimum temperature for phytoplankton growth and MAXTEMP is the maximum temperature. OPT, MAXTEMP, and A are all parameters that are different for different phytoplankton species.

The description of phytoplankton growth in the equations by use of the constant stoichiometric approach is a simplification, because the phytoplankton growth is in reality a two-step process. The first step is uptake of nutrients and the second step is growth of phytoplanktons (increase of the biomass). The more correct description can be formulated mathematically by the following equations:

$$\text{Uptake rate } P = dPA/dt = PA * \text{maxupp} * (PS/(k_p + PS)) *$$

$$((\text{PAMAX} - PA)/(\text{PAMAX} - \text{PAMIN}))$$

$$\text{Parallel for uptake rate } N \qquad (2.5)$$

$$\text{Uptake rate } C = dCA/dt = CA * \text{maxupc} * (CS/k_c + CS) *$$

$$((\text{CAMAX} - CA)/(\text{CAMAX} - \text{CAMIN})) *$$

$$((L/KL + L) - \text{RESP if } L < L_1$$

$$L \text{ is use; if } L_2 > L > L_1$$

$$L_1 \text{ is used;}$$

$$\text{if } L > L_2 \quad L_1 + L_2 - L \text{ is used for } L \qquad (2.6)$$

PA, NA, and CA are state variables that cover the amount of phosphorus, nitrogen, and carbon in the form of phytoplankton expressed as mg P, N, or C per liter of water. Notice that the unit is mg in 1 L of water. Maxupp, maxupn, macups, k_p, k_n, k_c, PAMAX, PAMIN, NAMAX, NAMIN, CAMAX, CAMIN,

KL, L_1, and L_2 are all parameters. PAMAX, PAMIN, NAMAX, NAMIN, CAMAX, and CAMIN are, however, known fairly well. They are the phytoplankton concentration times, respectively, 0.025, 0.005, 0.12, 0.05, 0.6, and 0.4 with good approximations. It is of course more difficult to calibrate the two-step growth equations than the CS approach due to the higher number of parameters in the NC equations, although the approximate knowledge that is available to the six parameters PAMAX, PAMIN, NAMAX, NAMIN, CAMAX, and CAMIN facilitates the calibration slightly. Notice that the uptakes of phosphorus, nitrogen, and carbon are, according to the equations, dependent on both the concentrations of the nutrients in the water and on the concentrations of nutrients in the cells.

The closer the nutrient concentrations in phytoplanktons are to the minimum, the faster is the uptake. When a nutrient concentration, on the other hand, has reached the maximum value, the uptake stops. The carbon uptake opposite the uptake of phosphorus and nitrogen is dependent also on light as shown by a Michaelis–Menten's expression that includes the light prohibition. Finally, RESP covers the respiration. Only carbon is of course involved in the respiration.

The growth process is quantified by the following equation:

$$\text{Growth} = g_{rmax} * \text{phytoplankton} *$$
$$(\min((PA - PAMIN)/(PAMAX - PAMIN))$$
$$((NA - NAMIN)/(NAMAX - NAMIN)),$$
$$((CA - CAMIN)/(CAMAX - CAMIN))) \tag{2.7}$$

g_{rmax} is a parameter on line with the corresponding parameter in Equation 2.2. Equation 2.7 indicates that the higher the nutrient concentrations are compared with the minimum levels the faster is the growth.

The phytoplankton growth model based on this approach has four state variables: PA, NA, CA, and phytoplankton. They are all assumed to apply the unit mg/L. As the minimum and maximum values are presumed to be a parameter times the phytoplankton concentration, they are also expressed in mg/L.

The two-step description is of course more difficult to calibrate and validate and use in generally, which of course raises the question when should the two-step description be applied instead of the easier applicable constant stoichiometric approach? The model experience has revealed that the need for the two-step description is increasing with the shallowness and the eutrophication of the aquatic ecosystem; see Figure 2.12 (see further details in Jørgensen and Fath 2011). It is definitely recommended for shallow very eutrophied aquatic ecosystems to apply the two-step

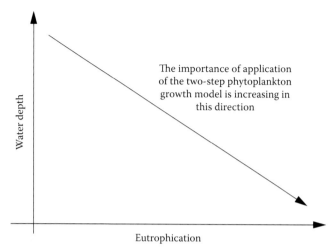

FIGURE 2.12
Need for the two-step description of phytoplankton growth increases with the nutrient concentration and decreases with the depth.

FIGURE 2.13
(See color insert.) Lake Taihu suffers from eutrophication and frequent bloom of blue-green algae.

description, while it is hardly needed for deep mesotrophic or oligotrophic aquatic ecosystems.

Figures 2.13 and 2.14 show the eutrophication in Lake Taihu and Dianchi Lake in China, where the eutrophication problems are significant due to the high density of population.

FIGURE 2.14
(See color insert.) Eutrophication in Dianchi Lake, China.

2.12 Solutions to the Eutrophication Problem

It is usually necessary to apply a wide spectrum of methods to solve the eutrophication problems, because the nutrients have many sources. The amount of nutrients discharged with the wastewater can be reduced by several wastewater treatment methods, see Chapter 15. Nutrients discharged by drainage water including drainage water from agriculture that may have high nutrient concentration can be removed or reduced by a number of ecological engineering methods, for instance, wetlands; see these entries; see Chapter 17. In this context, it is important to consider that lakes do not always respond to changes in a linear fashion but have a complex response dynamics; see Chapter 17 and ILEC (2005). The need for structurally dynamic models is due to a shift in species composition associated with the complex response dynamics; see Section 19.10.

It is possible to reduce the nutrient concentrations in the aquatic ecosystems by use of environmental technological methods, ecotechnological methods, cleaner technology, and environmental legislation. There are a number of cases where a consequent environmental strategy has solved the problems, partially or completely (see Figures 2.15 and 2.16), but there are also many examples of insufficient environmental management, which has led to only a partial solution or no solution at all of the eutrophication problem. From the experience, it can be concluded that quantitative nutrient balances including all sources of nutrients is the best starting point for a good environmental management strategy, because the nutrient balances show clearly which source is important to eliminate or reduce, and they facilitate a comparison of the costs for various management strategies. For further details about selection of an environmental management strategy see Chapter 14, Jørgensen (2000), and Jørgensen et al. (2004).

FIGURE 2.15
(See color insert.) Western lake in Hangzhou, China. By a management that has considered all the sources to the eutrophication problems, it has been possible to obtain a reasonable water quality in spite of a previous hypereutrophication. Many of the tool boxes that are mentioned in Chapter 14 have been applied.

FIGURE 2.16
(See color insert.) Lake Konstanz (Germany), where the management of the eutrophication has consequently considered all sources and has reduced the eutrophication significantly. The phosphorus concentration in 1980 was more than $80\,mg/m^3$ while it is about $13\,mg/m^3$ today.

2.13 Reservoirs and Lakes as Complex Systems and They Require an Integrated Management Plan

Due to their dynamic characteristics in space and time, reservoirs and lakes are complex system. The understanding of this complexity is fundamental from the management point of view since actions for control of water quality or the recovery of these ecosystems depends upon the basic knowledge about the functioning and the response of the reservoir or lake to the forcing functions (Figure 2.17).

The knowledge about basic features of different reservoir and lake types is important for the design and application of management procedures and technology. Shallow reservoirs (average depth <5 m) represent a separate category not much different from shallow lakes but different from deep reservoirs (Straskraba and Tundisi 1999). Two situations are typical for shallow water bodies: macrophyte domination and phytoplankton domination. The switching between the two states depends on water transparency and fish populations and the limiting nutrient concentration (further details see Scheffer 1998). Compared to shallow reservoirs, deep reservoirs or lakes are characterized by the scarcity or less dominance of macrophyte vegetation. The role of bottom sediments that is very important from the qualitative and quantitative point of view in shallow reservoirs and lakes adds new complexity features to these ecosystems; very complex limnological events such as seiches, reversed flows, and deep currents are observed in operations such as water pumping for irrigation, hydroelectricity generation, or other mechanical and technical activities.

Cascades of reservoirs add new complexity features for the management. The depth of the outlet of the upstream reservoirs determines the hydrodynamics and advection processes for the downstream reservoirs. Reservoirs upstream in the cascade retain phosphorus and suspended material and export nitrogen to the next reservoir in the cascade. In any operation of management of reservoirs, either a single reservoir or reservoirs in a cascade, the interaction of the ecosystem with the watersheds is fundamental; therefore, the knowledge of the inputs of the watersheds to the reservoirs must be known.

Impacts of nitrogen and phosphorus from nontreated wastewater, toxic metals from industrial plants, and pesticides and herbicides from the watersheds to the reservoirs and lakes must be known and export coefficients should be developed for each watershed. It is strongly recommended to set up a mass balance for all relevant pollutants including the nutrients. Soil uses, the vegetation cover, the degree of urbanization, and industry locations, the declivity of the watershed, and the drainage system should be known in this context (Jørgensen et al. 2005).

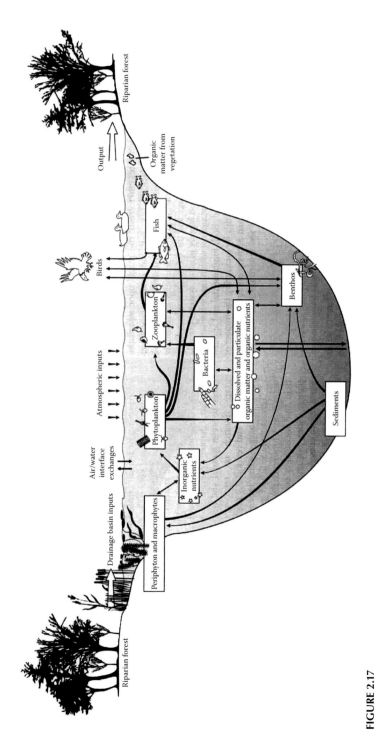

FIGURE 2.17

Major pathways of the energy and nutrient flows between the drainage basin, atmosphere, and lake or reservoir. (Modified from Kalff, J., *Limnology*, Prentice Hall, Upper Saddle River, NJ, 592pp., 2003.)

References

Armengol, J.; Garcia, J.G.; Comerna, M.; Romero, M.; Dolz, J.; Moura, M.; Han, B.H.; Vidal, A.; and Simek, K. 1999. Longitudinal process in Canyon type reservoirs: The case of SAU (N.E. Spain). In: Tundisi, J.G. and Straskraba, M. (Eds.), *Theoretical Reservoir Ecology and Its Application*. IIE, BAS, Backhuys Publishers, Leiden, the Netherlands, pp. 313–345, 583pp.

Cavalcanti, I.F.A. and Kousky, V.E. 2009. Parte I—Sistemas meteorológicos que afetam o tempo na América do Sul: Frentes frias sobre o Brasil. In: Cavalcanti, I.F.A.; Ferreira, N.J.; SILVA, M.G.A.J.; and Silva Dias, M.A.F. (Eds.), *Para Entender Tempo e Clima*. Oficina de Textos, São Paulo, Brazil, pp. 135–147.

Ford, D.E. 1990. Reservoir transport processes. In: Thornton, K.W.; Kimmel, B.L.; and Payne, F.E. (Eds.), *Reservoir Limnology: Ecological Perspectives*. Wiley Interscience, New York.

IIEGA/SVMA—PMSP. 2010. *Manual de Gerenciamento de Bacias Hidrográficas*. Cubo Multimídia, São Carlos, Brazil, 158pp.

ILEC. 2005. *Managing Lakes and Their Basins for Sustainable Use*. ILEC. Kusatsu, Japan, 142pp.

Imberger, J. and Patterson, J.C. 1990. Physical limnology. In: Wu, T. (Ed.), *Advances in Applied Mechanics*, Vol. 27. Academic Press, Boston, MA, pp. 303–455.

Jørgensen, S.E. 2000. *Principles of Pollution Abatement*. Elsevier, Oxford, U.K., 520pp.

Jørgensen, S.E. 2011. *Handbook of Ecological Models Used in Ecosystem and Environmental Management*. CRC Press, Boca Raton, FL, 620pp.

Jørgensen, S.E. and Faith, S. (Eds.). 2011. *Encyclopedia of Ecology*. ICES, Bremen, Germany.

Jørgensen, S.E.; Löfler H.; Rast, W.; and Straskraba, M. 2005. *Lake and Reservoir Management*. Developments in Water Sciences, Vol. 54. Elsevier, Amsterdam, the Netherlands, 502pp.

Jørgensen, S.E. and Svirezhev, Y.M. 2004. *Towards a Thermodynamic Theory for Ecological Systems*. Elsevier, New York, 366pp.

Junk, W.J. 2006. Flood pulsing and the linkages between terrestrial, aquatic and wetland systems. *Verh. Internat. Verein. Limnol. Stuttgart* 29:11–38.

Kalff, J. 2003. *Limnology*. Prentice Hall, Upper Saddle River, NJ, 592pp.

Kimmel, B.L. and Groeger, A.W. 1984. Factors controlling phytoplankton production in lakes and reservoirs: A perspective. In: *Lake and Reservoirs Management*, EPA 440/5/84 001.U.S. EPA, Washington, DC, pp. 277–281.

Kimmel, B.L.; Lind, O.T.; and Paulson, L.J. 1990. Reservoir primary production. In: Thornton, K.W.; Kimmel, B.L.; and Payne, F.E. (Eds.), *Reservoir Limnology: Ecological Perspectives*, Wiley Interscience, New York, pp. 133–194, 246pp.

Likens, G.E. 1992. The ecosystem approach: Its use and abuse. In: Kinne, O. (Series Ed.), *Excellence in Ecology*. Ecology Institute, Oldendorf/Luke, Germany, 166pp.

Margalef, R.; Plawas, D.; Armengol, J.; Vidal, A.; Prat, N.; Guiset, A.; Toja, J.; and Estrada, M. 1976. *Limnologia de los embases españoles. Dirección General de obras Hidraulicas*. MOPU, Madrid, Spain.

Matsumura-Tundisi, T. and Tundisi, J.G. 2005. Plankton richness in a eutrophic reservoir (Barra Bonita Reservoir, SP, Brazil). *Hydrobiologia (The Hague), Aquatic Biodiversity II* 542:367–378.

Paerl, H.W. and Huisman, J. 2008. Blooms like it hot. *Science* 320:57–58.

Paula Lima, W. and Zakia, M.J.B. 2001. Hidrobiologia de Matas Ciliares. In: Rodrigues, R.R. and Leitão Filho, H. (Eds.), *Matas Ciliares: Conservação e recuperação*. EDUSP, FAPESP, São Paulo, Brazil, pp. 33–44, 320pp.

Revenga, C.; Murray, S.; Abramovitz, J.; and Hammond, A. 1998. *Watersheds of the World*. World Resources Institute, Washington, DC, 164pp.

Rodrigues, R.R. and Leitão Filho, H. (Eds.). 2001. *Matas Ciliares: Conservação e recuperação*. EDUSP, FAPESP, São Paulo, Brazil, 320pp.

Scheffer, M. 1998. *Ecology of Shallow Lakes*. Chapman & Hall, London, U.K., 357pp.

Straskraba, M. and Tundisi, J.G. 1999. Reservoir ecosystem functioning: Theory and application. In: Tundisi, J.G. and Straskraba, M. (Eds.), *Theoretical Reservoir Ecology and Its Application*. IIE, BAS, Backhuys Publishers, Leiden, the Netherlands, pp. 565–597, 583pp.

Talling, J.F. 2001. Environmental controls on the functioning of shallow tropical lakes. *Hydrobiologia* 458:1–8.

Talling, J.F. and Lemoalle, J.L. 1998. *Ecological Dynamics of Tropical Inland Waters*. Cambridge University Press, Cambridge, U.K.

Tundisi, J.G.; Forsberg, B.R.; Devol, A.; Zaret, T.; Matsumura-Tundisi, T.; Santos, A.; Ribeiro, J.; and Hardy, E. 1984. Mixing patterns in Amazon lakes. *Hydrobiologia* 108:3–15.

Tundisi, J.G. and Matsumura Tundisi, T. 1990. Limnology and eutrophication of Barra Bonita Reservoir, São Paulo State, Southern Brazil. Arch. *Hydrobiol. Beith. Ergebn. Limnol. Alemanha* 33:661–676.

Tundisi, J.G. and Matsumura Tundisi, T. 1995. The Lobo-Broa ecosystem research. In: Tundisi, J.G.; Bicudo, C.E.M.; and Matsmura Tundisi, T. (Org.), *Limnology in Brazil*. Academia Brasileira de Ciências/SBL, Rio de Janeiro, Brazil, pp. 219–244.

Tundisi, J.G. and Matsumura Tundisi, T. 2008. Biodiversity in the neotropics: Ecological, economic and social values. *Braz. J. Biol. (Impresso)* 68:913–915.

Tundisi, J.G. and Matsumura-Tundisi, T. 2010. Potential impacts of changes in the Forest law in relation to water resources. *Biota Neotropica* 10(4):68–75.

Tundisi, J.G.; Mtasumuratundisi, T.; and Abe, D.S. 2008. The ecological dynamics of Barra Bonita (Tietê River, SP, Brazil) reservoir: Implications for its biodiversity. *Braz. J. Biol. (Impresso)*, 68:1079–1098.

Tundisi, J.G.; Matsumura Tundisi, T.; Arantes Jr., J.D.; Tundisi, J.E.M.; Manzini, N.F.; and Ducrot, R. 2004. The response of Carlos Botelho (Lobo, Broa) reservoir to the passage of cold fronts as reflected by physical, chemical and biological variables. *Braz. J. Biol., São Carlos* 64:177–186.

Tundisi, J.G.; Matsumura-Tundisi, T.; and Sidagis Galli, C. 2006. *Eutrophication: Causes, Consequences and Technologies for Control and Management*. Brazilian Academy of Sciences, International Institute of Ecology, IANAS, São Paulo, Brazil, 532pp.

Vollenweider, R.A. 1987. Scientific concepts and methodologies pertinent to lake research and lake restoration. *Schweiz. Z. Hydrol.* 37:53–84.

Wetzel, R.G. 1990. Reservoir ecosystems: Conclusions and speculations. In: Thorton, F.; Kimmel, B.L.; and Payne, F.F. (Eds.), *Reservoir Limnology—Ecological Perspectives*. John Wiley & Sons, New York, pp. 227–238, 246pp.

Yang, C.T. 2003. Reservoir sedimentation. In: Yang, C.T. (Eds.), *Sediment Transport: Theory and Practice*. Krieger Publishing Company, Melbourne, FL, pp. 267–315.

3

Physical Processes and Circulation in Lakes and Reservoirs

3.1 Introduction

A lake or reservoir is a body of water located at a certain latitude, longitude, and altitude occupying depressions in a watershed or is a result of damming a river. Lakes and reservoirs are permanently interacting at the water–air interface by exchanging heat and gases (oxygen, nitrogen, carbon dioxide, and so on) with the atmosphere. They are also subject to forcing functions such as solar radiation, advection (inflows and outflows), and wind stress. The net input to the lake or reservoir varies seasonally, and it is dependent on the meteorological conditions in the watershed. The balance between the fluxes, the wind stress at the surface, and the inflows and outflows change continuously. In general, these hourly or daily variations are superimposed on the seasonal changes.

3.2 Physical Processes

Figure 3.1 shows the main characteristics of the heat transfer and mechanical energy across a lake or a reservoir interface. Therefore, in lakes and reservoirs, mixing phenomena results from the external inputs of energy, such as solar radiation, and wind magnitude and direction and outflows and inflows. Complicated patterns of vertical structure are observed in these natural or artificial aquatic ecosystems. Mixing is a dynamic process that is the result of the wind stress or other disturbing forces (such as inflows and outflows) against the "potential energy gradient" (sic, Imberger and Patterson 1990, p. 34), which is a result of the radiation. For reservoirs, the situation is even more complicated due to outflows at different levels and the input of several tributaries that disturb the system.

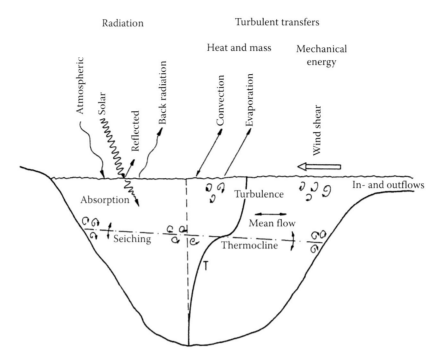

FIGURE 3.1
Main processes of heat transfer and mechanical energy across a lake or a reservoir. (From Bloss, S. and Halemann, D.R.F., Effect of wind mixing on the thermocline formation in lakes and reservoirs, Report 249, MIT, Cambridge, MA, 147pp., 1979.)

Figure 3.2 shows the classical stratification pattern that occurs in lakes or reservoirs and the three layers that characterize the stratification system.

The depth of the first layer the epilimnion varies with latitude, longitude, or altitude. The depths of metalimnion and hypolimnion (see Figure 3.2) are dependent upon the volume and depth of the lake and the degree of stratification. Since the density of water varies with temperature (see Figure 3.1), the energy requirements to destratify a 1°C difference at 25°C is one order of magnitude higher than to destratify a 1°C difference at 5°C (Figure 3.3).

The energy available to warm the water of a lake or reservoir comes from solar radiation; therefore, it is expected that the seasonal behavior of this forcing function influences the degree of stratification and the stability of the water column. The thermal structure changes, therefore, with the seasonal behavior of the forcing functions that are decisive for the stratification and mixing processes. Lake D. Helvécio located in eastern Brazil in the Rio Doce valley can be used as an illustration of the complex behavior. This lake has very little wind influence. During the beginning of spring and summer, (August/March) the lake stratifies due to surface heating and practically no wind. Thermal structure shows a gradient from 30°C or 32°C

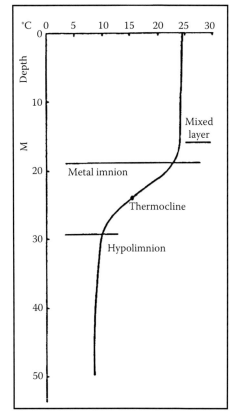

FIGURE 3.2
Classical stratification figure in a lake.

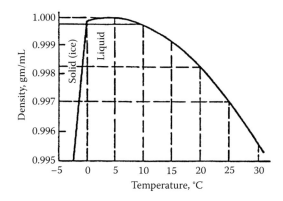

FIGURE 3.3
Relationship of density of water with temperature. (From Bloss, S. and Halemann, D.R.F., Effect of wind mixing on the thermocline formation in lakes and reservoirs, Report 249, MIT, Cambridge, MA, 147pp., 1979.)

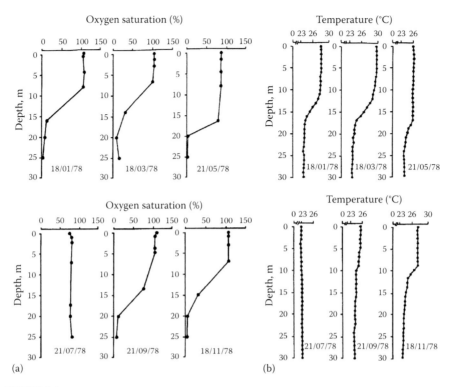

FIGURE 3.4
Seasonal cycle of stratification and circulation in a monomictic tropical lake—D. Helvécio Lake of Eastern Brazil (Lat 19°46′43.89″ S, Long 42°36′16.39″O). (a) Oxygen saturation (%) and (b) temperature (°C). (From Tundisi, J.G. and Saijo, Y. (Eds.), *Limnological Studies at the Rio Doce Valley Lakes*, Brazilian Academy of Sciences, University of São Paulo, São Paulo, Brazil, 5213pp., 1997.)

at the surface to 22°C to 23°C at the bottom (hypolimnion) of the lake (maximum depth 30 metrics). When the air temperature and the solar radiation start to fall (from March, July, and to winter), the water temperature gradually changes until in the winter time (July), where there is only one continuous vertical structure of 23°C from surface to bottom (see Figure 3.4) (Tundisi and Saijo 1997).

These thermal variations correspond to the period of 1 year and there is not much variability of stratification from year to year. Hypolimnetic temperature and thermocline depths are very similar from year to year. The surface layer or epilimnion responds to the surface fluxes and/or to the wind stress. Barbosa and Padisak (2002) and Imberger and Patterson (1990) called attention to the importance of the dynamic, diurnal reorganization of the surface layer in the upper strata of the epilimnion. This has an influence on the vertical distribution of density and on the distribution of nutrients, elements, substances, and phytoplankton.

Cycles of importance in the thermal structure of lakes and reservoirs are annual cycles, seasonal changes in temperature as a consequence of change is solar radiation, wind air temperature, and inflows or outflows. Diurnal cycles have a period of 24 h and correspond to heating and cooling periods, respectively, in daytime and nighttime. Other cycles that are typical for many lakes and reservoirs are the cycles related to the passage of cold fronts or warmer fronts as a result of the major weather systems. They are cycles with 5–10 day periods (synoptic cycles) (Tundisi et al. 2008).

When a lake or reservoir undergoes a cycle of stratification and circulation in the course of 1 year, it is called monomictic. When there are two circulations yearly, this is denoted as a dimictic system. Many circulations with permanent mixing are characteristics of what is named a polymictic system. A lake or reservoir permanently stratified is a meromictic ecosystem. Meromixis is due to salinity intrusions to deep water or the accumulation of organic matter at the bottom water below the thermocline in an ecosystem generally protected from wind. The knowledge concerning mixing and stratification in lakes and reservoirs is based on the vertical profile of temperature and its temporal and spatial variations. However, the distribution of the physical or chemical and biological variables follows the vertical temperature profile and the density profile (see Figure 3.5).

Therefore, a complex set of states follows the stratification and mixing processes in lakes under the influence of energy and wind. Figure 3.6 shows the complexity of mixing processes in lakes caused by energy flow variations and wind.

Temperature and density are not uniform in the surface layer especially when surface heating is very strong and the wind stress is decreasing with time. Diurnal thermocline occurs close to the surface by a depth of a few

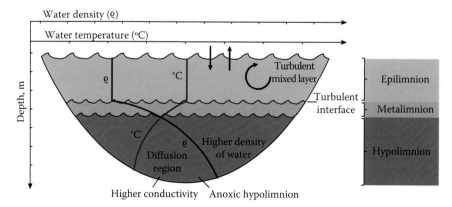

FIGURE 3.5
(See color insert.) Vertical structure of a stratified lake and its physical and chemical features (Original Degani and Tundisi 2011.).

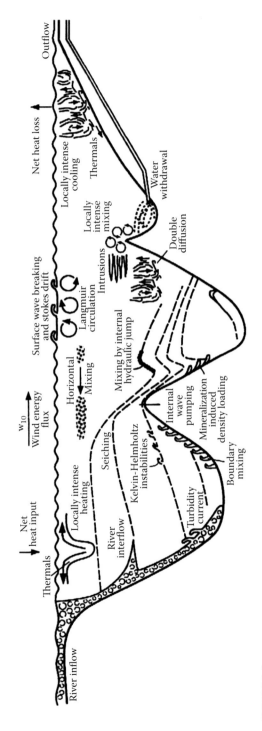

FIGURE 3.6

Complex mixing processes in lakes. (From Imboden, D.M. and Wuest, A., Mixing mechanisms in lakes, in: Lerman, A., Imboden, D., and Gat, J. (Eds.), *Physics and Chemistry of Lakes*, Springer-Verlag, Berlin, Germany, pp. 83–138, 1995.)

TABLE 3.1

Characteristic Timescales of Mixing Processes in Lakes

Timescale	Process	Examples
Seconds and minutes	Surface waves	
	Turbulent overturns	Thorpe (1977) and Dillon (1982)
	Stability oscillation in stratified water	
	Langmuir circulation	Leibovich (1983)
Hours	Wind setup	Spigel and Imberger (1980)
	Internal waves	
	Diurnal mixed layer	Imberger (1985)
	Lateral convection due to differential heating/cooling	Horsch and Stefan (1988)
	Convective turbulence	Lombardo and Gregg (1989)
	Turbidity currents	Lambert (1988)
	Inertial modes	
Days	Mixing due to storms	Imboden et al. (1988)
	Basin modes (topographical waves)	Saylor et al. (1980)
Weeks and months	Annual stratification cycle	
	Basin-wide exchange due to horizontal density gradients	Wüest et al. (1988)
	Internal-wave damping in regularly shaped basin	Mortimer (1974), Csanady (1974)
	Thermal bar	
Years	Meromixis	Sanderson et al. (1986), Steinhorn (1985)

Source: Imboden, D.M. and Wuest, A., Mixing mechanisms in lakes, in: Lerman, A., Imboden, D., and Gat, J. (Eds.), *Physics and Chemistry of Lakes*, Springer-Verlag, Berlin, Germany, pp. 83–138, 1995.

centimeters or meters. This thermal microstructure is followed by a pleustonic and neustonic organization in the vertical axis.

The mixing phenomena and stratification in lakes and reservoirs may cover a wide range of temporal and spatial scales as shown in Table 3.1.

For reservoirs, the patterns of mixing and circulation tend to be more complex (see Figure 3.7) due to the specific features of these artificial ecosystems, retention time, depth of the outflows, depth of the inflows, and currents near the mouth of tributaries.

Tundisi et al. (1990) showed for Barra Bonita reservoir (see Figure 3.8), São Paulo State, Brazil, how the daily needs of hydroelectricity govern the deep current flux in this reservoir due to variation in hydroelectricity production.

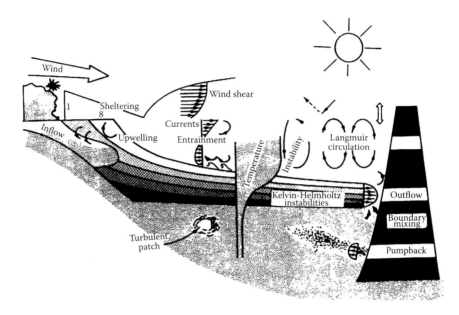

FIGURE 3.7
Patterns of circulation in reservoirs. (From Thornton, F.W. et al., *Reservoir Limnology: Ecological Perspectives*, John Wiley & Sons, New York, 246pp., 1990.)

FIGURE 3.8
(See color insert.) Bara Bonita reservoir, S. Paulo State, Brazil.

3.3 Potential Energy and the Turbulent Kinetic Energy

The potential energy of a lake or reservoir indicates the amount of work that is required to mix a stratified water column against the force of gravity:

$$PE = mgH = \int_{o}^{Z_m} gzA(z)\rho(zt)dz$$

m is the total mass of the lake and reservoir (kg)
g is the acceleration due to gravity $(m \cdot s^2)$
H is the height of the center of mass of the reservoir or lake 1 (m)
Z_m is the maximum elevation (m)
z is the elevation above the lake or reservoir bottom (m)
A(z) is the horizontal area of the reservoir or lake at elevation Z (m^2)
$\rho(z, t)$ is the reservoir or lake density at elevation z and time t (kg/m^3)

Potential energy is stored energy that a lake or reservoir has due to its configuration (morphometry, volume, mean depth, maximum depth, altitude, latitude, and longitude). This potential energy can be converted into kinetic energy. Turbulent kinetic energy (TKE) is the energy input from wind or inflows, which can produce partial or complete vertical mixing of lakes and reservoirs. TKE, for example, can change the thermocline and eventually result in complete vertical mixing. TKE from winds or inflows shows seasonal variations. For example, in natural lakes in Rio Doce (they are tropical lakes), TKE from the inflows dominates in summer, while TKE from wind is negligible all year round. At Lobo Broa reservoir (see Figure 3.9)

FIGURE 3.9
(See color insert.) Broa reservoir, S. Paulo State, Brazil.

in São Paulo State, TKE from inflows dominates during summer (December to March), and TKE from wind dominates during winter months (June to September; Tundisi and Matsumura Tundisi 1995).

3.4 Transport Process in Lake and Reservoir

Mixing and stratification patterns are followed by the transport of various substances and organisms between the different strata of water or at the interface of water and sediments. The transport processes are advection, molecular diffusion, turbulence, turbulent diffusion, convection, dispersion, or entrainment.

Advection—This is the transport due to the directional motion of a fluid. Inflows and outflows and wind shear at the air–water interface are examples of advective processes in lakes and reservoirs. The inflowing water intrudes horizontally into a stratified lake transporting substances and organisms and thereby changing the chemical composition of the stratified lake as shown by Tundisi and Saijo (1997) for Lake D. Helvecio, southeastern Brazil, and Rio Doce valley.

Molecular diffusion—This is a transport process determined by the concentration gradient. It follows Fick's first law, rate = $D*dc/dl$, where D is the diffusion coefficient and dc/dl the gradient (c is concentration and l the distance). Molecular diffusion occurs, for example, from sediments to water, transferring elements or substances according to a concentration gradient at their interfaces.

Turbulence—This is the motion generated by a forcing function such as wind over the surface of a lake or reservoir, producing regions of eddies (or rotating areas of fluid). Turbulence flows can be irregular, diffusive, and spatially varying and dissipative (depends on a continuous source of energy to be permanent). Density stratification inhibits turbulence and mixing. In lakes and reservoirs, turbulence can be generated by wind, inflows, outflows, convection, and boundaries Thornton et al. (1990).

Turbulent diffusion—This is the transport of substances and elements through diffusion processes induced by turbulence. A large turbulence diffusion coefficient replaces the molecular diffusion coefficient. For details about this process see Imberger and Patterson (1990).

Convection—This is a vertical transport process induced by density instabilities. This is a buoyancy-induced flow occurring when a fluid becomes unstable due to density differences. This process can be measured by applying the Brünt Vaisalla Equation (Harris 1986).

Dispersion—The advection of fluid at different speeds and at different positions produces dispersion, which is common at the mouth of tributaries of lakes and reservoirs and at the riverine regions of reservoirs.

Entrainment—This process occurs at the boundary and interface between turbulent and nonturbulent regions of a lake or reservoir. The transport by advection advances into the unstirred layer and sharpens thereby the gradients (Ford et al. 1980, Lewis 1983).

3.5 Stratification and the Circulation of Lakes and Reservoirs and the Ecological Processes

The changes that a lake or a reservoir undergoes during daily, seasonal, or synoptic cycles have an impact on the water quality of these ecosystems and on the spatial and temporal organization of the biological communities from bacteria to fishes. The atmospheric heating and cooling processes impact the mixing (by adding or removing heat) and change water density. The wind force and wind shear transmit energy to the water body and produce surface waves, circulation currents, and turbulence. These transport mechanisms, both the vertical and the horizontal ones, have a strong interference with the biogeochemical cycles and with the dispersion of organisms. Circulation currents within a lake or reservoir are water movements controlled by external and friction forces or large-scale movement of water such as the Coriolis force generated by wind (TKE) or by other circulation processes (Bloss and Halemann 1979, p. 54; see the description of the Langmuir Cells; Matsumura Tundisi and Tundisi 2005).

These mixing processes can transport nutrients, concentrate suspended matter, and transport planktonic organisms such as bacteria and cladocera. Turbulent eddies can concentrate cyanobacteria blooms, and variations in the horizontal density of water can also concentrate organisms, toxic substances, and suspended matter.

The interaction between the dynamics of the mixing processes, the stratification pattern and the temporal and horizontal distribution of organisms, and the production and decomposition of organic matter govern the magnitude of the responses of the chemical and biological variables. Inflows can influence the lake or reservoir water quality by introducing suspended material with high oxygen demand, nutrients, and bacteria. The intrusion of eutrophic water from an upper lake or reservoir by a river can stimulate phytoplankton blooms. Excess bacteria in the intrusion water may harm the water quality in lakes and reservoirs used for recreation. Decrease of the euphotic zone by intrusion of inflowing water with high suspended matter concentration occurs in many lakes and reservoirs, particularly where deforestation of the watersheds is intensive. Interflows, underflows, or outflows have interference on water quality of lakes and reservoirs. Controlling these transport processes by monitoring is fundamental for the management of the lake and reservoir.

The knowledge of the circulation and transport process across the horizontal axis and the vertical boundaries of lakes or reservoirs are of fundamental importance for the management of phytoplankton blooms and of low oxygen concentration and for the success of recovery projects. Manipulation of retention time and controlling stratification and vertical density currents may be very useful methods to reduce phytoplankton blooms in reservoirs and lakes and thereby control the eutrophication. The knowledge of microstratification processes is useful to understand the complexity of the organization of the biological communities such as bacteria, phytoplankton, and fishes at very small spatial scales (a few centimeters). This may therefore provide new insights about the management of lakes and reservoirs at these scales.

From the ecological and management points of view, some processes are fundamental regarding the mixing patterns of lakes and reservoirs. Imberger and Patterson (1990) highlighted these processes as follows:

1. *Seasonal behavior*—Comparative studies on the seasonal cycle of mixing and stratification in lakes and reservoirs at different latitudes is relevant for the better knowledge of the forcing functions that govern the stratification and mixing processes. Shallow lakes and reservoirs respond very fast to forcing function such as rainfall or wind. The knowledge of their seasonal behavior and their responses to these forcing functions could be very important in this context.

2. *Surface fluxes and horizontal transport processes in lakes and reservoirs*—Lakes and reservoirs under the strong influence of inflows from upstream rivers exhibit horizontal gradients in chemical and biological variables as well as gradients in physical variables such as conductivity or water temperature. The advection process intrudes water with different properties or qualities, interfering with decomposition processes and the general rates of bacterial activity.

3. *Outflows and inflows*—Outflows at different levels at the dam site in reservoirs or at the discharge gates of lakes can influence the water quality upstream and downstream. Monitoring of the water quality at the outflows is useful for management of downstream reservoirs or rivers. The water quality of the inflows changes of course quantitatively and qualitatively depending on the water quality of the lake or reservoir. The knowledge of the load introduced is of fundamental importance for the nutrient balance of the aquatic system, which is crucial for the environmental management.

4. *Mixing below the surface layer*—The mixing efficiency generated below the surface layer by turbulent patches should be known in order to identify areas of nutrient input into the euphotic layer or areas of accumulation of organic matter in the reservoir or lake. Accumulation of organic matter by the inflow of more dense drainage water was identified in tropical lakes (Tundisi et al. 1984) and reservoirs (Tundisi and Matsumura Tundisi 1995).

5. *Upwelling*—With the increase of surface wind stress, there is a longitudinal water movement with the isopycnals surfacing at the upwind and deepening at the downwind end. (Bloss and Harlemann 1979, Imberger and Patterson 1990). Vertical microstructures of different water temperatures, densities, and nutrient concentrations develop. Water enriched by these upwelling processes enhances primary production of phytoplankton. Upwelling occurring in lakes and reservoirs can promote the development of patches of phytoplankton as discussed by Tundisi et al. (2008) for Barra Bonita reservoir is São Paulo State, Brazil.

3.6 Classification of Lakes

Lakes have their origins by a wide variety of natural processes. A large number of lakes were formed between 15,000 and 6,000 years before present; therefore, they originated in the late Pleistocene. Hutchinson (1957) identified six major types of lakes originated from the following processes: *glacial lakes, tectonic lakes, coastal, riverine,* and *volcanic* and *lakes with miscellaneous origins.*

Figure 3.10 illustrates some of the important lake formation processes and Table 3.2 gives an overview of the six processes reviewed in the next sections. The table indicates the number of lakes and the lake areas formed by the six processes. As seen in the table, 85% of the lake areas are originated in glacial and tectonic processes.

3.7 Reservoirs

Man's activity produced many reservoirs for thousands of years. These reservoirs of artificial origin have an important role in water storage for several purposes as listed in Chapter 2. Main differences between artificial lakes and natural lakes are also described in Chapter 2.

3.8 Lake Morphometry and Lake Forms

Lakes vary widely in morphometric features and shapes, depending on their origin and the mechanisms that gave origin to them. Lake forms can be *circular, subcircular, elliptical, rectangular, triangular,* and *dendritic.*

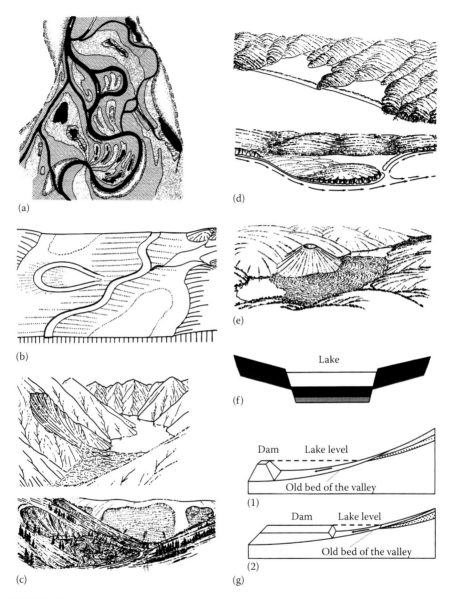

FIGURE 3.10
Mechanisms of lake formation and lake origin (Cole, 1983). (a) Various patterns of lakes and river floodplains, (b) formation of lakes in a horseshoe shape, (c) lakes formed by displacement of sediment by dams, (d) coastal lakes formed by dams, (e) volcanic lakes, (f) lakes formed by tectonic movement, and (g) lakes formed by sediment deposition: (1) profile of a deep lake and (2) profile of a shallow lake. (Modified from Welcome, R., River fisheries, FAO Fisheries Technical Papers 262, Rome, Italy, 1985; Horne, A.J. and Goldman, C.R., *Limnology*, 2nd edn, McGraw-Hill, New York, 1994; Wetzel, R.G., *Limnology: Lake and River Ecosystems*, Academic Press, San Diego, CA, 1006pp., 2001; Cole, G.A., *Textbook of Limnology*, 3rd edn., C.V. Mosby Company, St. Louis, MO, 402pp., 1983; Tundisi, J.G. and Matsumura-Tundisi, T., *Limnologia*, Oficina de Textos, São Paulo, Brazil, 632pp., 2008.)

TABLE 3.2

Overview of Lake Formation by the Six Reviewed Processes

Origin	Number of Lakes[a]	Total Lake Area (km²)	Percent Total Area
Glacial	3,875,000	1,247,000	50
Tectonic	249,000	893,000	35
Coastal	41,000	60,000	2
Riverine	531,000	218,000	9
Volcanic	1,000	3,000	<<1
Miscellaneous	567,000	88,000	4
Total	5,264,000	2,509,000	100

Source: Modified from Meybeck, M., Global distribution of lakes, in: Lerman, A. and Gat, J. (Eds.), *Physics and Chemistry of Lakes*, Springer Verlag, Berlin, Germany, pp. 1–32, 1995.

[a] Approximate value. Other total quoted is 8.4×10^6 lakes $>0.01 \text{ km}^2$.

TABLE 3.3

Morphometric Parameter of a Lake

Surface Area (km²)	A
Volume (m³ or km³)	V
Maximum length (m or km)	l_m
Maximum width (m or km)	b_m
Maximum depth (m)	Z_m
Mean depth (m)	Z
Relative depth (%)	Z_r
Length of shoreline (m)	S
Shoreline development	D_s
Volume development	D_v

The main morphometric characteristics of lakes are (see Table 3.3)

- *Maximum depth*: Lake depths range from a few meters to 1.800 m. Lake Baikal in Siberia is the deepest lake known with 1.741 m. The second deepest lake is Lake Tanganyika with a maximum depth of 1.470 m.
- *Mean depth*: This is the relationship between the volume (V) and the area (A) of a lake, which is V/A.
- *Length (l)*: This is the distance between the farthest points on the shore of a lake.
- *Area*: A lake's area is the extent of its surface in m² or km².
- *Volume*: This is the volume of water contained in a lake, a pond, or a reservoir. It can be calculated by dividing the lake to a number of horizontal strata and estimating the volume of each stratum. A lake volume is measured in cubic meters (m³) or cubic kilometers (km³). The lake with the greatest volume is the Caspian Sea 79,319 km³ (Hutchinson 1957).

- *Shore length*: It is used to describe a lake. Shoreline length is given in meters. Index of shoreline development is defined as the ratio of the shoreline to the length of a circumference of a circle of the same area as the lake. The shoreline gives the borderline between the lake and its environment and is therefore a measure of the lake's openness.

Area—A lake's area is the extent of its surface in m^2 or km^2.

Volume—The volume of water contained in a lake, a pond, or a reservoir. This is determined by measuring the area of each contour of the lake, finding the volume between the planes of successive contour, and summing the volume.

Maximum length—This is the shortest distance between the two most remote points on the lake shore.

Maximum width—This is the maximum distance between shores at right angles to the maximum length.

Maximum depth—This is read directly from the bathymetric map and represents the deepest point of the lake.

Mean depth—This is calculated by dividing volume/area. This is a very important morphometric feature of a lake or reservoir.

Relative depth—This is defined by the ratio of maximum depth (in meters) to the mean diameters of the lake: $Z_R = Z_r = (z_m \sqrt{\pi}/20 \sqrt{A})$, where A is the area of the lake, in km^2.

Length of shoreline—This is the length of the shore of the lake.

Shoreline development—This is the measure of the degree of irregularity of the shoreline. It is given by the formula $D_s = (s/2\sqrt{A\pi})$. This ratio gives an index of the potential importance of littoral influences on a lake or a reservoir (Timms 1992).

Volume development—This index is used to characterize the form of the basin. The volume development compares the shape of the basin to an inverted cone with a height equal to Z_m (maximum depth) and a base equal to the lake's surface area.

References

Barbosa, F.A.R. and Padisak, J. 2002. The forgotten lake stratification pattern atelomixis and its ecological importance. *Verh. Int. Verein. Limnol. Stuttgart* 28:1385–1395.

Bloss, S. and Halemann, D.R.F. 1979. Effect of wind mixing on the thermocline formation in lakes and reservoirs. Report 249. MIT, Cambridge, MA, 147pp.

Cole, G.A. 1983. *Textbook of Limnology*, 3rd edn. C.V. Mosby Company, St. Louis, MO, 402pp.

Csanady, G.T. 1974. Spring thermocline behavior in Lake Ontario during IFYGL. *J. Phys. Oceanogr.* 4:425–445.

Dillon, T.M. 1982. Vertical overturns: A comparison of Thorpe and Ozmidov length scales. *J. Geophys. Res.* 87:9601–9613.

Ford, D.E.; Johnson, M.C.; and Monismith, S.G. 1980. Density inflows in Degray Lake, Arkansas. In: Carstens, T. and McClimans, T. (Eds.), *Proceedings of the Second International Symposium on Stratified Flows*. International Association for Hydraulic Research, Tapir, Trondheim, Norway, pp. 977–987.

Harris, G.P. 1986. *Phytoplankton Ecology: Structure, Function and Fluctuation*. Chapman & Hall, London, U.K.

Horne, A.J. and Goldman, C.R. 1994. *Limnology*, 2nd edn. McGraw-Hill, New York.

Horsch, G.M. and Stefan, H.G. 1988. Convective circulation in littoral water due to surface cooling. *Limnol. Oceanogr.* 33:1068–1083.

Hutchinson, G.E. 1957. A treatise on limnology. *Geography, Physics and Chemistry*. John Wiley & Sons, New York, Vol. 1, 1015pp.

Imberger, J. 1985. The diurnal mixed layer. *Limnol. Oceanogr.* 30:737–770.

Imberger, J. and Patterson, J. 1990. Physical limnology. In: Wu, T. (Ed.), *Advances in Applied Mechanics*. Academic Press, Boston, MA, Vol. 27, pp. 303–475.

Imboden, D.M.; Stotz, B.; and Wüest, A. 1988. Hypolimnic mixing in a deep alpine lake and the role of a storm event. *Verh. Int. Verein. Limnol.* 23:67–73.

Imboden, D.M. and Wuest, A. 1995. Mixing mechanisms in lakes. In: Lerman, A.; Imboden, D.; and Gat, J. (Eds.), *Physics and Chemistry of Lakes*. Springer-Verlag, Berlin, Germany, pp. 83–138.

Lambert, A. 1988. Records of riverborne turbidity currents and indications of slope failure in the Rhone Delta of lake Geneva. *Limnol. Oceanogr.* 33:458–468.

Leibovich, S. 1983. The form and dynamics of langmuir circulations. *Annu. Rev. Fluid Mech.* 15:391–427.

Lewis, J.R.W. 1983. A revised classification of lakes based on mixing. *Can. J. Fish. Aquat. Sci.* 40:1779–1787.

Lombardo, C.P. and Gregg, M.C. 1989. Similarity scaling of ε and x in a convecting surface boundary layer. *J. Geophys. Res.* 94:6273–6284.

Matsumura-Tundisi, T. and Tundisi, J.G. 2005. Plankton richness in an eutrophic reservoir (Barra Bonita), S.P. Brazil. *Hydrobiologia* 542:367–378.

Meybeck, M. 1995. Global distribution of lakes. In: Lerman, A. and Gat, J. (Eds.), *Physics and Chemistry of Lakes*. Springer Verlag, Berlin, Germany, pp. 1–32.

Mortimer, C.H. 1979. Strategies for coupling data collection and analysis with dynamic modeling of lake motion. In: Graf, W.H. and Mortimer, C.H. (Eds.), *Hydrodynamics of Lakes*. Elsevier, Amsterdam, the Netherlands, pp. 183–222.

Sanderson, B.; Perry, K.; and Pederson, T. 1986. Vertical diffusion in meromictic Powell Lake, British Columbia. *J. Geophys. Res.* 91:7647–7655.

Saylor, J.H.; Huang, J.C.K.; and Reid, R.O. 1980. Vortex modes in southern Lake Michigan. *J. Phys. Oceanogr.* 10:1814–1823.

Spigel, R.H. and Imberger, J. 1980. The classification of mixed-layer dynamics in lakes of small to medium size. *J. Phys. Oceanogr.* 10:1104–1121.

Steinhorn, I. 1985. The disappearance of the long term meromictic stratification of the Dead Sea. *Limnol. Oceanogr.* 30:451–462.

Thornton, F.W.; Kimmel, B.L.; and Payne, F.E. 1990. *Reservoir Limnology: Ecological Perspectives*. John Wiley & Sons, New York, 246pp.

Thorpe, S.A. 1977. Turbulence and mixing in a Scottish loch. *Philos. Trans. R. Soc. Lond. Ser. A* 286:125–181.

Timms, B.V. 1992. *Lake Geomorphology*. Gleneagles Publishing, Adelaide, Australia, 180pp.

Tundisi, J.G.; Forsberg, B.R.; Devol, A.; Zaret, T.; Matsumura-Tundisi, T.; Santos, A.; Ribeiro, J.; and Hardy, E. 1984. Mixing patterns in Amazon lakes. *Hydrobiologia* 108:3–15.

Tundisi, J.G. and Matsumura Tundisi, T. 1990. Limnology and eutrophication of Barra Bonita Reservoir, São Paulo State, Southern Brazil. *Arch. Hydrobiol. Beith. Ergebn. Limnol. Alemanha* 33:661–676.

Tundisi, J.G. and Matsumura-Tundisi, T. 1995. The Lobo-Broa ecosystem research. In: Tundisi, J.G.; Bicudo, C.F.M.; and Matsumura-Tundisi, T. (Eds.), *Limnology in Brazil*. Academia Brasileira de Ciências, Sociedade Brasileira de Limnologia, Rio de Janeiro, Brazil, pp. 219–244.

Tundisi, J.G. and Matsumura-Tundisi, T. 2008. *Limnologia*. Oficina de Textos, São Paulo, Brazil, 632pp.

Tundisi, J.G.; Matsumura-Tundisi, T.; and Abe, D.S. 2008. Ecological dynamics of Barra Bonita reservoir: Implications for its biodiversity. *Braz. J. Biol.* 68(4):1079–1098.

Tundisi, J.G. and Saijo, Y. (Eds.). 1997. *Limnological Studies at the Rio Doce Valley Lakes*. Brazilian Academy of Sciences, University of São Paulo, São Paulo, Brazil, 5213pp.

Welcome, R. 1985. River fisheries. FAO Fisheries Technical Papers 262. Rome, Italy.

Wetzel, R.G. 2001. *Limnology: Lake and River Ecosystems*. Academic Press, San Diego, CA, 1006pp.

Wüest, A.; Imboden, D.M.; and Schurter, M. 1988. Origin and size of hypolimnic mixing in Urnersee, the southern basin of Viewaldstättersee (Lake Lucerne). *Schweiz. Z. Hydrol.* 50:40–70.

4

Rivers as Ecosystems

4.1 Physical Characteristics: Horizontal Gradients

In contrast to lakes, reservoirs, or wetlands, rivers are characterized by a unidirectional flow where the horizontal movement is the main forcing function. Rivers also have a strong interaction with the watershed receiving allochthonous material such as organic and inorganic suspended matter, leaves, fruits, and aquatic insects. The unidirectional flux controls the sediment deposition and the bottom structure. The physical characteristics that are important in rivers are the width and depth of the river channel, the current velocity and the roughness of the substrate, and the degree of meandering of the river. The declivity of the river is also important, because it determines the pattern of current velocity including the downstream transportation of the suspended material and organic matter (Barila et al. 1981, Drago and Amsler 1981, Drago 1989, 1990).

The flow velocity of the river (in m^3/s) is dependent upon the hydrological regime and periods of rainfall and dryness. The organization of spatial heterogeneity and the microhabitats where fauna and flora of the river are established are dependent upon the substrate and discharge. The transport of organic and inorganic suspended materials depends upon the soil erosion in the watersheds. The size of the transported particles is related to the velocity of the water and the morphometric characteristics of the river. During transportation along the river channel, organic matter and inorganic suspended material are processed in such a way that downstream there is generally more fine particulate material and dissolved organic material (Allan 1995). The spatial scale in rivers ranges from some millimeter (individual particles) to the entire drainage network that can reach 10^5 m or even much more. This spatial scale represents a gradient of morphologic features upstream, current velocities, physicochemical conditions of the waters, therefore, they are also important gradients for the biota as well. Tributaries add to the complexity of the rivers; therefore, the discharge generally increases downstream. The deposition of the material and the size of the sediment particles and their chemical composition are important characteristics for the biota (Richey et al. 1991). Rivers, streams, and creeks are classified according to their

TABLE 4.1

River Classification

Size of the River	Average Discharge (m³/s)	Drainage Area (km²)	Width (m)	River Order
Very large rivers	>10,000	>106	1–500	>10
Large rivers	1,000–10,000	100,000–10⁶	800–1500	7–11
Rivers	100–1,000	10,000–100,000	200–800	6–9
Streams	1–10	100–1,000	8–40	3–6
Creeks	<0.1	<10	1–8	2–5

Source: Modified from Chapman, D. (Ed.)., *Water Quality Assessments*, UNEP, UNESCO, WHO, Chapman & Hall, London, U.K., 585pp., 1992; Tundisi, J.G. and Matsumura Tundisi, T., *Limnologia*, Oficina de Textos, São Paulo, Brazil, 632pp., 2008.

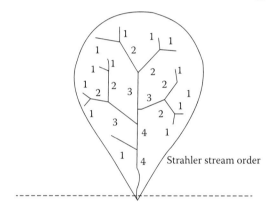

FIGURE 4.1
Strahler stream order. (Modified from Petts, G.E. and Amoros, C., *Fluvial Hydrosystems*, Chapman & Hall, London, U.K., 322pp., 1996.)

discharge, drainage area, and width. Chapman (1992) has given a table with a river classification (Table 4.1).

Small rivers and creeks are classified as first order. Small rivers that merge as tributaries of large rivers are second-order rivers. Large rivers are in the 8th or 10th order. The following is the classical figure of stream orders as given by Strahler (1957) (Figure 4.1).

4.2 Biogeochemical Cycles

The rivers transport dissolved and particulate material, both organic and inorganic. Among the factors that influence these substances, elements, and particulate material, the regional hydrogeochemistry is an important source (see also Chapters 10 and 11). Rain and drainage are a quantitative and qualitative source of chemical inputs to the river. Due to the variable nature of rock that is related to the geology and soil composition, the river chemistry is variable and depends upon these regional characteristics. Thus, the rock composition influences the total dissolved solids of the rivers and this is also dependent on the runoff (Meybeck 1976, 1982). Gibbs (1967) proposed a global scale scheme showing the relationships of total dissolved solids, the chemical composition of the rock substrate ($CaHCO_3$ or $NaCl$), and the rainfall. Therefore, water in rivers transports suspended inorganic matter with major elements such as Ca, Na, Mg, K, Fe, Al, and Si; dissolved ions such as Ca^{++}, Na^+, Mg^{++}, K^+, HCO_3^-, and Cl^-; nutrients (dissolved and particulate) such as N, P, and Si; particulate suspended organic matter; gases (O_2, N_2, Co_2); and heavy metals (Allan 1995). The watersheds are sources of inputs of these chemicals and elements to the rivers from point and nonpoint sources (Richey et al. 1980, 1986, 1991) (Figure 4.2).

Besides the natural composition of the river water, there are other contributions from the watersheds as point and nonpoint sources, originating from human activities. Pollutants and contaminants, such as pesticides, herbicides, and fertilizers from agricultural activities and discharges of industries, are added to the river waters making their water chemistry extremely complex; as a result the monitoring of river waters is also a very complex task (Amoros et al. 1987). The contribution of atmospheric contaminants either as dissolved in rainfall or as particulate matter should also be considered especially in urban or metropolitan regions where air pollution is high (Bayley and Li 1992, Kalff 2003).

Among dissolved gases in river water, dissolved oxygen, carbon dioxide, and nitrogen are important components (see in Chapter 10 about the calculations of the concentrations). Due to the constant turbulence and flow of waters, diffusion of dissolved oxygen and carbon dioxide are near saturation. When excess organic matter is added to the river water (such as the wastewater of untreated sewage), the consumption of oxygen increases and the concentration of dissolved oxygen decreases below 50% saturation or even less. Thus, the amount of organic pollution is a source of biological oxygen demand, regulating to a certain extent the concentration of oxygen in river waters. Diurnal changes (24h) of the oxygen and carbon dioxide concentrations are a consequence of day and night activities in the rivers. There is photosynthetic activity by submerged macrophytes or periphyton that increases the O_2 concentration in the water during daytime and the plants require oxygen for respiration during the night. In addition, the microbial activity decreases the dissolved oxygen concentration and increases the CO_2 concentration. The fluctuation of dissolved gases depends upon the water temperature, the amount of dissolved and particulate organic matter, and the biological activity (production/respiration of organic matter). Quantification can easily be made by the use of the equations presented in Chapter 10. The horizontal variations in dissolved oxygen in river waters are due to the discharge of organic matter as point and nonpoint sources and the capacity of the river to recover its oxygen from air through the turbulence process.

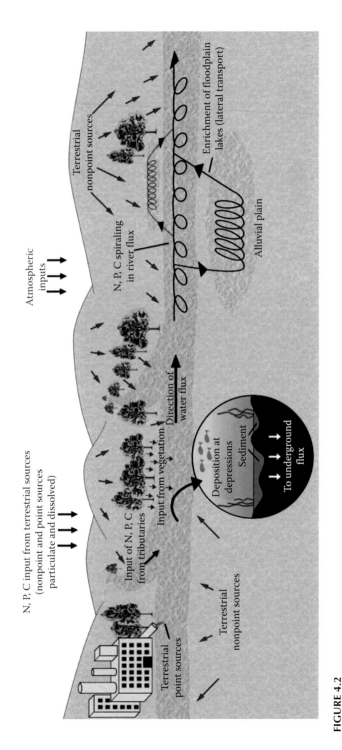

FIGURE 4.2
(See color insert.) Sources of point and nonpoint inputs of nutrients and organic matter to the river; N, P, C spirals in river and floodplain lakes. (Credit to Degani and Tundisi, 2011.)

The major dissolved components of river water and the biogeochemical cycles are thus dependent on the local/regional hydrogeochemical characteristics of the substrate and the interactions of the aquatic and terrestrial biota that interact with the nutrient cycles and the ionic composition (Spitzy and Leenheer 1991). Furthermore, the riparian forests and vegetations contribute particulate organic matter such as leaves, wood, and insect remains and are sources of nitrogen, phosphorus, and carbon to the river (Lewis and Saunders 1989). Organic matter in suspended particles is processed by bacteria, protozoa, insects, and fishes, and this accelerates the decomposition process and promotes other sources of nutrients in the river water (Armour et al. 1991, Allan 1995).

The particulate organic material is a major source of dissolved organic phosphorus and nitrogen. Dissolved organic carbon from terrestrial sources also inputs the rivers from the watersheds (Richey 1982). The concentration of nutrients is fundamental for the growth of periphyton and macrophytes in the rivers (Allan 1995) or phytoplankton in regions of low current velocity or bays (Talling et al. 2009). All these processes are highly dependent on pH and the redox potential. The determination of these important factors for the biogeochemical processes and their influence on the processes are presented in Chapter 11.

Rivers also transport sediments in varying quantities. Some of the large rivers are the largest suspended matter transportation systems of the world (Paredes et al. 1983). The channel morphology and typology contribute to the rate of transportation of sediments and the accumulation in regions with low current velocity (Degens et al. 1983, Paolini et al. 1983, Callow and Petts 1992a,b). Carbon, phosphorus, and nitrogen have biogeochemical cycles in rivers that include and encompass the input of allochthonous materials from the watersheds, the decomposition of autochthonous sources such as the dead biomass of organisms, and the microbial community and their decomposition process. Besides transportation downstream, nutrients can be retained in rivers by sediment accumulation in regions of low currents; this accumulation enhances production of biomass of periphyton, developing a "microbial carpet" consisting of algae, bacteria, and particulate and organic matter. Provided that there is sufficient light at the bottom sediment of the river, this "microbial carpet" may have an important role in the food chain by being the food for bottom dwelling fishes or macroinvertebrates (Tundisi and Matsumura-Tundisi 2008). Phosphorus, nitrogen, and carbon are removed from the water by bacteria and cyanobacteria, macrophytes, periphyton, and phytoplankton. The efficiency of this removal is dependent upon the water temperature and the oxidation/reduction (redox) potential of the water. Adsorption and desorption of phosphorus in particles of organic and inorganic composition can also occur (Kempe and Richey 1991, Newbold 1992).

The watersheds also add to the river water and hydrographic network varied toxic substances, organic materials (such as pesticides and herbicides from agricultural lands) or toxic metals such as lead (Pb), cadmium (Cd), and mercury (Hg) that originate from industrial activities. Nontreated domestic wastewater is an important source of nitrogen, phosphorus, and carbon in urban rivers enhancing algae growth and macrophyte growth, therefore producing eutrophication of river waters, especially, in countries of Africa, Latin America, and southeast Asia.

Hydrological cycles, bottom sediment composition, redox potential, and biota have an important role in the spirals of nutrients in the rivers (Paolini et al. 1983, Paolini 1990). The plant and animal community modify and interfere with the nutrient cycling by assimilating or excreting phosphorus, nitrogen, and carbon. Fishes excrete ammonia, increasing its concentration in the river water. Aquatic macrophytes remove nutrients or heavy metals from the river water. Migration of animals can influence inputs and outputs of phosphorus and nitrogen. For example, the use of riparian forests of rivers by aquatic birds for nitrification can be a source of phosphorus and nitrogen to the rivers (Decamps et al. 1987). Animals that dwell in water and in the terrestrial systems and ecotones such as capybaras (*Hydrochoerus hydrochoeris*) can have a strong role in adding or removing nutrients to the river water (Tundisi and Matsumura-Tundisi 2008).

All in all, there is a complex cycling and network of biogeochemical processes in rivers. The processes are decisive for the biogeochemical water quality that determines the life conditions for the aquatic biota of rivers (Kempe and Richey 1991).

4.3 Aquatic Biota of Rivers

Rivers have a permanent unidirectional flow that poses many specific challenges to the rivers' biota. Thus, the following factors are fundamental to the organisms that are river dwellers:

- *Current velocity* and associated physical forces—fauna and flora of rivers have to adapt to the water flux and to the drifting; this drifting transports organisms, eggs, and larvae downstream, away from their optimum region for survival and reproduction.
- *Water flux* near the sediment—laminar or turbulent flux.
- *Substrate*—quality of substrate influences the abundance and diversity of organisms. Sand, rocks, fine clay particles, leaves, and vegetation remains are some of the organism diversity and distribution.
- *Water temperature*—water temperature of rivers changes diurnally, monthly, or seasonally and this influences development ratios, distribution of organisms, their reproduction, and survival. In rivers where vegetation cover is dense, water temperature has smaller variations. When groundwater discharges into rivers, advection

currents with lower temperatures may be influencing life cycles and reproduction. Water temperature of the rivers is also related to the concentration and saturation of dissolved oxygen.

- *Dissolved oxygen*—the concentration of dissolved oxygen has a fundamental role in the distribution, survival, and physiology of lotic fauna and flora. The fish fauna of the rivers is determined by the distribution and concentration of dissolved oxygen; this concentration is dependent upon the water temperature, the discharge of organic wastewaters, the turbulence that promotes reoxygenation, and the oxidation of dissolved organic matter.
- *Interactions with other organisms*—Predator–prey relationships, symbiosis, and parasitism are other factors that can influence the aquatic biota in rivers.

Several organisms such as invertebrates have body shapes adapted to running waters, and with this they are able to withstand currents.

A *river zonation* was proposed by several authors such as Illies and Botosaneau (1963), Macan (1961), and Hawkes (1975). This zonation was based on the physical characteristics of the rivers, the horizontal distributions of organisms, and the associations of these physical characteristics with the density of the benthic fauna and biotic indicators such as benthic algae or macroinvertebrates. Hawkes (1975) considers that it is very important to classify the river zones according to the fish fauna.

The two main divisions proposed by Illies and Botosaneau (1963) are

- Rhithron—zone of high current velocity; substrate with stones, rocks, and fine sand; average yearly temperature 20°C.
- Potamon—zones of low current velocity, laminar flux predominant, annual average water temperature exceeding 25°C, organic substrate with fine particulate matter and low oxygen concentration.

Other classifications include the subzones epi-, meta-, and epirhithron (Illies and Botosaneau 1963) and crenon divided into eucrenon (sources) and hypocrenon (headwaters) as regions above rhithron.

These classifications are important to identify conservation of river zones and ecological studies. The best association is probably that of organisms such as families of aquatic insects and the physical characteristics of the river.

The river continuum concept was introduced by Vannote et al. (1980) based on the physical characteristics, the type of organic particulate material, and the type of benthic or macroinvertebrates present in the gradients of organic and inorganic matter of the substrate. This concept can be applied to small rivers and creeks of temperate and tropical rivers but probably does not apply to large rivers where the complexity of the relationship of river and floodplain is much greater (Tundisi and Matsumura-Tundisi 2008).

4.4 Energy Flux and Food Chains

The energy flux in rivers is dependent upon the abundance and diversity of primary producers and other components of the food chain. This is related to the physical conditions of the rivers such as current velocities and substrates (Horne and Goldman 1994). The contribution of allochthonous material to the food chain can be very significant from the quantitative and qualitative point of view. The microbial loop and the protozoan community can have a strong role in the energy flux and the transference of organic matter to other levels of the food chain. The invertebrate fauna, consisting of several consumers in running waters and composed of shredders, suspension feeders, collectors, grazers, and predators, has a strong role in processing the allochthonous material and maintaining a food chain that develops from bacteria, protozoans, to top carnivores such as alligators (Décamps et al. 1987, Tundisi and Matsumura-Tundisi 2008).

The riverine fishes have also varied food habits from piscivores to herbivores, detritivores, omnivores, and benthic invertebrates (Allan 1995) with several specialized anatomical and physiological characteristics, especially related to the diet. There are extremes on that. For example, in Rio Negro of the Amazon watershed, Goulding (1981) demonstrated that despite low organic dissolved matter in the river, the biomass and diversity of fish fauna was very high. This is due to the feeding habits of the fishes that feed on detritus, invertebrates, and fruits and leaves from the inundated vegetation.

4.5 Large Rivers

As Margalef (1960, 1990) pointed out "the large river has to be understood as a complex system, very important in its interaction with the watershed utilizing external energy and in a permanent and continuous exchange with the terrestrial system. A large river interferes with the ecological dynamics of the whole watershed" (see Margalef 1990).

In these large rivers, as in all river ecosystems, ecological, limnological, hydrological, and geomorphological knowledge have to be integrated to understand the complex mechanism of functioning and to find solutions to reduce, minimize, or avoid impacts on the water quality, aquatic biota, and limnological features (Ward and Stanford 1983, Ward 1989).

Large rivers are an important ecosystem of great ecological, economical, and social significance. In a volume describing the ecological dynamics of 13 river systems—Nile, Orange, Orange-Vaal, Volta Zaire, and Zambezi (Africa); Colorado and Machenzie (North America); Amazon, Parana, and Uruguay (South America); Murray–Darling (Australian); and Mekong (southeast Asia)—edited by Davies and Walker (1986), the authors empha-size the geographical, climatological, hydrological, and ecological features of these large ecosystems; endemism; biogeographical unity; community char-acteristics; floodplain–river interaction; and zonation.

The diversity of each one of these large rivers from several approaches, from the watershed to the aquatic biota and fisheries, suggests that these river systems can be effective ecosystems for testing ecological theories. Margalef (1997) considers large river systems, such as the Amazon River with its complex processes of water level fluctuation, large floodplain areas, and exchanges of nutrients and energy, as an active *center of evolution* due to their dynamic character (Salo et al. 1986).

Considering the needs for conservation and recovery of large river sys-tems, it is important to emphasize the following points:

1. They represent a vast supply of freshwater responsible for trans-portation of large amounts of suspended material to the oceans (see Table 4.2).

2. Multiple and varied uses of water are important features of large rivers (from supply of freshwater to human populations; to biomass production, irrigation, and fisheries; to navigation; to religious and cultural importance).

3. In large rivers, the relationships between evolutions and succes-sion are fundamental and occur under the influence of periodic perturbations(Legendre and Demers 1984, Junk et al. 1989). The study of these ecosystems can be useful to detect and describe pos-sible regulating factors (Fisher 1983, Merona 1990).

4. The tributaries of large rivers play an important role in the supply of organic and inorganic material, nutrients, being in the watershed, an element of spatial heterogeneity able to stimulate diversity.

5. The fish community of large rivers is diverse and of varied abundance representing an important protein resource for human populations. The management of the large rivers systems is a complex task due to their size, dynamics, hydrological, and ecological diversity along the main axis. A possible approach to manage the large river systems is to focus on subbasins: First, build up knowledge about them (Agostinho et al. 2007) and later propose measures to recover and conserve the specific characteristics of each watershed or subbasin (Tundisi and Matsumura-Tundisi 2008, Tundisi et al. in preparation). Large rivers are also international river systems such as the Nile (Dumont 2009), Paraná (Bonetto et al. 1989), or the Amazon (Sioli 1975). Therefore, their management has to consider international relationships too.

Table 4.3 shows the drainage area, discharge volume, and carbon fluxes of important world rivers in South and North American continents. Table 4.4 shows

TABLE 4.2

Absolute and Relative TDS and Bicarbonate and DIC Transport from Continents

Continent	Discharge[a] (km³/year)	TDS (ppm)	Load (10¹⁵ g/year)	Load (%)	HCO₃ (ppm)	Load (10¹⁵ g/year)	DIC (ppm)	Load (10¹⁵ g/year)	Load (%)
Europe	2,800	182	0.510	12.6	95	0.266	18.7	0.052	12.1
Asia	12,200	142	1.732	42.9	79	0.964	15.5	0.189	43.9
N. America	5,900	142	0.838	20.7	68	0.401	13.3	0.078	18.1
S. America	11,100	69	0.407	10.1	31	0.344	6.1	0.068	15.8
Africa	3,400	121	0.411	10.2	43	0.146	8.5	0.029	6.7
Australia	2,400	59	0.142	3.5	31.6	0.076	6.2	0.015	3.4
Total	37,700		4.040	99.8		2.197		0.431	

Source: Kempe, S., Pettine, M., and Callwet, G.: *Biogeochemistry of Major World Rivers*. 355pp. 1991. Copyright Wiley-VCH Verlag GmbH & Co. KGaA.; Kempe, S., Carbon in the freshwater cycle, in Bolin, B., Dejens, E.T., Kempe, S., and Fetner, P., Eds., *Global Carbon Cycle*, John Wiley, New York, pp. 317–342, 1979.

TDS, total dissolved solid; DIC, dissolved inorganic carbon.

TABLE 4.3

Drainage Area, Total Discharge Volume, and Carbon Fluxes of Prominent World Rivers in South and North American Continents

River/Station	Area ($\times 10^6$ km²)	Volume (km³/year)	TDS ($\times 10^6$ t/year)	TSS ($\times 10^6$ t/year)	DOC ($\times 10^6$ t/year)	POC ($\times 10^6$ t/year)	TOC ($\times 10^6$ t/year)	DIC ($\times 10^6$ t/year)
South American								
Amazon/Obidos	4.69	5,780	2990.0[5]	900.0[5]	19.1[3]	13.0[3]	31.0[3]	31.7[K]
Orinoco/Cuid Bolivar	1.0	1100	30.5[5]	150.0[5]	4.5[5]	2.0[5]	6.6[K]	1.7[5]
Paraná/Paraná-Sta. Fe	2.8	470	38.3[5]	80.0[5]	5.9[5]	1.3[5]	7.2[5]	3.0[5]
Uruguay/Salto Grande	0.24	145	6.0[5]	11.0[5]	≈0.5[5]	≈0.1[5]	≈0.6[5]	≈3.6[5]
Magdalena	0.26	215	20.0[5]	2220.0[5]	n.d	n.d	n.d	n.d
S. Francisco	0.63	120	n.d	6.0[5]	n.d	n.d	n.d	n.d
Estimate of continental total		11,039	551	1927	44.2	24.1	66.9	58.9
North America								
Mississippi/Bel Chase	3.22	410	142.0[K]	296.0[K]	3.5[4]	0.8[4]	3.6[4]	11.6[K]
Columbia/The Dales	0.67	182	21.0[K]	14.0[K]	0.5[L]	n.d	0.6[K]	2.6[K]
St Lawrence/Quebec	1.15	413	70.3[C]	5.1[4]	1.55[4]	0.31[4]	1.87[4]	7.02.[4]
Makenzie/Arctic Red	1.81	249	43.7[4]	n.d	1.3[4]	1.8[4]	3.1[4]	4.9[4] (TIC)
Yukon/Pilot Station	0.84	210	34.2[4]	n.d	0.9[4]	0.3[4]	1.2[4]	4.0[K]
Other U.S. rivers to W. Coast	n.d	33	n.d	n.d	0.12[L]	0.02[L]	0.14[L]	n.d
Other U.S. rivers to E. Coast and Gulf of Mexico	n.d	179	n.d	1.8[L]	0.51[L]	1.5[L]	n.d	n.d
Estimate of continental total		5,840	1241	1831	33.8	14.6	41.8	120.9

Source: Degens, E.T. et al., (Eds.), *Biogeochemistry of Major World Rivers*, SCOPE/ICSU/UNEP, John Wiley & Sons, New York, 355pp., 1991.

TABLE 4.4

Size of the Population in Large European River Basins

River Basin Name	Inhabitants (10^6 People)	Population Density (People/2)
Baltic Sea drainage area	*c.* 90	*c.* 58
Vistula	22.1	112
Rhine	41.4	184
Rhone	8.1	84
Arno	2.1	253
Tiber	4.5	265
Po	15.5	232
Adige	1.2	98
Danube	80.8	99

Source: Welcomme, R.L., *Fisheries Ecology of Floodplain Rivers,* Longmans, London, U.K., 371pp., 1979; Compiled from Helmer (1989); Pattine et al. (1985).

TABLE 4.5

General Features of the South American Rivers

River	Discharge (m^3/s)	Area ($\times 10^6$ km^2)	Length (km^2)	Runoff (1/s/km^2)	TDS M.T.R ($\times 10^6$ t/year)	TSS M.T.R ($\times 10^6$ t/year)
Amazon	175,000	6.3	6577	28.0	290	900
Paraná	15,000	2.8	4000	5.3	38.3	80
Orinoco	36,000	1.0	2150	32.7	30.5	150
São Francisco	3,760	0.63	2900	6.0	—	6
Magdalena	6,800	0.26	1316	26.5	20	220
Uruguay	4,600	0.24	—	16.0	6(?)	11(?)

Sources: Depetris, P.J., *Limnol. Oceanogr.*, 21(5), 736, 1976; Ducharne, D., Informe tecnico de biologia pesquera (Limnologia). *Projeto para El Desarollo de La Pesca Continental.* Inderena, FAO, Publ. 4DP/Col.71552/4, Bogota, Colombia, 1975; Furch, K., Water chemistry of the Amazon Basin: The distribution of chemical elements among freshwaters, in: Sioli, H. (Ed.), *The Amazon: Limnology and Landscape Ecology of a Mighty Tropical River and Its Basin*, Dr. W. Junk Publishers, Dordrecht, the Netherlands, pp. 167–200, 1984; Milliman, J.D. and Meade, R.H., *J. Geol.*, 91, 1, 1983; Meybeck, M., *Hydrol. Sci. Bull.*, 21, 265, 1976; Paolini, J. et al., Hydrochemistry of the Orinoco and Caroní Rivers, in: Degens, E.T., Kempe, S., and Soliman, H. (Eds.), *Transport of Carbon and Minerals in Major World Rivers, Part II*, Mitt. Geol. Paläont. Inst. Univ. Hamburg, Vol. 55, pp. 223–236, 1983; Paredes, J. et al., São Francisco River: Hydrological studies in the dammed lake of Sobradinho, in: Degens, E.T., Kempe, S., and Soliman, H. (Eds.), *Transport of Carbon and Minerals in Major World Rivers, Part 2*, Mitt. Geol-Paläont. Inst. Univ. Hamburg, SCOPE/UNEP Sonderband, Vol. 55, pp. 193–202, 1983; Depetris, P.J. and Paolini, J.E., Biogeochemical Aspects of South American Rivers: The Paraná and the Orinoco, in: Degens, E.T., Kempe, S., and Richey, J.E., (eds.), *Biogeochemistry of Major World Rivers*, SCOPE, Wiley, New York, pp. 105–125, 1991.
TDS M.T.R, total dissolved solid mass transport rate; TSS M.T.R, total suspended solid mass transport rate.

TABLE 4.6

Average Solute Concentrations of Major African Rivers

River	Precipitation (mm)	Runoff (mm)	R/P (%)	TDS (mg/L)	TSS (mg/L)	Transport 10⁶ t		TSS/TDS
						TDS	TSS	
Zaire	1520	338	22	28	37	36.6	48	1.39
Niger	1140	124	11	67	127	14.0	25.4	1.86
Nile	510	47	9	318	54	11.8	2	0.18
Senegal	650	48	7	42	196	0.4	1.9	2.44
Orange	380	15	4	140	57	1.6	0.7	0.44
Zambezi	1020	157	15	113	90	25.2	20	0.80
Gambia	1100	219	20	17	19.5	0.08	0.09	1.10

Source: Martins and Probst (1991) to Martins, O. and Probst, J.L., Biogeochemistry of major African rivers: Carbon and mineral transport, in Degens, E.T., Kempe, S., and Rickey, J.E., Eds., *Biogeochemistry of Major World Rivers*, SCOPE 42, John Wiley & Sons, Chichester, U.K., pp. 127–154, 355pp, 1991.

the size of populations in important rivers and large rivers of European river basins. Tables 4.5 and 4.6 show the general features of South American and African rivers, respectively.

4.6 River Fisheries

Welcomme (1979, 1985, 1990) discussed the importance of river fisheries and the application of a modeling procedure and mathematical analysis that followed two methods to manage the fishery:

1. A range of mathematical analysis describing the dynamics of populations that depends on the stocks and their characteristics (Petrere 1996)
2. The determination of the magnitude of the fisheries' resources related to the environment by using parameters that reflect the morphometry of the system and their trophic characteristics

Both approaches were applied to African rivers and South American rivers (Petrere 1983, Bayley 1991). The fisheries in the large rivers of the African and South American continents are important ecological and economic resources and clear examples of the importance of hydrological, ecological, geomorphological, and biological processes in promoting a large and varied fish biomass (Oldani et al. 1992). The fish communities in these large rivers are very complex, and Figure 4.3 shows the number of species of fishes present in different river systems plotted in relation to the basin areas.

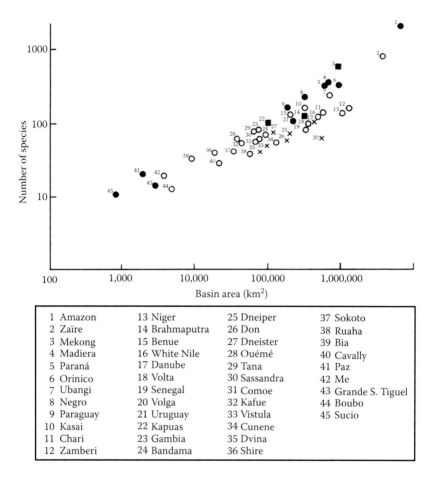

1 Amazon	13 Niger	25 Dneiper	37 Sokoto
2 Zaïre	14 Brahmaputra	26 Don	38 Ruaha
3 Mekong	15 Benue	27 Dneister	39 Bia
4 Madiera	16 White Nile	28 Ouémé	40 Cavally
5 Paraná	17 Danube	29 Tana	41 Paz
6 Orinico	18 Volta	30 Sassandra	42 Me
7 Ubangi	19 Senegal	31 Comoe	43 Grande S. Tiguel
8 Negro	20 Volga	32 Kafue	44 Boubo
9 Paraguay	21 Uruguay	33 Vistula	45 Sucio
10 Kasai	22 Kapuas	34 Cunene	
11 Chari	23 Gambia	35 Dvina	
12 Zamberi	24 Bandama	36 Shire	

FIGURE 4.3
Number of species of fishes in different watersheds plotted in relation to the areas. (From Welcomme, R.L., *Fisheries Ecology of Floodplain Rivers*, Longmans, London, U.K., 371pp., 1979.)

There are vast basins with several different geomorphological characteristics. The biogeographical factors indicate that groups of species are located in different regions of the system. For example, Welcomme (1979) describes several habitats of river–floodplain systems consisting of main *channels, tributaries, streams, flooded grasslands, lagoons and depressions, lakes, semipermanent channels in dry season, large larges, floodplain pools also in dry season, and backwaters connected to the main channel in dry season.*

Abujamra et al. (2009) described similarly varied habitat range for fish species of the high Paraná floodplains (Revista de Ictiología 1999).

According to Lowe McConnell (1999) the "river–floodplain" systems in tropical latitudes provide a great heterogeneity of habitats and high biodiversity.

The characteristics of flood pulses and alternating dry and flood periods are fundamental for the integrity of the ecosystem and high fish diversity. The availability of flood with several resource and diverse quality of food is another source of this biodiversity (Salo et al. 1986). For example, for the high Parana river system, Hahn et al. (2004) described 5 species of herbivores, 12 species of insectivores, 10 species of detritivores, 5 species of omnivores, 20 species of piscivores, and 2 species of planktivores.

Since there are several differences in the floodplain morphology and characteristics of large rivers, including the biology and ecology of fish population, a certain degree of dissimilarities in terms of species diversity, physiology, and composition of communities occur. The differences in the way that nutrients and energy are transferred throughout the food chain in the backwaters and the floodplain point out possible different approaches in the management of the fisheries in large rivers.

Despite the differences in the energy flow, Welcomme (1979) considers similarities in the ecology of fish communities of the large rivers. These are provided by the large fish biomass and fish diversity along the main axis and in the tributaries. Floodplains and large river watersheds are submitted to changes due to human impacts. The agricultural exploitation of the flood-plain, flood control structures, excess fisheries that deplete stocks and biodiversity, introduction of exotic species, and introduction of barrages upstream in the tributaries that affect the floodplain are the main problems (Bonetto et al. 1989, Bechara et al. 1996, 1999).

4.7 Small Creeks and Streams

Small creeks and streams play a relevant role in the watershed due to their function of collecting organic material for transportation downstream. Small, well-preserved creeks are also important ecosystems for maintaining a good water quality (Tundisi and Matsumura-Tundisi 2010) and a high fish fauna diversity with different feeding habitats and physiological niches (Buckup 1999, Charamaschi et al. 1999).

4.8 Ecological, Economical, and Social Importance of the Rivers: The Hydrosocial Cycle

Small streams and large rivers are fundamental components of the land-scape and watersheds in all continents and regions. The supply of protein (fisheries), the use of water for irrigation, and recreation are some of the most important ecological services promoted by small streams and creeks; in many towns around the world, these small streams are part of the urban landscape and elements of recreation and leisure activities. They also serve as a heterogeneous landscape component in an otherwise homogeneous urban environment. Large rivers, such as the Amazon, the Yang-Tse, the Nile, Zambezi, Paraná, and Ganges are important references since ancient times, and in many of these large rivers, floodplains have a central role in connecting watersheds and promoting gene flux and biomass and nutrient fluxes (Welcomme 1979). These large rivers are important sources of fisheries, have intensive uses of navigation and some of these rivers such as the Ganges are also sites for religious ceremonies and cultural practices. Therefore, these large rivers have, besides an economic value due to their many ecological services, cultural and social values. In all rivers of any dimension and discharge, a hydrosocial cycle is attached to them. This hydrosocial cycle is a strong link between the river as ecosystem and the human population (Tundisi 1990).

Contamination and pollution, deforestation and removal of the riparian forest, and vegetation and dam construction remove or disrupt this hydrosocial link and thereby cause a loss of important cultural and ecological services. River restoration from small streams to large rivers will be an important step in the future, since recovery of river ecosystem will be essential in many regions of the world. Therefore, a systemic approach to research and management of rivers has to be considered and this involves an important and fundamental degree of conceptualization of the relationship between the river and their watershed (Boom et al. 1992, Bormann and Likens 1979).

4.9 Human Impacts on River Ecosystems

The rivers are affected by a number of human activities that are developed in the watersheds. These are as follows:

- Wastewater disposal (nontreated or treated insufficiently) from domestic and industrial effluents
- Increase of suspended material from erosion in the watersheds (deforestation, road construction, and agricultural activities)
- Dam construction and operation (Ruggles and Watt 1975, Petts 1984, Petts 1990, Bonetto et al. 1989)
- Excess fisheries
- Introduction of exotic species of fishes and other organism (mollusks and macrophytes)

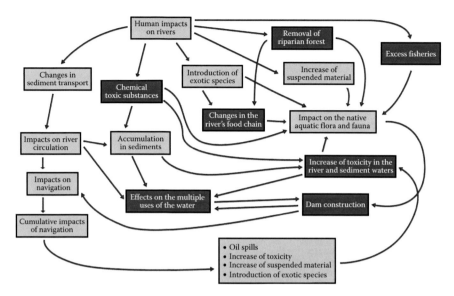

FIGURE 4.4
(See color insert.) Human impacts on rivers systems: a synthesis.

- Discharge of chemical toxic substances by point or nonpoint sources
- Navigation and construction of waterways
- Disposal of treated or partially treated wastewater (will inevitably discharge nutrients, nitrogen, and phosphorus, which will enhance the eutrophication)
- Deforestation and removal of the riparian forest and vegetation (Van der Hoek 1987)

These activities (see Figure 4.4) change the river flow, produce areas of lower current and sedimentation, accumulate residues and toxic substances, and disrupt the hydrosocial cycle by impacting the river's ecosystem services.

References

Abujamra, F.; Agostinho, A.A.; and Hahn, N.S. 2009. Effects of flood regime on the body condition of fish of different trophic guilds in the Upper Paraná floodplain. *Braz. J. Biol.* 69(Suppl. 2):469–479.

Agostinho, A.A.; Gomes, L.C.; and Pelicice, F.M. 2007. *Ecologia e Manejo de Recursos Pesqueiros em Reservatórios do Brasil*. Eduem, Maringá, Brazil, 501 pp.

Allan, J.D. 1995. *Stream Ecology. Structure and Function of Running Waters*. Chapman & Hall, London, U.K., 388 pp.

Amoros, C. et al. 1987. A method for applied ecological studies of fluvial hydrosystems. *Regul. Rivers* 1:17–36.

Armour, C.L.; Duff, D.A.; and Elmore, W. 1991. The effects of liverstock grazing on riparian and stream ecosystems. *Fisheries* 16:7–11.

Barila, T.Y.; Williams, R.D.; and Stauffer, J.R. 1981. The influence of stream order and selected stream-bed parameters on fish diversity in Raystown Branch, Susquehanna River Drainage, Pennsylvania. *J. Appl. Ecol.* 18:125–131.

Bayley, P.B. 1991. The flood-pulse advantage and the restoration of river-floodplain systems. *Regul. Rivers* 6:75–86.

Bayley, P.B. and Li, H.W. 1992. Riverine. In: Calow, P. and Pettes, G.E. (Eds.), *The Rivers Handbook*, Vol. 1. Blackwell Scientific Publications, Oxford, U.K., pp. 251–281.

Bechara, J.A.; Domitrovic, H.A.; Flores Quintana, C.; Proux, J.P.; Jacabo, W.; and Gavilán, G. 1996. The effect of gás supersaturation of fish health below Yacyretá Dam (Paraná River, Argentina). In: Leclerc, M.; Capra, H.; Valentín, S.; Boudreault, A.; and Côté, Y. (Eds.), *Proceedings of the Second International Symposium of Habitat Hydraulics. Ecohydraulics 2000*, Vol. A. INRS-Eua, Québec, Canadá, pp. A3–A12.

Bechara, J.A.; Sanchez, S.; Roux, J.P.; Terraes, J.C.; and Quintana, C.F. 1999. Variaciones Del factor de condición relativo de La ictiofauna Del rio Paraná águas abajo de La represa de Yacyretá, Argentina. In: *Revista de Ictiología–Estudios dedicados a la Biología Pesquera em El área de Yacyretá*, Vol. 7. Número Especial 1999. Instituto de Ictiología del Nordeste Facultad de Ciencias Veterinarias; Universidad Nacional Del Nordeste, Corrientes, Argentina, pp. 75–90.

Bonetto, A.A.; Wais, J.R.; and Castello, H.P. 1989. The increasing damming of the Parana Basin and its effect on the lower reaches. *Regul. Rivers* 4:333–346.

Boon, P.J.; Calow, P.; and Petts, G.E. (Eds.). 1992. *River Conservation and Management*. John Wiley, Chichester, U.K.

Bormann, F.H. and Likens, G.E. (Eds.). 1979. *Pattern and Process in a Forested Ecosystem*. Springer-Verlag, New York.

Buckup, P.A. 1999. Sistemática e biogeografia de peixes de riacho. In: Caramaschi, E.P.; Mazzoni, R.; and Peres-Neto, P.R. (Eds.), *Ecologia de peixes de riachos*. Série Oecologia Brasiliensis, Vol. VI. PPRE-UFRJ, Rio de Janeiro, Brazil, pp. 91–130.

Callow, P. and Petts, G.E. (Eds.). 1992a. *The Rivers Handbook*, Vol. 1. Blackwell Scientific Publications, Oxford, U.K., 526pp.

Callow, P. and Petts, G.E. (Eds.). 1992b. *The Rivers Handbook*, Vol. 2. Blackwell Scientific Publications, Oxford, U.K., 523pp.

Chapman, D. (Ed.). 1992. *Water Quality Assessments*. UNEP, UNESCO, WHO. Chapman & Hall, London, U.K., 585pp.

Charamaschi, E.P.; Mazzoni, R.; and Peres Neto, P.R. 1999. *Ecologia de peixes de riachos*. Series Ecologia Brasiliensis, Vol. 6. Rio de Janeiro, Brazil.

Davies, B.R. and Walker, K.F. (Eds.). 1986. *The Ecology of River Systems*. Monographieae Biologicae, Vol. 60. Dr. W. Junk Publishers, Dordrecht, the Netherlands, 793pp.

Décamps, H.; Joachim, J.; and Lauga, J. 1987. The importance for birds of the riparian woodlands within the alluvial corridor of the River Garonne, S.W. France. *Regul. Rivers* 1:301–316.

Degens, E.T.; Kempe, S.; and Richey, J.E. (Eds.). 1991. *Biogeochemistry of Major World Rivers*. SCOPE/ICSU/UNEP, John Wiley & Sons, New York, 355pp.

Degens, E.T.; Kempe, S.; and Soliman, S. (Eds.). 1983. *Transport of Carbon and Minerals in Major World Rivers, Part 2*. Mitt. Geol-Paläont. Inst. Univ., Hamburg, SCOPE/UNEP Sonderband, 55, 535pp.

Depetris, P.J. 1976. Hydrochemistry of the Paraná River. *Limnol. Oceanogr.* 21(5):736–739.

Depetris, P.J. and Paolini, J.E. 1991. Biogeochemical aspects of South American rivers: The Paraná and the Orinoco. In: Degens, E.T.; Kempe, S.; and Richey, J.E. (Eds.), *Biogeochemistry of Major World Rivers*. SCOPE, Wiley, New York, pp. 105–125.

Drago, E.C. 1989. Morphological and hydrological characteristics of the floodplain ponds of the Middle Paraná River (Argentina). *Rev. Hydrobiol. Trop.* 22(3):183–190.

Drago, E.C.E. 1990. Geomorphology of large alluvial rivers: Lower Paraguay and Middle Parana. *Interciencia* 15(6):378–387.

Drago, E.C. and Amsler, M.L. 1981. Sedimentos suspendidos en el tramo medio del río Paraná: Variaciones temporales e influencia de los principales tributarios. *Rev. Asoc. Cienc. Nat. Litoral* 12:28–43.

Dumont, H.J. (Ed.). 2009. *The Nile: Origin, Environments, Limnology and Human Use*. Monographiae Biologicae. Springer, Berlin, Germany, 818pp.

Fisher, S.G. 1983. Successions in streams. In: Barnes, J.R. and Minshall, G.W. (Eds.), *Stream Ecology. Application and Testing of General Ecological Theory*. Plenum Press, New York, pp. 7–27.

Furch, K. 1984. Water chemistry of the Amazon Basin: The distribution of chemical elements among freshwaters. In: Sioli, H. (Ed.), *The Amazon: Limnology and Landscape Ecology of a Mighty Tropical River and Its Basin*. Dr. W. Junk Publishers, Dordrecht, the Netherlands, pp. 167–200.

Gibbs, R. 1967. The geochemistry of the Amazon River System, 1. The factors that control the salinity and the composition and concentration of the suspended solids. *Geol. Soc. Am. Bull.* 78:1203–1232.

Goulding, M. 1981. Man and fisheries on the Amazon frontier. In: Dumont, H.J. (Ed.), *Developments in Hydrobiology*, Vol. 4. Dr. W. Junk Publisher, The Hague, the Netherlands, 121pp.

Hahn, N.S.; Fugir, R.; and Andrian, I.F. 2004. Trophic ecology of fish assemblages. In: Thomaz, S.M.; Agostinho, A.A.; and Hahn, N.S. (Eds.), *The Upper Paraná River and Its Floodplain: Physical Aspects, Ecology and Conservation*. Backhuys Publishers, Leiden, the Netherlands, pp. 247–269.

Hawkes, H.A. 1975. River zonation and classification. In: Whiton, B.A. (Ed.), *River Ecology. Studies in Ecology*, Vol. 3. Blackwell Scientific Publication, Oxford, U.K., pp. 312–374, 725pp.

Horne, A. and Goldman, C. 1994. *Limnology*. Mc Graw Hill, New York, 576pp.

Illies, J. and Botosaneau, I. 1963. Problems et methods de la classification et la zonation ecologique des eaux continentales consideres surtant du. Point de view faunistique. *Milt. Int. Verein Theor. Angew. Limnol.* 12:1–57.

Junk, W.J.; Bayley, P.B.; and Sparks, R.E. 1989. The flood pulse concept in river-floodplain systems. *Can. Spec. Publ. Fish. Aquat. Sci.* 106:110–127.

Kalff, J. 2003. *Limnology*. Prentice Hall, Upper Saddle River, NJ, 592pp.

Kempe, S. 1979. Carbon in the freshwater cycle. In: Bolin, B.; Dejens, E.T.; Kempe, S.; and Fetner, P. (Eds.), *Global Carbon Cycle*. John Wiley, New York, pp. 317–342.

Kempe, S.; Pettine, M.; and Callwet, G. 1991. Biogeochemistry of European rivers. In: Degens, E.T.; Kempe, S.; and Richey, J.E. (Eds.), *Biogeochemistry of Major World Rivers*. SCOPE, John Wiley & Sons, New York, pp. 169–211, 355pp.

Kempe, S. and Richey, J.E. (Eds.). 1991. *Biogeochemistry of Major World Rivers*. SCOPE/ICSU/UNEP, John Wiley & Sons, New York, 355pp.

Legendre, L. and Demers, S. 1984. Towards dynamic biological oceanography and limnology. *Can. J. Fish. Aquat. Sci.* 41:2–19.

Lewis, W.M. Jr. and Saunders III, J.F. 1989. Concentration and transport of dissolved and suspended substances in the Orinoco River. *Biogeochemistry* 7:203–240.

Lowe McConnel, R.H. 1999. *Estudos Ecológicos das Comunidades de Peixes Tropicais.* Vazzoler, A.E.A.; Agostinho, A.A.; and Connimgham, P. (Translators). University of S. Paulo, S. Paulo, Brazil.

Macan, T.T. 1961. A review of running water studies. *Verh. Intern. Verein Theor. Angew. Limol. Stuttgant* 14:587–602.

Margalef, R. 1960. Ideas for a synthetic approach to the ecology of running waters. *Int. Revue Ges. Hydrobiol.* 45:133–153.

Margalef, R. 1990. Ecosistemas fluviales, transporte horizontal y dinamica sucesional evolutiva. *Interciencia* 15(6):334–336.

Margalef, R. 1997. Our biosphere. In: Kinne, O. (Ed.), *Excellence in Ecology.* Ecology Institute, Germany.

Merona, B. 1990. Amazon fisheries: General characteristics based on two case-studies. *Interciencia* 15(6):461–475.

Meybeck, M. 1976. Total minerals dissolved transport by world major rivers. *Hydrol. Sci. Bull.* 21:265–284.

Meybeck, M. 1982. Carbon, nitrogen and phosphorus transport by world rivers. *Am. J. Sci.* 282:401–450.

Milliman, J.D. and Meade, R.H. 1983. Worldwide delivery of river sediments to the oceans. *J. Geol.* 91:1–21.

Newbold, J.D. 1992. Cycles and spirals of nutrients. In: Callow, P. and Petts, G.E. (Eds.), *Rivers Handbook,* Vol. 1. Blackwell Scientific Publications, Oxford, U.K., pp. 379–408, 526pp.

Oldani, N.O.; Iwaszikiw, J.M.; Padín, O.H.; and Otaegui, A. 1992. Fluctuaciones de la abundancia de peces en el Alto Paraná (Corrientes, Argentina). Serie Técnico-Cientíca. Publicaciones de La Comisión Administradora Del Río Uruguay, Vol. 1, pp. 43–55.

Paolini, J. 1990. Carbono Organico disuelto y particulado en grandes rios de la América Del Sur. *Interciencia* 15(6):358–366.

Paolini, J.; Herrera, R.; and Németh, A. 1983. Hydrochemistry of the Orinoco and Caroní Rivers. In: Degens, E.T.; Kempe, S.; and Soliman, H. (Eds.), *Transport of Carbon and Minerals in Major World Rivers, Part. II.* Mitt. Geol-Paläont. Inst. Univ., Hamburg, Vol. 55, pp. 223–236.

Paredes, J.; Paim, A.J.; Da Costa-Doria, E.M.; and Rocha, W.L. 1983. São Francisco River: Hydrological studies in the dammed lake of Sobradinho. In: Degens, E.T.; Kempe, S.; and Soliman, H. (Eds.), *Transport of Carbon and Minerals in Major World Rivers, Part 2.* Mitt. Geol-Paläont. Inst. Univ., Hamburg, SCOPE/UNEP Sonderband, Vol. 55, pp. 193–202.

Petrere, M. 1983. Yield per recruit of the tambaqui Colossoma Macropomum Cuvier in the Amazon state. *Brazil Journ. Fish Biol.* 22:133–144.

Petrere, M. Jr. 1996. Fisheries in large tropical reservoirs in South America. *Lakes Reservoirs: Res. Manage.* (Carlton South) 2:111–133.

Petts, G.E. 1984. *Impounded Rivers.* Wiley, Chichester, U.K., 326pp.

Petts, G.E. 1990. Regulation of large rivers: Problems and possibilities for environmentally-sound River development in South America. *Interciencia* 15(6):388–395.

Petts, G.E. and Amoros, C. 1996. *Fluvial Hydrosystems*. Chapman & Hall, London, U.K., 322pp.

Revista de Ictiología. 1999. Estudios dedicados a la Biología Pesquera em El área de Yacyretá, Vol. 7. Némero Especial 1999. Instituto de Ictiología del Nordeste Facultad de Ciencias Veterinarias; Universidad Nacional Del Nordeste, Corrientes, Argentina.

Richey, J.E. 1982. The Amazon River system: A biogeochemical model. In: Degens, E.T. (Ed.), *Transport of Carbon and Minerals in Major World Rivers, Part 1*. Mitt. Geol-Paläont. Inst. Univ., Hamburg, SCOPE/UNEP Sonderband, Vol. 58, pp. 356–378.

Richey, J.E.; Brock, J.T.; Naiman, R.J.; Wissmar, R.C.; and Tallard, R.F. 1980. Organic carbon: Oxidation and transport in the Amazon River. *Science* 207:1348–1351.

Richey, J.E.; Meade, R.H.; Salati, E.; Devol, A.H.; Nordin, C.F.; and dos Santos, U. 1986. Water discharge and suspended sediment concentrations in the Amazon River: 1982–1984. *Water Resour. Res.* 22:756–764.

Richey, J.E.; Victoria, R.L.; Salati, E.; and Forsberg, B.R. 1991. The biogeochemistry of a major river system: The Amazon case study. In: Degens, E.T.; Kempe, S.; and Richey, J.E. (Eds.), *Biogeochemistry of Major World Rivers*. SCOPE/ICSU/UNEP, John Wiley & Sons, New York, pp. 57–74, 355pp.

Ruggles, C.P. and Watt, W.D. 1975. Ecological changes due to hydroelectric development on the St. John River. *J. Fish. Res. Board Can.* 32:161–170.

Salo, J.; Kalliola, R.; Hakkinen, I.; Makinen, Y.; Niemela, P.; Puhakka, M.; and Coley, P.D. 1986. River dynamics and the diversity of Amazon lowland forest. *Nature* 322:254–258.

Sioli, H. 1975. Tropical river. The Amazon. In: Wnitton, B.A. (Ed.), *River Ecology*. Blackwell, London, U.K., pp. 461–491.

Spitzy, A. and Leenheer, J. 1991. Dissolved organic carbon in rivers. In: Degens, E.T.; Kempe, S.; and Richey, J.E. (Eds.), *Biogeochemistry of Major World Rivers*. SCOPE/ICSU/UNEP. John Wiley & Sons, New York, pp. 213–232, 355pp.

Talling, J.F.; Sinuda, F.; Taha, O.E.; and Sobhy, E.M.H. 2009. Phytoplankton: Composition, and productivity. In: Dumont, H.J. (Ed.), *The Nile: Origin, Environment, Limnology and Human Use*. Monographicae Biologicae. Springer, Berlin, Germany, pp. 431–462, 818pp.

Tundisi, J.G. 1990. Ecology and development. Perspective for a better society. In: Kawanabe, H.; Ogushi, T.; and Higashi, M. (Eds.), *Physiology and Ecology Japan*. Vol. 27, special number, pp. 93–130, 205pp.

Tundisi, J.G. and Matsumura-Tundisi, T. 2008. *Limnologia*. Oficina de Textos, São Paulo, Brazil, 632pp.

Tundisi, J.G. and Matsumura-Tundisi, T. 2010. Impacto of changes in the forest code in Brazil on the water resources. *Biota Neotropica* 10(4):67–76.

Tundisi, J.G.; Saraiva, A.; Matsumura Tundisi, T.; and Campanholi, F. How many more dams in the Amazon? (in preparation).

Van der Hoek, D. 1987. The input of nutrients from arable lands on nutrient poor grassland and their impact on the hydrological aspects of nature management. *Ekologia* 6:313–323.

Vannote, R.L.; Minshall, G.W.; Cummins, K.W.; Sedell, J.R.; and Crushing, C.E. 1980. The river continuum concept. *Can. J. Fish. Aquat. Sci.* 37:130–137.

Ward, J.V. 1989. The four dimensional nature of lotic ecosystems. *J. N. Am. Benthol. Soc.* 8:2–8.

Ward, J.V. and Stanford, J.A. 1983. The serial discontinuity concept of lotic ecosystems. In: Fontaine, T.D. and Bartell, S.M. (Eds.), *Dynamics of Lotic Ecosystems.* Ann Arbor Science, Ann Arbor, MI, pp. 29–42.

Welcomme, R.L. 1979. *Fisheries Ecology of Floodplain Rivers.* Longmans, London, U.K., 371pp.

Welcomme, R.L. 1985. River fisheries, FAO Fisheries Department. Publication number 262, Summary, Report.

Welcomme, R.L. 1990. Status of fisheries in South American Rivers. *Interciencia* 15(6):337–345.

5

Estuaries and Coastal Lagoons as Ecosystems

5.1 Introduction

An estuary according to Ketchum (1951) "is defined as a body of water in which the river water mixes with and is measurably diluted by sea water." Or according to Pritchard (1967) "an estuary is a semi enclosed coastal body of water which has free connection with the open sea and within which sea water is measurably diluted by fresh water derived from land drainage."

A coastal lagoon is defined as a shallow lake connected directly with the sea or through a river or outlet. Lagoons are separated from the ocean by offshore bars or marine islands. Coastal lagoons may have a free connection with the sea that changes with the tide or are inundated by seawater at irregular intervals. Kjerfve (1994) defines a coastal lagoon as a "shallow coastal water body separated from the ocean by a barrier, connected at least intermittently to the ocean by one or more restricted inlets and usually oriented coastal parallel." In estuaries and coastal lagoons, there is a salinity gradient. As seawater enters the estuary, freshwater from rivers or land drainage is also discharged into the estuary. Salinity variations and salinity gradients therefore are common features in estuaries with a varying degree depending upon the tide and the freshwater volume that enters the estuary. The patterns of dilution of the seawater entering the estuary varies from estuary to estuary: they depend on the tidal amplitude, the extent of evaporation of the water, the volume of freshwater, and the number of tributaries that contribute to the estuary (Mc Lusky 1989). Estuaries can be classified as positive, negative, or neutral. In a positive estuary, freshwater runoff is greater than evaporation and the estuary has a vertical salinity gradient, with seawater entering at the bottom of the estuary and flowing upward. In a negative estuary, evaporation exceeds the freshwater entering the estuary. Therefore, the salinity of these estuaries increases toward the sea. If the freshwater input into the estuary equals the evaporation, a static salinity regime occurs and the estuary is denoted as a neutral one.

Coastal lagoons may be highly stratified, well mixed, and predominantly brackish or hypersaline, depending on the various inputs of freshwater and marine water and outputs such as evaporation. In some semiarid coastal regions, hypersaline coastal lagoons occur.

5.2 Classification and Zonation of Estuaries

Figure 5.1 shows the several types of estuary shaped by geomorphology processes.

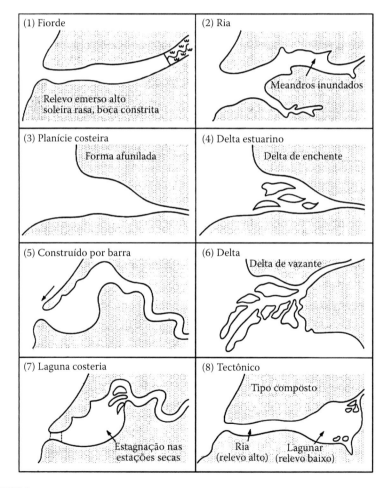

FIGURE 5.1
Estuarine types and physiographic characteristics. (Modified from Miranda, L.B. et al., *Princípios de Oceanografia física de estuários*, EDUSP, São Paulo, Brazil, 414pp., 2002.)

Estuaries can be classified according to the geomorphological type or the salinity stratification. The *first classification* on a geomorphological basis presents five different types of estuaries:

1. Coastal plain
2. Bar estuaries
3. Rias
4. Delta estuary
5. Tectonic estuaries

These estuaries are dependent upon sedimentation or tectonic processes or originate from glaciers such as fords.

The classification of estuaries taking into account the salinity stratification is useful for establishing the characteristics of the estuaries of the coastal plain geomorphology. Following are four types of classifications:

1. Saline intrusion estuaries Type A, with moderate or strong vertical stratification
2. Moderately or partially stratified estuary Type B
3. Vertically well mixed and laterally stratified Type C
4. Completely well mixed estuary Type D, with the absence of halocline

The transition between the different types is dependent upon the freshwater input, the geomorphology, the current velocity, and the depth of the estuary.

Estuaries therefore present a degree of horizontal and vertical salinity gradient that is variable; this gradient is fundamental for the complexity of the physicochemical processes and for the distribution of organisms.

Generally, the zonation of the estuary reflects the balance of freshwater and saline water entering the estuarine environment. A sequence that is common to occur is as follows:

1. *Head*—predominant river currents and freshwater, limited salt intrusion in this region, salinity <5%
2. *Upper reaches*—zone of mixing of freshwater and saline water, salinity 5%–18%
3. *Middle reaches*—tides influence currents, mud deposits with some sandy deposition at deeper regions, salinity 18%–25%
4. *Lower reaches*—currents strongly influenced by tides, sand deposits with some mud deposits in protected bays or enclosures with low circulation, salinity 25%–35%
5. *Mouth*—the region near the coastal water, salinity >30%, currents from tidal origin predominant

The estuary and coastal lagoons are dynamic systems and the gradient of environmental conditions may suffer strong variations in temporal and spatial scales due to the peculiarity of each estuarine or coastal lagoon system (latitude, longitude, morphometry, freshwater input, and tidal height).

5.3 Environmental Factors

5.3.1 Tides

The tide is the most important factor that controls the basic environmental characteristics of estuaries and coastal lagoons.

5.3.2 Waves

In some estuaries and shallow coastal lagoons, wave action may be an important factor stirring up sediment, changing and altering the morphometry of the ecosystem. Tidal bores—a wave of translation can occur with waves greater than 4 m (Emery et al. 1957).

5.3.3 Currents

Tides are the chief cause of current in estuaries. Tidal height, river discharge, and current velocity are related depending on the degree of closure of bays and the estuary morphometry. Excessive evaporation causes slow currents.

5.3.4 Temperature

This factor is dependent on the temperature of the seawater entering the estuary, the river water temperature, and the solar heating that, together with low velocity winds, produces a high temperature at the surface with a thermal gradient. Tundisi and Matsumura-Tundisi (2001) reported values of 36°C of surface temperature at the mangrove backwaters in the Laguna's Region of Cananéia (25° South Lat.) in Brazil. In fjord-type estuaries in northern latitudes, which lack complete exchange with the sea, the overlying brackish on freshwater act like the glass of a greenhouse heating up the subsurface water (Emery et al. 1957). In some northern latitudes, the formation of ice sheets during winter is an important factor.

5.3.5 Salinity

Salinity is an important forcing function in estuaries and coastal lagoons. Estuaries and coastal lagoons may be classified as marine, brackish, or hypersaline. This can change with the season. The salinity gradient and the changes during a tidal cycle or during periods of high rainfall may have a strong effect on the salinity variation of the estuary, with effects on the survivorship, reproduction, and distribution of organisms (Bassindale 1943).

The composition of the estuarine or coastal lagoon water can differ from the seawater to the freshwater entering their ecosystems. The effects of freshwater and evaporation combined, leads to differential composition of salts as already mentioned. In estuaries, in general, there is a higher ratio of carbonate and sulfate to chloride and of calcium to sodium than in seawater due the composition of the runoff water. These changing ratios may be an important factor in the distribution and composition of aquatic flora and fauna of estuaries.

5.3.6 Oxygen, Carbon Dioxide, pH, and Biogeochemical Cycles

The oxygen content, the CO_2 and pH are dependent upon the characteristics of the freshwaters and the seawaters entering the estuary. Besides this, the dynamics of the nictimeral cycles of tides and runoff and activities of the organisms interfere with the dissolved oxygen, carbon dioxide, and pH. The amount of organic matter in estuaries with origin, for example, from the mangrove vegetation surrounding the estuary, decreases the oxygen concentration of the estuarine water. In some backwaters anoxic bottom water is found due to accumulation of remains of decomposing mangrove leaves (Tundisi and Matsumura-Tundisi 2001). Changes in the pH during the diurnal cycle also occur. Increasing runoff dilutes alkalinity. Respiration of bottom organisms increases CO_2 and decreases pH in the shallow regions of the estuary where there is reduced circulation.

Nutrient cycles are, in general, similar to these of shallow coastal water but the runoff from land drainage may yield a difference in the cycles. Streams and freshwater tributaries draining to the estuary contribute to the phosphorus and nitrogen cycles. Anoxic water where the circulation is low and the concentration of organic matter high in the bottom water is a major source of dissolved inorganic phosphate to the surface or near surface waters. Many organisms such as bottom dwelling fishes, mollusks, or crustaceans can also contribute to changes in the classical pattern of the biogeochemical cycles due to bioturbation of the sediment (Emery et al. 1957, Mc Lusky 1989).

The discharge of suspended material to the estuary is another input to the biogeochemical cycles of estuaries and coastal lagoons. Being regions of high primary productivity and with a biomass of fishes, aquatic plants, macrophytes, and planktonic organisms, estuaries produce considerable amount of particulate organic matter that by decomposition are source of inputs of nutrients (Figure 5.2 and Table 5.1).

Estuaries and coastal lagoons accumulate large concentrations of inorganic and organic nutrients in particulate and dissolved forms. If there is a

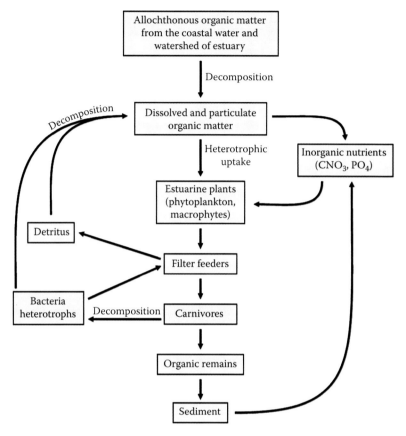

FIGURE 5.2
Cycle of organic matter in an estuary.

TABLE 5.1

Concentration of Organic Carbon in Natural Waters

Concentration (mg/L) of Organic Carbon	River	Estuary	Coastal Sea	Open Sea Surface	Deep	Sewage
Dissolved (DOC)	10–20 (50)	1–5 (20)	1–5 (20)	1–1.5	0.5–0.8	100
Particulate (POC)	5–10	0.5–5	0.1–1.0	0.01–1.0	0.003–0.01	200
Total	15–30 (60)	1–10 (25)	1–6 (21)	1–2.5	0.5–0.8	300

Source: After Head, P.C., Organic processes in estuaries, in: Burton, J.D. and Liss, P.S. (Eds.), *Estuarine Chemistry*, Academic Press, London, U.K., pp. 54–91, 1976.

Figures in brackets represent extreme values.

strong and intensive land use in the watershed of an estuary, these concentrations increase approximately proportional to the land use contribution. In estuaries with high degree of urbanization and discharge of nontreated domestic wastewater, eutrophication can occur impairing estuary waters for several uses such as recreation, tourism, or aquaculture. Examples of eutrophication of estuaries are found in the Baltic Sea. The Thames Estuary in London (but the eutrophication is controlled better now) and many estuaries in tropical or subtropical coastal regions (see impacts on estuaries) have high concentrations of nutrients that are interfering with nitrogen, phosphorus, and carbon cycles. Figure 5.2 shows the cycle of organic matter in a typical estuary. This cycle can be enhanced, accelerated, or strongly modified when mangrove vegetation surrounds the estuary and becomes a source of inputs of carbon, nitrogen, phosphorus, and humus substances as mentioned earlier (Tundisi 1969, Lacerda et al. 2008).

Coastal lagoons are generally more complex systems than estuaries as regard the environmental factors and the connection with the sea. As described by Esteves et al. (2008), coastal lagoons are strongly affected by the geophysical characteristics that interfere with their hydrological balance. These geophysical features are utilized to identify different types of coastal lagoons: lentic microtidal (permanent connection with the sea) or lentic nontidal (without a permanent connection with the sea). The second criterion of classification considers the lagoon origin. In general, the neotropical coastal lagoons were originated from the flooding of coastal lowland areas due to the rise in the sea level during the late Quaternary (Caldas et al. 2006). Lagoons are formed by the deposition of sediments and change of barriers due to tidal action; coastal lagoons present a vast heterogeneity of morphometric, geomorphological, and ecological dynamics creating a large scale of ecological gradients and microhabitats, enhancing aquatic and terrestrial biodiversity.

5.4 Aquatic Biota of the Estuaries

Since the estuaries and coastal lagoons present environmental gradients, the distribution, zonation, diversity, and colonization of the estuary by the aquatic flora and fauna depends on these gradients. The survival of organisms in an estuary or a coastal lagoon is determined by the wide range of the fluctuations that occur in environmental factors such as salinity, current velocity, turbidity, dissolved oxygen concentration, and nutrient cycles.

The survival and success of colonization of organisms in an estuary therefore is depending on the interaction of several environmental factors such as the ones described previously (Day 1951, Emery et al. 1957,

Castanares et al. 1969, Tundisi 1970). The distribution of organisms in estuaries falls in the following categories:

1. *Oligohaline organisms*—This is the majority of freshwater organisms that can tolerate only freshwater with salinities up to 0%–5%.
2. *Estuarine organisms*—These are organisms that are truly estuarine and can tolerate salinity ranges between 5% and 18%.
3. *Euryhaline marine organisms*—These are salinity tolerant species of organisms living in the coastal water reaching the areas of estuarine salinity between 18% and 25%. Few of these organisms can survive at salinities of 5%.
4. *Stenohaline marine organism*—These are marine organisms with a limited range of distribution in the estuary, only in the mouth of the estuary at salinities of 25%.
5. *Migrating organisms*—These are animals that spend part of their life cycle in estuaries and migrate to freshwater rivers or the sea. (Salmon, *Salmo solar*, and eel, *Anguilla anguilla*).

Day (1951) classifies the fauna of estuaries as

1. Freshwater organisms
2. Stenohaline marine organisms
3. Euryhaline marine organisms
4. Estuarine organisms
5. Migratory organisms

Figure 5.3 shows a "reciprocal biological gradient" of species with optimum tolerance in freshwater and brackish saline water.

The estuary community is adapted to the variations in environmental factors in estuaries, which is a permanent feature of the estuarine environment. The organisms, therefore, are adapted to fluctuation. Euryhalinity is a physiological characteristic of many estuarine components of the community. There are several hypotheses to explain the origin and permanence of the estuarine flora and fauna. The wide tolerance of estuarine organisms to charging environmental conditions may explain their ability to establish in several estuaries of the world. Since the dispersal of estuarine community is difficult via the sea due to ecological competition with marine organisms, the colonization of estuaries by introduced species is in many instances made by ships. One of the most known episodes of colonization of introduced species is the introduction of *Venus mercenaria* from the east coast of the United States to Southampton estuary during the World War II (Raymont 1963).

The osmotic concentration is regulated by many organisms throughout special physiological adaptations. Tundisi and Matsumura-Tundisi (1968) describe how several species of planktonic copepods were distributed at

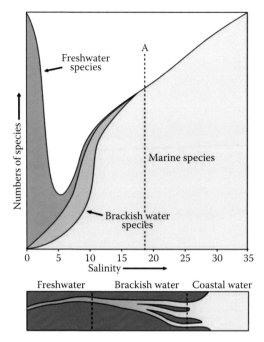

FIGURE 5.3
(See color insert.) Reciprocal biological gradient in estuaries. (Modified from Emery, K.O. and Stevenson, R.E., Estuaries and lagoons, in Hedgpeth, J.W., Ed., *Treatise on Marine Ecology and Paleoecology*, Geological Society of America, Boulder, CO, Memoir 67, Chapter 23, pp. 673–750, 1227pp, 1957.)

different regions of the Laguna Region of Cananeia due to their different tolerance to the salinity gradients. The effects of salinity on organisms may furthermore differ at the different stages of the life cycle.

The maintenance of the stock of planktonic and benthonic components of the estuarine community at regions of optimum or near optimum is dependent on strategies for migration, physiological adaptation, and tolerance to freshwater, brackish water, or saline water. As shown by Lance (1962), Tundisi and Matsumura-Tundisi (1968), and Tundisi (1970), vertical migration associated with salinity tolerances is one of the mechanisms that maintain the population near their best environmental conditions for survival and tolerance.

In estuaries, the benthic community plays an important role from the point of view of energy transfer and food chain. Shallow estuaries with fast currents near the bottom are regions of spatial heterogeneity, which is enhancing diversity of species. The benthic fauna of estuaries is composed of polychaetes, shrimps, and crabs, living near the bottom sediments. The rich organic matter of the sediment is very important as food source for these invertebrates. Many bird species also are estuarine dwellers and therefore contribute to the recycling of organic matter and energy transfer (Horne and Goldman 1994).

In coastal lagoons, salinity fluctuation is an important factor and can be a driving force in the biodiversity patterns and fluctuation. If the spatial and temporal effects of salinity are discontinuously distributed in coastal lagoons, salinity can promote regional biodiversity due to diversification of the species compositions. Biodiversity in coastal lagoons is dependent on size, volume, and morphometry of these ecosystems, as well as the extent of environmental gradients (Esteves et al. 2008).

The varying nature of coastal lagoons that is due to tidal, wind stress river input, precipitation and evaporation balance, and salinity influences make these systems unique compared with island freshwater ecosystems or adjacent coastal or marine waters (Caliman et al. 2010).

Despite this variable characteristic coastal lagoons can have endemic species. This is probably due to their environmental and ecological gradients. These gradients promote the isolation of the populations character displacement and act as a driving force for evolutionary divergence (Esteves et al. 2008).

5.5 Freshwater Inflow into Estuaries

The freshwater inflows to estuaries are important sources of nutrients and interfere with the salinity gradient, the faunal distribution, and the primary production. River-dominated estuaries have low salinity ranges and tend to have higher concentrations of nutrients. In contrast, lagoons influenced by coastal water have higher salinities and lower nutrient concentrations. (Palmer et al. 2011). Species diversity in estuaries in the Texas coast increased with increasing salinity according to Palmer et al. (2011). For the lagoons region and estuary of Cananéia, similar results are obtained by Teixeira et al. (1969). Tundisi and Matsumura-Tundisi (2001) demonstrated that freshwater inflow regulated primary production of phytoplankton, distribution of planktonic species, and concentration of dissolved and particulate material.

In general, duration and frequency of the freshwater inflows in the estuary affects salinity, sediment conditions, and particulate material distribution and interferes with species composition, nutrient recycling, sustainability, habitat changes, diversity, and biomass (Palmer et al. 2011).

5.6 Primary Production in Estuaries and Coastal Lagoons

In shallow estuaries and coastal lagoons, primary productivity is sustained by a biomass of aquatic macrophytes, periphyton, and phytoplankton. The nutrient input from the continental watersheds supplies almost continuously phosphorus, nitrogen, and other elements to the primary producers. Internal sources also contribute—from the sediment stock of nutrients and the decomposition processes in the water and sediment. The annual range of total primary production for coastal lagoons is within 200–400 g C/m² year. This range is comparable to that for estuaries and coastal waters of medium productivity (Nixon 1982). In lagoons where phytoplankton production is reduced, primary production is sustained by benthic algae and sea grasses. Tundisi (1969) found a value of 1 g C/m² day similar to that given by Nixon (1982) for coastal lagoons; with the contribution of sea grasses and microphytobenthos, the primary production at this region can however be much higher. Table 5.2 shows the primary production (net primary production and total production) in estuaries and coastal lagoons as compared with other marine systems (Knoppers 1994).

Estuaries and coastal lagoons are efficient ecotone systems between land and sea, modifying organic and inorganic matter, recycling elements and nutrients, and accumulating organic matter and biomass.

The permanent pulse of freshwater and coastal waters into the estuaries and in coastal lagoons connected to the sea is an enrichment system that enhances productivity and carbon fixation by primary producers. It also promotes, periodically, biogeochemical changes in the system by supplying or removing elements, nutrients, and metals.

Because of the high turnover of organic matter and the permanent input of nutrients from the continental waters, estuaries are ecosystems with high productivity. Development of fisheries, aquaculture, and harvest of crustaceans and mollusks in estuaries supports the local economy. Tourism and recreation are of course also sources of income for the inhabitants in estuaries regions. On the other land, estuaries accumulate impacts from the continental watersheds due to extensive land use, discharge of nontreated domestic wastewater, and industrial output. Estuaries can therefore be subject to eutrophication, which can reduce the ecosystem services of estuaries.

TABLE 5.2

Net Areal Primary Production and Total Production of Coastal Lagoons in Comparison to Other Marine Systems of the World

System	Area (10⁻⁶ km²)	Net Production (g C/m² year)	Total Production (10⁻¹² kg C/year)
Ocean	332	125	41.5
Upwelling	0.4	500	0.2
Shelf	33	183	4.1
Estuaries	1.4	300	0.4
Lagoons	0.3	300	0.1

Source: Knoppers, B., Aquatic primary production in coastal lagoons, in: Kjerfve, B. (Ed.), *Coastal Lagoons Processes*, Elsevier, Amsterdam, the Netherlands, Chapter 9, pp. 243–286, 577pp., 1994.

Estuaries are, however, also regions of high navigation activities due to the development of harbors and installations for receiving goods transported by sea. Apart from several impacts resulting from these activities estuaries can become the "entrance door" for the introduction of exotic species such as *Limnoperna fortunei* in the estuary of La Plata River, South America. This species has become a very successful invader of brackish and freshwater in South America and is today a big economical threat in the whole La Plata Basin due to its rapid geographic expansion and highly competitive efficiency (Oliveira et al. 2010).

5.7 Anthropogenic Impacts on Estuaries

In several estuaries in tropical and temperate regions, increasing development and human activities in the estuarine and continental watersheds contributed significantly to the alternations in environmental conditions. Nitrogen and phosphorus loads due to untreated wastewater have increased considerably in many estuaries that become eutrophic with changes in the structure and function of biotic communities. But there are many other impacts that have adversely affected estuaries and need a strong action to reverse the degradation of the water quality (Kennish 2004). A nonexhaustive list of impacts compiled from several sources and known from estuaries in several tropical and temperate regions is presented as follows:

- Pollution due to heavy industrialization and several human activities: excessive watershed development, toxic metal decompositions, organic contaminants, volatile organics.
- Habitat degradation resulting from shore changes, including the removal of mangrove vegetation in tropical regions.
- Nutrient load: phosphorus, nitrogen, and carbon.
- Changes in water quality due to urban, agricultural, and industrial runoff and increase of pesticides in water and sediments.
- Input of suspended material reducing the transparency affects primary production.
- Dredging of estuarine channels changes the structure and morphology of the estuary.
- Soil erosion in the continental watersheds due to agricultural development and industrial construction affects the estuarine fauna and flora.
- One large impact on some tropical estuaries is the development of aquaculture that increases suspended matter, produces inputs of nitrogen and phosphorus, and thereby accelerates the eutrophication.
- Tourist activities and recreation in estuaries can affect water quality and sediment chemistry significantly.

References

Bassindale, R. 1943. A comparison of the varying conditions of the tees and severn estuaries. *J. Anim. Ecol.* 12(1):1–10.

Caldas, L.H.D. et al. 2006. Geometry and evolution of Holocene transgressive and regressive barriers on the semi-arid coast of NE Brazil. *Geo-Mar. Lett.* 26(5):249–263.

Caliman, A. et al. 2010. Temporal coherence among tropical coastal lagoons: A search for patterns and mechanisms. *Braz. J. Biol.* 70(Suppl. 3):803–814.

Castanares, A. et al. 1969. *Coastal Lagoons: A Symposium.* Universidad Nacional Autónoma de México, de México, Mexico.

Day, J.H. 1951. The ecology of South African estuaries. Part 1. A review of estuarine conditions in general. *Trans. R. Soc. S. Afr.* 33:53–91.

Emery, K.O. and Stevenson, R.E. Estuaries and lagoons. In: Hedgpeth, J.W. (Ed.), *Treatise on Marine Ecology and Paleoecology.* Geological Society of America, Boulder, CO, Memoir 67, Chapter 23, pp. 673–750, 1227pp.

Esteves F.A. et al. 2008. Neotropical coastal lagoons: An appraisal of their biodiversity, functions threats and conservation management. *Braz. J. Biol.* 68(Suppl. 4):967–981.

Head, P.C. 1976. Organic processes in estuaries. In: Burton, J.D. and Liss, P.S. (Eds.), *Estuarine Chemistry.* Academic Press, London, U.K., pp. 54–91.

Horne, A.J. and Goldman, C.R. 1994. *Limnologia.* MC Graw Hill International Editions, New York, 576pp.

Kennish, M.C. (Ed.). 2004. *Estuarine Research, Monitoring and Resource Protection.* CRC Press, Boca Raton, FL, 297pp.

Ketchum, B.H. 1951. The flushing of tidal estuaries. *Sew Ind. Wastes* 23(2):198–209.

Kjerfve, B. (Ed.). 1994. *Coastal Lagoon Processes.* Elsevier, Amsterdam, the Netherlands, 577pp.

Knoppers, B. 1994. Aquatic primary production in coastal lagoons. In: Kjerfve, B. (Ed.), *Coastal Lagoons Processes.* Elsevier, Amsterdam, the Netherlands, Chapter 9, pp. 243–286, 577pp.

Lacerda, L.D.; Milisani, M.M.; Sera, D.; and Maria, P.L. 2008. Estimating the importance of natural and anthropogenic sources on N and P emission to estuaries along the Ceará State Coast NE Brazil. *Environ. Monit. Assess.* 141:149–164.

Lance, J. 1964. The salinity tolerance of some estuarine planktonic crustaceans. *Biol. Bull.* 127(1):108–118.

Mc Lusky, D.S. 1989. *The Estuarine Ecosystem.* Chapman & Hall, New York, 215pp.

Miranda, L.B. et al. 2002. *Princípios de Oceanografia física de estuários.* EDUSP, São Paulo, Brazil, 414pp.

Nixon, S.W. 1982. Nutrients, primary production and fisheries yields in coastal lagoons. *Oceanol. Acta* 5:357–371.

Oliveira, M.D. et al. 2010. Modeling the potential distribution of the invasive golden mussel *Limnoperna fortune* in the upper Paraguay River system using limnological variables. *Braz. J. Biol.* 70(Suppl. 3):831–840.

Palmer, A.T. et al. 2011. The role of freshwater inflow in lagoons, rivers and bays. *Hydrobiologia* 667:49–67.

Pritchard, D.W. 1967. What is an estuary: Physical view point. In: Lauff, G.H. (Ed.), *Estuaries.* American Association for the Advancement of Science, Washington, DC, pp. 3–5.

Raymont, J.E.G. 1963. *Plankton and Productivity in the Oceans.* Pergamon Press, Oxford, U.K., 660pp.

Teixeira, C.; Tundisi, J.G.; and Santoro, Y.L. 1969. Plankton studies in a mangrove environment. VI. Primary production, zooplankton, standing-stock and some environmental factors. *Int. Revue Ges. Hydrobiol.* (Alemanha) 53(2):289–301.

Tundisi, J.G. 1969. Plankton studies in a mangrove environment: Its biology and primary production. In: *Lagunas Costeras. Mem. Simp. Int. Lagunas Costeras,* Vol. 28/30. Mexico City, Mexico, UNAM, UNESCO, pp. 485–494.

Tundisi,. J.G. 1970. O plâncton estuarino. *Cont. Inst. Ocean. USP S. Paulo* 19:1–22. *Sec. Ocean. Biologica.*

Tundisi, J.G. and Matsumura-Tundisi, T. 1968. Plankton studies in a mangrove environment. V. Salinity tolerances of some planktonic crustaceans. *Bolm Inst. Oceanogr.* (São Paulo) 17(1):57–65.

Tundisi, J.G. and Matsumura-Tundisi, T. 2001. The lagoon region and estuary ecosystem of Cananeia, Brazil. In: Seeliger, U. and Kjerfve, B. (Eds.), *Ecological Studies—Coastal Marine Ecosystems in Latin America.* Springer-Verlag, Berlin, Germany, Vol. 144, pp. 119–130.

6

Wetlands

6.1 Introduction: The Importance of Wetlands

Wetlands are a major feature of the landscape in many parts of the world. They are among the most important ecosystems on Earth as they are the kidney of the landscape. Up to the mid-nineteenth century, wetlands were often given a sinister image. As a consequence of this view and the need for more agriculture land, wetlands have disappeared at alarming rates. They were drained and turned into agriculture land, which has resulted, particularly in the industrialized countries, in a massive pollution threat of pesticides and nutrient discharge by agricultural activities—a pollution that the wetlands and other natural ecosystems (ditches, trees, wind shelterbelts, ponds, forests, and so on) would otherwise eliminate. Today, the lack of many different small or big ecosystems as a pattern in the landscape has been a disaster for the abatement of the nonpoint pollution from agriculture. The various natural ecosystems are crucial for the health of the landscape.

6.2 Ecosystem Services by Wetlands

Today, we have realized the importance of wetlands and their role in landscape. The roles of wetlands are manifold and the following list of ecosystem services offered to the society of wetlands could easily be prolonged:

1. Production of rice
2. Grazing
3. Production of proteins in general
4. Flood control
5. Enhanced cycling of nutrients
6. Purification of drainage water and even wastewater

7. Buffer zones between aquatic and terrestrial ecosystems

8. Production of sphagnum

9. Peat mining

10. Conservation of high biodiversity

11. Bird reserve

12. Fishery

It is possible to calculate the economic value of wetlands by the services that they are offering the society. Costanza et al. (1997) found that the services offered by wetlands amounted to about 15,000 dollars/ha year. Wetlands produce a biomass corresponding to $18\,MJ/m^2$ year, but if we calculate the eco-exergy (ecological exergy or work capacity, which is biomass and information; see Jørgensen 2012 and Jørgensen and Fath 2011), it would be in the order of $45,000\,GJ/ha$ year (see the details of these calculation in Jørgensen 2010). With an energy (exergy) price of 1 EURO cent/MJ, the value would be 450,000 EURO/ha or about 550,000 dollars/ha, more than 30 times the value indicated by Costanza et al. (1997). The exergy value considers, however, the total content or work energy and indicates therefore the entire value of wetlands included all possible utilized or nonutilized services, while Costanza et al. only consider the values of the services that we actually are utilizing.

6.3 Types of Wetlands and Wetland Processes

There is a wide spectrum of different wetlands with different properties. The list of wetland types and their properties are presented in Table 6.1. For a more comprehensive overview, see Jørgensen (2009).

All these types of wetlands play a major role in nature and in different landscapes. They have slightly different properties, functions, and roles as can be seen in Table 6.1, but they are all very important for the health of the landscape and as buffer zones for other aquatic ecosystems; see the wetland at Brazilian reservoir in Figure 6.1.

Let us exemplify the importance of wetlands further by focusing on the importance of wetland type number two, mangrove wetlands:

1. Physical protection of coast (for instance against Tsunamis)

2. Very high productivity

3. Important nesting areas for mussels, shrimps, crayfish, lobsters, and fish

4. Can be used for the treatment of wastewater but it is very important not to overload them

TABLE 6.1

Types of Wetlands and Their Characteristics and Properties

Salt marshes: the water is saline. This type of wetland is important for the protection of coasts but is not considered in this volume about inland aquatic ecosystems

Mangroves: These are forested wetlands that dominate the intertidal zone of generally tropical coastal landscapes. They grow between land and sea from nearly freshwater to marine conditions. They have a very high biodiversity

Freshwater marshes: These are produced by flooding and dominated by herbaceous plants such as cattails and reeds. They tend to occur along rivers and lakes, producing wildlife and human food. They are very productive ecosystems

Forested wetland (or swamps): These are dominated by trees and shrubs and consist of spongy, soft, and wet soil. Often very important as temporary storage for floodwaters, thereby reducing the peak flows to downstream areas

Peatlands (or mires): These are characterized by deep accumulation of incompletely decomposed organic matter or peat. Eighty seven percent of the world's 4 mill km^2 peatlands can be found in the boreal and subarctic areas. The peat consists often of a deep anaerobic layer, often acidic and nutrient poor. *Fens* are peatlands under the influence of geogenous waters, while *bogs* are ombrogenous peatlands

Floodplains: These are riparian systems that interact with streams. The hydrology, soil characteristics, flora, fauna, and biogeochemistry of the floodplains are determined by the streams. The feedbacks from the floodplains influence the same list of characteristics in the streams

Tundra: These are characterized by constraints of low temperatures, strong winds, low precipitation, and repeated freezing and thawing on soils and water bodies. Permafrost is found in arctic tundra but is less frequently found in alpine tundra. Tundra is very sensitive to climate changes

Ponds: These are small shallow lakes and the properties are therefore similar to lakes, although they are more open and may therefore change faster, when they are exposed to changing forcing functions

Ditches: These are small nonflowing or slowly flowing streams. They have very variable water quality because of the relatively low volume, but they are often important buffer zones in the landscape

Riparian wetlands: These are characterized by intensive ecological interactions between aquatic and terrestrial ecosystems and are therefore very important buffer zones that are crucial for the maintenance of a good water quality in streams. They are also important as buffer zones for the water budget. Vegetation may be herbal or trees or a mixture of the two possibilities

Constructed wetlands, surface wetlands, and subsurface wetlands: These are constructed by application of ecological engineering principles to solve a pollution problem, either associated with wastewater or with drainage water. See also Figure 6.2

FIGURE 6.1
(See color insert.) Wetland at a Brazilian reservoir.

Similar lists can be made for all the wetland types, although the list of course is dependent on the local conditions, it should be underlined that wetlands are very frequently indispensable buffer zones for flooding and important buffer zones for pollution.

6.4 Constructed Wetlands

Due to the increasing interest to have wetlands in the landscape to cope with the agricultural pollutants and due to the possibilities to treat wastewater by wetlands, there has been an increasing interest to construct wetlands, which has resulted in the development of models to design constructed wetlands. The model facilitates construction of wetlands and is able to give a better design, because a model can consider all the influencing factors at the same time. It can also be used to assess the natural wetland area needed for a well-defined water treatment task, whatever it be, drainage water or wastewater. The design of a constructed or a natural wetland is obtained within 20 min

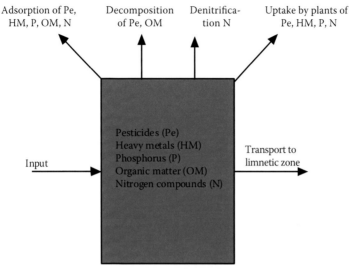

Adsorption of Pe, HM, P, OM, N

Decomposition of Pe, OM

Denitrification N

Uptake by plants of Pe, HM, P, N

Pesticides (Pe)
Heavy metals (HM)
Phosphorus (P)
Organic matter (OM)
Nitrogen compounds (N)

Input

Transport to limnetic zone

Transition zone (ecotone)

FIGURE 6.2
Processes that wetlands utilize to reduce the concentration of pesticides, heavy metals, phosphorus, organic matter, toxic organic compounds, and nitrogen compounds. Wetlands may be constructed or they may be ecotones (transition area between two ecosystems) between terrestrial and aquatic ecosystems for instance agriculture land and a lake. Other ecotones than wetlands, for instance small ponds and ditches, are also able to improve the water quality, but wetlands offer usually higher removal efficiency.

by use of the model, named SubWet, see Jørgensen and Fath (2011). Figure 6.2 gives an overview of some of the most important processes that wetlands, natural or constructed, are applying for purification of drainage and wastewater (see point 6 on the list of the services offered by wetlands):

1. Removal of nitrogen by nitrification and denitrification
2. Decomposition of organic matter including toxic organic compounds, for instance, pesticides
3. Adsorption of phosphorus compounds into the wetland soil
4. Uptake of nitrogen and phosphorus by plants
5. Adsorption of heavy metal ions into the soil
6. Uptake of toxic organic compounds and heavy metals by the plants

By use of models, it is possible to design a wetland that is able to treat the wastewater or drainage water to obtain a defined water quality with given concentrations of ammonium-N, nitrate-N, organic-N, total phosphorus, biological oxygen demand measured over a period of five days (BOD_5), pesticides, and several of the most toxic heavy metals. The model answers with other words the question: how can we design a wetland (area and flow pattern) to utilize the six processes listed before to obtain a defined water quality? Figure 6.3 shows a constructed wetland and Table 6.2 gives characteristic empirical efficiencies for wetlands with different plant density. They can be used if it is not possible to get the information needed to use SubWet or other models for the design. It is strongly recommended to apply a model whenever it is possible, because a model can consider many factors

FIGURE 6.3
(See color insert.) Constructed wetland in Tanzania. A high density of plants are selected in the design phase to ensure a high treatment efficiency. The selected plants species are local wetland plants.

TABLE 6.2

Typical Removal Efficiencies in $g/24\,h\,m^2$ of BOD_5 Removal, Nitrogen Removal, and Phosphorus Removal for a Constructed or Natural Wetland for Different Plant Densities

	Removal Efficiencies		
Plant Density t/ha	**BOD_5**	**Nitrogen**	**Phosphorus**
2	5.0	1.0	0.15
5	7.5	1.6	0.22
10	13	2.1	0.31
15	20	3.0	0.45
20	24	3.6	0.54
25	25	3.8	0.57

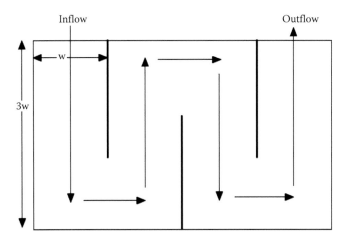

FIGURE 6.4
Flow pattern selected for a constructed wetland in Tanzania is shown. The pattern is selected in the design phase to ensure higher efficiency.

simultaneously and include also the interactions among the factors. Figure 6.4 shows how it is possible to increase the linear flow rate (m/24h) to obtain a better utilization of the edge and corners of the wetland.

For natural wetlands that are used for treatment of wastewater or drainage water, the same considerations as used for constructed wetlands are valid. SubWet can also be used here to design the wetland and find the area that is needed to treat properly a defined volume of wastewater. Figures 6.5 and 6.6 show the use of tundra in northern Canada for the treatment of waste.

The model SunWet has been applied to design 13 wetlands in Africa, including a design of a wetland for the teachers' college in Iringa, Tanzania. The wetland should treat the wastewater from 2500 students, totally 50 m³/24 h. Even before the construction of the wetland, from the design it was found that 625 m² of wetland was needed to obtain a BOD₅ reduction from about 125 mg/L to between 25 and 30 mg/L and a reduction of the total nitrogen content from about 42 to about 10 mg/L. The recommended horizontal flow rate was found to be 9 m/24 h with 2% suspended matter in the untreated wastewater. Furthermore, it was found that a proper utilization of the wetland requires that four pathways should be used. As the area foreseen for the wetland was 25 m × 25 m, it meant that the flow width should be 6.25 m. A good utilization of the wetland area requires that four pathways should be used. As the area foreseen for the wetland was 25 m × 25 m, it meant that the flow width should be 6.25 m. The results obtained by the use of the model before the construction and the observed values are compared in Table 6.3. The forcing functions applied for the model are listed in Table 6.4.

As seen from the application of the model for this case study, the validation of the model is very acceptable.

FIGURE 6.5
(See color insert.) Discharge of wastewater for the treatment on natural tundra in northern Canada.

FIGURE 6.6
(See color insert.) Typical tundra applied for wastewater treatment in northern Canada.

TABLE 6.3

Comparison of Model Results and Observed Values

	Model Results	Observed Results
Concentrations (treated water)		
BOD_5 (mg/L)	27	26
Nitrate-N (mg/L)	0.40	0.50
Ammonium-N (mg/L)	9.2	9.1
Organic N (mg/L)	0.45	0.50
Total N (mg/L)	10.1	10.1
Total P (mg/L)	5.2	5.0
Removal efficiencies (%)		
BOD_5	79	79
Nitrate-N	98	97
Ammonium-N	23	24
Organic-N	96	96
Total N	74	74
Total P	35	38

TABLE 6.4

Forcing Functions

Temperature: 30°C
Flow rate: $50\,m^3/24\,h$
BOD_5 untreated water: 123 mg/L
Nitrate-N untreated water 18 mg/L
Ammonium-N untreated wastewater 12 mg/L
Organic nitrogen-N untreated wastewater 12 mg/L

Total P untreated wastewater 8 mg/L
Particulate matter 20 mg/L

6.5 Natural Wetlands

The processes that explain the removal of pollutants by wetlands from wastewater or drainage water have been listed in Section 6.4. However, there are a number of processes of importance for the sustainability of natural wetland ecosystems. They are the basic for all services that wetlands offer our society, including the ability to remove pollutants. The most important of these crucial, additional processes are listed in Table 6.5.

TABLE 6.5

Important Processes for the Sustainability of Natural Wetlands, in Addition
to the Processes Listed Earlier and Summarized in Figure 6.1

Decomposition—mainly microbiologically of detritus/dead organic matter
A very substantial photosynthesis during spring and summer
A very comprehensive network of many species of insects, birds, plants, fish, and amphibians recycle the essential elements and recycle the energy
Both adaptation and selection are continuously operating and ensure a high biodiversity in these ecosystems

For natural wetlands that are not used for treatment of wastewater or drainage water, other issues are of interest for the environmental management
of the wetlands. The following relevant questions would be significant to
answer in this context, and the answers would require assessment of several
ecological questions:

1. Which species are characteristic for the wetland? Are the species
   normal for wetland in the region?
2. What is the spatial distribution of the species?
3. What is the diversity, included in the biodiversity? Does the diversity give the wetland a wide spectrum of buffer capacities?
4. Are the wetland density of plants and the wetland area able to
   "absorb" flooding and a tsunami?
5. Are the animals characteristic for the type of wetland in the region
   represented in the wetland or are some typical species that would be
   important for the functioning of the wetland missing?

To answer these five issues, it is recommended to get information about the
following ecological questions:

1. What is the plant density and area of the wetland? The answer is com-
   pulsory to find if a model—ecological as well as hydrological—should
   be applied but is of course also important if empirical data should be
   applied (see Table 8.2).
2. Which are the dominant plant and animal species? Would it be beneficial to harvest the plants and thereby remove a significant amount
   of nutrients annually?
3. Determine the species diversity. The simple number of species may
   be applied. The number of species is important not for the resilience
   or buffer capacity of the wetland but for the spectrum of these stability concepts. The higher the species diversity, the higher is the probability to meet a wider spectrum of disturbances including new and
   unexpected disturbances (see Jørgensen [2012]).

4. Determine the respiration/biomass ratio and the production/biomass ratio—two of E.P. Odum's attributes. The information would be able to tell use if the wetland is in balance and moreover, if the wetland is a sink or a source of carbon emission. Wetlands may often emit methane, which has about 23 times higher greenhouse effect than carbon dioxide. It is therefore important to determine the methane emission and carbon dioxide emission and sequestration. Although methane has a much stronger greenhouse effect than carbon dioxide, methane has a half-life time of about 7 years in the atmosphere, which means that the long-term effect of methane is less than that of carbon dioxide with the same greenhouse effect. Here, it is also advised to apply models that make it possible to consider both methane and carbon dioxide, their greenhouse effects, and their retention time in the atmosphere.

5. Find the most characteristic buffer capacity. The buffer capacities are correlated with eco-exergy, which is a thermodynamic indicator (see Chapter 18 about Ecological Indicators). It would therefore be possible indirectly to determine the buffer capacity by determination of the eco-exergy of the wetland. It is a matter of knowing the biomass density of the major plant and animal species; see Chapter 18 for the details of these calculations.

6. How is the distribution of the dominant plant species? Answer to this question makes it possible to obtain a spatial overview of the wetland and to distinguish between different uses of the wetlands to solve different problems or offer different ecosystem services.

7. Make at least a conceptual model of the most important energy flows in the wetland and determine the concentration of the species that are included in the flow diagram. The information will be useful to obtain a better answer to the questions mentioned in point 6.

In summary, wetlands are extremely important ecosystems in the landscape. Many wetlands—smaller or larger—as pattern in the landscape are very important for a healthy and sustainable nature.

References

Costanza, R. et al. 1997. The value of the world's ecosystem services and natural capital. *Nature* 387:252–260.

Jørgensen, S.E. 2009. *Introduction to Ecological Modelling*. WIT, Southampton, U.K., 210pp.

Jørgensen, S.E. 2010. Ecosystem services, sustainability and thermodynamic indicators. *Ecol. Complex.* 7:311–314.

Jørgensen, S.E. 2012. *Fundamentals of Systems Ecology*. Taylor & Francis Group, Boca Raton, FL, 320pp.

Jørgensen, S.E. and Fath, B. 2011. *Fundamentals of Ecological Modelling*, 4th edn. Application in Environmental Management and Research. Elsevier, Amsterdam, the Netherlands, 390pp.

Mtisch W.J.; Wang, N.; Deal, R.; Wu, X.; and Zuwerink, A. 2005. Using ecological indicators in a whole-ecosystem wetland experiment. In: Jørgensen, S.E.; Costanza, R.; and Fu-Liu, X. (Eds.), *Handbook of Ecological Indicators for the Assessment of Ecosystem Health*, 1st edn. CRC Press, Boca Raton, FL, Chapter 9, pp. 213–237, 440pp.

7

Tropical Freshwater Ecosystems

7.1 Tropical Lakes and Floodplains of the South American Continent

7.1.1 Large Floodplains of the South American Continent

The two major watersheds of the South American continent are the Amazon watershed and the La Plata watershed. Other watersheds of importance are the Orinoco, Magdalena, and São Francisco River basin (Figure 7.1).

The Amazon watershed has an area of 7 million km² representing 40% of the South American territory and shared by seven countries: Brazil, Colombia, Ecuador, Bolivia, Guiana, Peru, and Venezuela. The Amazon watershed has a large number of tributaries and associated floodplains, wetlands, inundated savannas, and inundated tropical forests.

The drainage system of the Amazon basin was consolidated in its geomorphology in the Pliocene when there was a definitive close up of the connection with the Pacific Ocean and the drainage and flushing took the direction of the Atlantic Ocean (Sioli 1984).

The concentration of sediments in the Amazon watershed was accumulated during the last 500 millions of years, being mostly composed of fine clay. The process of transport and sedimentation of this particulate material is continuous and according to Bigarella (1973), 13,500 tons of sediments/s is transported by the Amazon River.

The transport, deposition, and sedimentation of this particulate material in the Amazon basin plays an important ecological and geomorphological role, since new site or regions of accumulation of sediments are formed and destroyed during a seasonal cycle (Oltman 1970).

The interaction between terrestrial and aquatic ecosystems in the Amazon region is the main ecological factor. It is the primary forcing function that regulates the regional biodiversity, the evolutionary processes, and even the human uses of the ecosystem. Large internal deltas and extensive floodplain regions form a complex network of channels, rivers, shallow

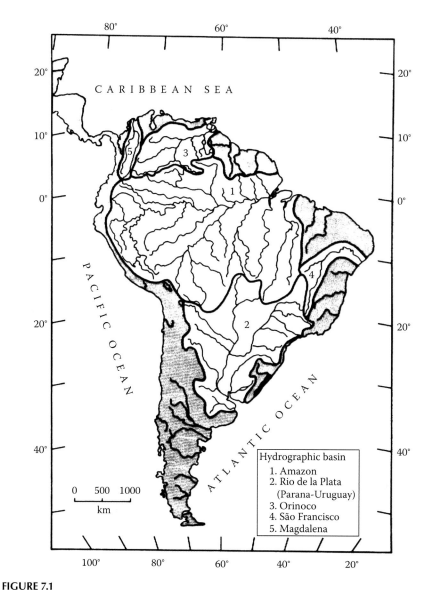

FIGURE 7.1
Major watersheds of the South American continent. (From Depetris, P.J. and Paolini, J.E.,
Biogeochemical aspects of South American Rivers: The Parana and the Orinoco, in: Degens,
E.T., Kempe, S., and Richey, J.E. (Eds.), *Biogeochemistry of Major World Rivers*, SCOPE 42, John
Wiley & Sons, New York, pp. 105–122, 355pp., 1991.)

lakes, and wetlands. The concept of the "pulse of inundation" was proposed by Junk and Weber (1997) in order to explain the periodical changes
in the biogeochemical cycles and in the biomass and species composition
of the terrestrial and aquatic biota. The chemical and physical environment
resulting from this periodic inundation promotes a chain of morphological,

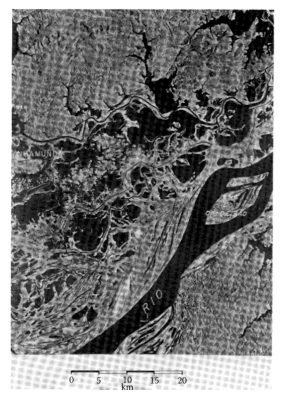

FIGURE 7.2
Floodplain morphology. (From Sioli, H., *The Amazon: Limnology and Landscape Ecology of a Mighty Tropical River and Its Basin*, Dr. W. Junk Publishers, Dordrecht, the Netherlands, p. 763, 1984.)

anatomical, physiological, geological, and ethological adaptations, with characteristic structures (Junk et al. 1989). Figures 7.2 through 7.4 shows the several types of fluvial dynamics associated with the Amazon River. Figure 7.3 also shows deposition and progressive erosions in inundated valleys (Sioli 1984).

Figures 7.5 and 7.6 show the water level fluctuation of the Amazon River. Floodplain lakes ("varzea lakes") are interconnected with temporary or permanent channels. The hydraulic energy removes and erodes river and creek banks, with effects on the biogeochemical cycles of carbon, nitrogen, and phosphorus or other elements. According to the inundation pulse, the interactions between terrestrial and aquatic ecosystems vary. Thereby, the response of the different species and groups of organisms, animals, and plants also vary.

Fisheries production is correlated positively with the inundation pulses and low water levels can produce large fish mortality, especially during the

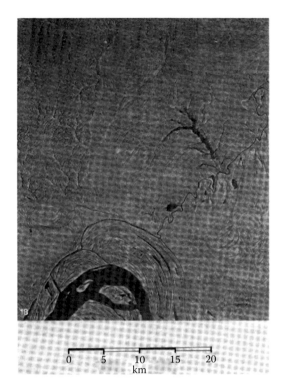

FIGURE 7.3
Progressive denudation of older alluvium. (From Sioli, H., *The Amazon: Limnology and Landscape Ecology of a Mighty Tropical River and Its Basin*, Dr. W. Junk Publishers, Dordrecht, the Netherlands, p. 763, 1984.)

"friagem" period (water turnover in lakes with anoxic hypolimnion). Junk (1997) described the various stages and types of inundation pulses that occur in different rivers (Table 7.1).

The rivers of the Amazon basin and the Amazon River are in the class of monomodal pulses, predictable and depending on the rainy season and dry periods.

The water from the inundation pulses accelerates the biogeochemical cycles, which is a fundamental forcing function for the different communities. The terrestrial vegetation of the inundated area is rapidly decomposing after the increase in water level adding nutrients that will stimulate the growth of phytoplankton, macrophytes, and periphytic algae (Junk and Bayley 2004). This aquatic vegetation is responsible for the increase of oxygen during daytime in the varzea lakes, since during the night the levels can vary between 0% and 20% saturation (Tundisi et al. 1984). Besides the nutrients transferred by the inundation water, the decomposition of the organic matter inputs contribute 80%–90% of the nutrient elements such as N, P, K, and Ca (Junk and Piedade 1997).

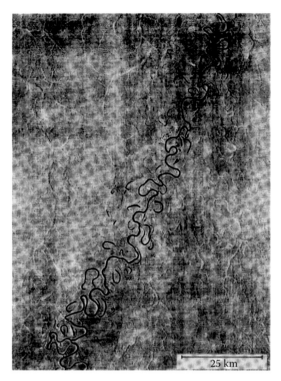

FIGURE 7.4
Section of Rio Juruá and its meanders and oxbow lakes. (From Sioli, H., *The Amazon: Limnology and Landscape Ecology of a Mighty Tropical River and Its Basin*, Dr. W. Junk Publishers, Dordrecht, the Netherlands, p. 763, 1984.)

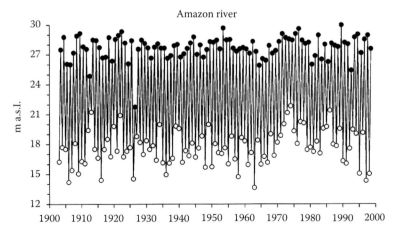

FIGURE 7.5
Water level of the Amazon River near Manaus. (From Junk, W.J. et al. (Eds.), *The Central Amazon Floodplain: Actual Use and Options for a Sustainable Management*, Backhuys Publishers, Leiden, the Netherlands, 584pp., 2000.)

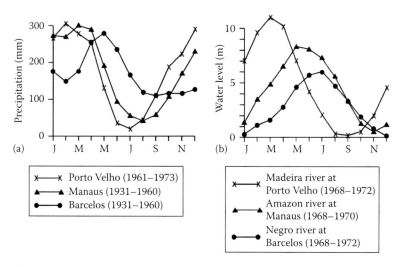

FIGURE 7.6
Monthly rainfall (a) and water level fluctuations (b) in the Madeira River at Porto Velho, the Amazon River at Manaus, and the Rio Negro River at Barcelos, corresponding to southern, central, and northern parts of the Amazon basin. (From Junk, W.J., Ecology of the varzea, floodplain of Amazonian white rivers, in: Sioli, H. (Ed.), *The Amazon: Limnology and Landscape Ecology of a Mighty Tropical River and Its Basin*, Dr. W. Junk Publishers, Dordrecht, the Netherlands, 1984.)

TABLE 7.1

Types of Inundation Pulses

Type of Inundation Pulse	Water Origin	River Type
Predictable monomodal	Precipitation/defrost	Large tropical rivers
Nonpredictable monomodal	Extensive rainfall multiannual not predictable	Rivers in semiarid regions
Polymodal predictable	Tidal influence	Rivers in tidal zones
Polymodal not predictable	Local heavy rains, defrost	Low order rivers, rivers in temperate regions

Source: From Junk, W.J., *The Central Amazon Floodplain: Ecology of a Pulsing System*, Ecological Studies, Vol. 126, Springer Verlag, Berlin, Germany, p. 526., 1997; Junk, W.J. and Nunes da Cunha, C., *Ecol. Eng.*, 24, 391, 2005.

The floodplain communities of primary producers have varied rates of primary production and biomass accumulation. Table 7.2 shows the primary production of the aquatic and terrestrial plant communities in the floodplain.

The inundation pulse, besides promoting the increase in primary production and the growth of the biomass, plays an important role in the

TABLE 7.2

Primary Production of Plant Communities in the Floodplain

	Biomass (in Dry Weight)	Primary Production (in ha/year)
Phytoplankton	10–40 kg	6 t
Periphyton	68 (?) kg	7.6 t
Aquatic annual plants	10–30 t	20–60 t
Territorial perennial plants and herbaceous aquatic vegetation	30–80 t	45–100 t
"Varzea" forests (inundation forests)	300–600 t	20–33 t

Source: Junk, W.J. and Piedade, M.T.F., Plan life in the floodplain with special reference to herbaceous plants, in: Junk, W.J. (Ed.), *The Central Amazon Floodplain: Ecology of a Pulsing System*, Ecological Studies, Vol. 126, Springer Verlag, Berlin, Germany, pp. 147–186, 1997.

aquatic animal communities. Aquatic invertebrates colonize wetlands with large populations. With short life cycles and asexual reproduction, their efficiency in occupying ecological niches in the floodplain is very high (Adis 1997).

Fishes, aquatic birds, amphibians, and mammals migrate from terrestrial habitats to the inundated areas, utilizing several mechanisms for feeding, taking advantage of the large and varied food offer. Physiological adaptations to these periods of inundation occur, for example, during the flood period, fishes accumulate fat that is used as energy source for extensive migrations and reproduction (Junk 1997, 2005).

Food chains show large changes of their connections and of their organization during and after the flooding period. The detritus food chain has an enormous quantitative and qualitative importance during these periods. Plant surfaces are sites for the growth of periphyton, bacteria, fungi, and protozoa, and this promotes short cuts in the food chain, making readily available the energy from the microbial loop to the fishes. The small creeks and streams in the inundated forests are collectors of detritus leaves, vegetation remains that are carried out downstream and are a source of inputs of carbon, nitrogen, and phosphorus to the mainstream. Bacteria and protozoa have an important role in the decomposition of this detritus as shown by Walker (1995).

The shallow permanent lakes of the inundated regions of the floodplain receive nutrients and suspended matter from the main river. These "varzea lakes" are thus enriched by the water entering them through natural channels. Several natural channels as a network transport these nutrients to the lakes, increasing the growth rate of plants and promoting large biomass growths. They are furthermore introducing a pulse that influences the food chain and the lakes water circulation. Tundisi et al. (1984) have shown how this pulse of river water entering the "varzea lakes" interferes with the mixing patterns of these aquatic ecosystems, changing their vertical structure. The enriched "varzea lakes" in the floodplain are nursery grounds for fishes and invertebrates. The high primary production of phytoplankton and periphyton and the high concentrations of detritus are rich sources of food for young fishes, invertebrate larvae, and shrimps. When the water level decreases, organic detritus, aquatic macrophytes, and planktonic organisms are drained to the rivers increasing the concentration of carbon, nitrogen, and phosphorus in the rivers. Figure 7.7 shows a simplified scheme of the main interactions between inundations, aquatic communities, and changing water levels. Table 7.3 describes the areas of floodplains

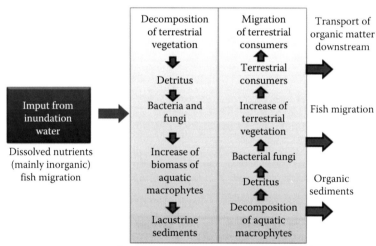

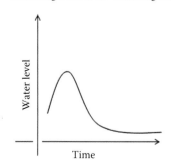

FIGURE 7.7
Simplified scheme of the main interaction of biological, chemical, and physical events that occur during the seasonal cycle inundation and low water level. (Modified and adapted from Junk, W.J., *The Central Amazon Floodplain. Ecology of a Pulsing System*, Ecological Studies, Vol. 126, Springer Verlag, Berlin, Germany, p. 526, 1997.)

TABLE 7.3

Areas of Wetlands and Floodplains in Tropical South America

Geographical Wetland Area	Extent (km² × 10³)	Hydrological Regime	Prevailing Vegetation
Along low order streams within Amazonian *term firme*	1000	Polymodal unpredictable	Forest
Andean foot zone	>100[ct]	Polymodal unpredictable	Forest
Border area Colombia/Venezuela/ Brazil	50	Polymodal unpredictable	Forest (Amazon Caatinga Rana), savanna[d]
Along the Amazon, and its bigger tributaries, mostly in			
Brazil	300[e]	Monomodal predictable	Forest, savanna
Eastern Bolivia	150	Monomodal predictable	Savanna, forest[f]
Pantanal	120	Monomodal predictable	Savanna, forest
Central Brazil ca.	100	Monomodal predictable	Savannas scattered in the carrado of the central Brazilian shield
Venezuelan llanos	80	Monomodal predictable	Savanna, forest
Bananal, Rio Araguaia, Brazil	65	Monomodal predictable	Savanna, forest[f]
South of Roraima			
West of Rio Branco and North of Rio Negro, Brazil Roraima lowlands	50	Monomodal predictable	Forest and shrubs, major portion of palms
Brazil and Rupununi lowlands, Guayana	33	Monomodal predictable	Savanna
Magdalena watershed in Colombia	20	Bimodal predictable	Forest, savanna[f]
A. *Uplands*			
High Andes	Unknown, but small	Monomodal	Saline swamps, salt pans, salt-tolerant savanna, or desert bogs
B. *Costal regions*			
Atlantic–Caribbean Coast	120	Daily bimodal[g]	Mangrove, tidal forest, and grassland
Amazon estuary	50	Daily bimodal[g]	Mangrove, tidal forest, and grassland

(continued)

TABLE 7.3 (continued)

Areas of Wetlands and Floodplains in Tropical South America

Geographical Wetland Area	Extent (km² × 10³)	Hydrological Regime	Prevailing Vegetation
Orinoco Delta	22	Daily bimodal[g]	Mangrove, tidal forest, and grassland
Magdalena Delta	5	Daily bimodal[g]	Mangrove, tidal forest, and grassland
Pacific Coast	?	Daily bimodal[g]	Mangrove, tidal forest, and grassland

Source: Junk, W.J. et al. (Eds.), *The Central Amazon Floodplain: Actual Use and Options for a Sustainable Management*, Backhuys Publishers, Leiden, the Netherlands, 584pp., 2000.

[a] Palm swamps *(marichales)* are not listed. Their total area is unknown. Probably not all of these wetlands arc truly permanently wet.

[b] Savannas usually include isolated tree groups and small forested areas.

[c] No better data available.

[d] In his vegetation map of die Venezuelan Amazon territory, reports on 23,000 km² of the Amazon Caatinga and Bana on Podzols and very small savanna patches (both llanos type and Amazonian savanna type).

[e] Estimate includes water surface. Water surface: forest: grassland = 1:1:1.

[f] Savanna: forest probably equal to 1:1.

[g] In part superimposed by monomodal flooding.

and wetlands in South America, and Figure 7.8 shows the distribution of floodplain, including wetlands, flooded savannas, and central wetlands in the South American Continent.

7.1.2 Biodiversity in the Amazon Region and Its Floodplains

According to Sioli (1984) and Margalef (1983, 1997), the Amazon region is considered an "active center of evolution" due to the dynamic character of the interactions between the aquatic and the terrestrial systems, and the fluctuating conditions between low and high water levels. As Junk (2005) emphasized, the inundation pulse integrates aquatic and terrestrial systems, with several dynamic exchanges between the hydrological cycle, the geomorphology, and the biota. Isolation of lakes and mixing during the seasonal cycle increase the gene flux and therefore are of high importance, quantitatively and qualitatively, for the adaptive radiation and the biodiversity. The spatial heterogeneity that promotes the biodiversity accordingly to Wilson (1998) can be measured in terms of large scales (thousands of kilometers) or small scales (meters or centimeters). These spatial scales are all represented in the Amazon region during high water and low water. Transition systems, such as flooded forests, temporary wetlands, and the "varzea" (the portion of land between lakes and during the low water level), provide changing ecological niches with several adaptations of animals and plants to these dry periods. The inundation periods and the transition regimes of water level disperse organisms and resistant eggs, promoting a

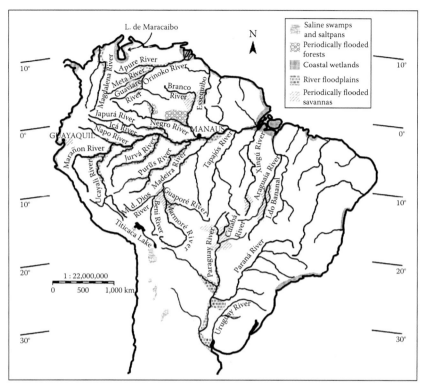

FIGURE 7.8
Distribution of floodplains, main rivers, and associated wetlands in the neotropics. (From Junk, W.J. et al. (Eds.), *The Central Amazon Floodplain: Actual Use and Options for a Sustainable Management*, Backhuys Publishers, Leiden, the Netherlands, 584pp., 2000.)

large-scale strategic process of survival adaptations for several plants and animals. The volume and diversity of organic detritus is very high in the Amazon. Therefore, the detritus food chain is of prime importance in all the floodplain aquatic ecosystems. The megadiversity of the Amazon biota is not only originated in the large Amazon River and its tributaries. Walker (1995) demonstrated the role of the small creeks, streams, and tributaries as providers of spatial heterogeneity and food niches that maintain a high biodiversity of bacteria, protozoa, mollusks, insects, and fishes. Figure 7.9 from Barthem and Goulding (2007) shows the extension of the seasonal inundated forests, the tidal forests, and the aquatic macrophyte growth that promotes a high biodiversity of habitats and substrates in a spatial scale of almost 2000 km. The three habitats exploited by fishes in the Amazon are the macrophyte stands, the open waters of the lakes during the inundation period, and the inundated forests. These forests offer an abundant and varied food diet for herbivore fishes consisting of fruits, leaves, vegetation remains, and seeds. This may explain how blackwater rivers that are periodically inundated but have low nutrient concentration and low primary

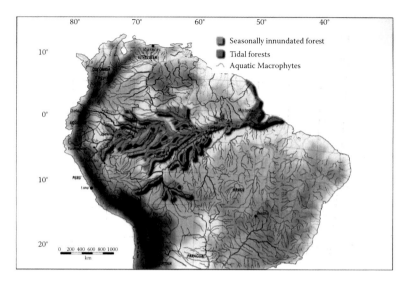

FIGURE 7.9
(See color insert.) Extension of the macrophyte growth in Amazon River and its tributaries. (From Barthem, R.C. and Goulding, M., *Um ecossistema inesperado. Amazônia revelada pela pesca*, ACA SCM, Lima, Peru, p. 241, 2007.)

production of aquatic plants maintain a large biomass of several species of herbivorous fishes (Goulding 1980, Barthem and Goulding 2007).

Several hypotheses were proposed in order to explain the origin of the Amazonian biodiversity. All these hypotheses give emphasis to the dynamic processes in the Amazonian ecosystem, consisting of a large-scale interaction of climatological, geomorphological, hydrological, and fluvial dynamics that promotes an opportunity for speciation and adaptive radiation of the biota. Several instances of isolation, barrier formation, barrier breakdown (due to the continuous alteration of geophysical structures) occur in the Amazon region with the seasonal cycle of inundation and low water level. According to Haffer (2008), these theories were developed around the barrier effects of rivers (temporary or permanent), *changes in the vegetation during periods of dryness and humidity, isolation of forest mosaics during dry periods*, and *large changes in spatial heterogeneity* (hypothesis of intermediate perturbation). Salo et al. (1986) called attention to the river dynamics and the diversity of Amazon lowlands as a function of changing spatial patterns of heterogeneity. Ward et al. (1999) emphasized the importance of ecotones and connecting in the floodplain as one of the main factors enhancing biodiversity.

According to Ayres (1986), the hypothesis that the river acts as barriers that triggered the speciation in the Amazon is the prevailing one. Preferences of habitat and diet are important factors that influence patters of speciation in the Amazon. The importance of streams in the wetland and floodplain as contributors to the biodiversity and influencing the community structure and diversity was stressed by Wantzen and Junk (2006). Streams and creeks in the floodplains and wetlands cover a wide area, and Junk and Furch (1985) estimated in central Amazonia a stream density of one stream per kilometer cross country. Their estimate is that 5% of the whole floodplain area is covered by streams. The environmental factors that potentially affect biodiversity of streams are substrate complexity, heterogeneity, substrate type and area, velocity, and altitude. Current and hydraulic conditions are important components of forcing functions that trigger the high biodiversity in the stream–wetland–floodplain complex. This is valid for benthic invertebrates and fish biodiversity (Junk 1997).

Seasonality in the water discharge of the streams, the pulsing systems of the streams, the pulsing systems of the streams, the gradient of water current sediment deposition, and suspended matter transportation are other factors that enhance biodiversity in the streams.

The flood pulse is a periodic disturbance factor and the level of disturbance that consists in flooding–drought periods depends on the duration and amplitude (Junk et al. 2000). The organisms are adapted to these changes. The speciation and gene flow along the river continuum and across the terrestrial aquatic systems is another factor that enhances biodiversity.

7.1.3 Fish Fauna of the Amazon Region and the Floodplains

The fish fauna of the Amazon was described by many authors such as Lowe McConnell (1984), Goulding (1980), and Goulding et al. (1988). These research efforts described the trophic relations and the diversity of fishes in blackwater rivers (Rio Negro). The water level fluctuation was pointed out as the main factor that influences the feeding behavior of the Amazon fishes. Food items described were insect larvae, crustaceans, mollusk, zooplankton, and algae.

These authors classified the fish fauna of the Amazon in three large groups as regards their good preferences:

1. Exclusively herbivorous fishes, in the adult stage of their life. Fruits and seeds are the main food items. Extensive fat reserves accumulate in these fishes during these feeding periods. Genuses: *Colossoma*, *Mylossoma*, *Myleris*, and *Brycon* are the main herbivore fishes.

2. Fishes that can feed on vegetation resources as fruits and seeds but that can also eat animals. These are species of *Serrasalmus* sp. ("piranha" fish) and *Pimelodus* sp.

3. Fishes that feed on fine detritus. Most important genera in this category are *Prochilodus*, *Semaprochilodus*, and *Curimatus*. These are fishes feeding in rivers with low nutrient concentrations. Goulding (1980) stressed the importance of the water level fluctuation for providing feeding niches for fishes. Seventy-five percent of the commercial fisheries are coming, according to this author, from the food chain developed in the inundated forests.

The physiology and the biochemistry of the Amazonian fishes were studied by Val et al. (1996) and Val and Almeida Val (1995). The high environmental variability of the Amazonian floodplains represented by hydrological fluctuations are two fundamental variables for the water temperature and the dissolved oxygen in the water.

About 2000 species, according to Menezes (1970), evolved in a highly variable environment therefore adjusting periodically their biochemical and physiological conditions to the fluctuations. As Walker and Henderson (1998) pointed out, the physiology of the Amazonian fishes can be considered an intermediate process between the ecology and the evolution of the fishes; this includes reproductive rates, life cycles, general metabolism, and behavior (Walker and Tyler 1984, Walker et al. 1991).

The survival at low concentrations of dissolved oxygen is one of the physiological characteristics that is most important for the Amazonian fishes: aerial respiration, metabolic depression, morphological adaptations, and general adaptations to low oxygen concentration are some of the physiological mechanisms that can meet the low oxygen concentration in water (Val et al. 1996).

The fluctuations of the dissolved oxygen concentration in Amazonian waters is dependent on the rate of photosynthesis by aquatic plants; the rate of decomposition of organic matter; and the circulation patterns in lakes, natural channels, and rivers (Tundisi et al. 1984). During the inundation period in some lakes, anoxy or hypoxy conditions that occur submit the fishes to a high physiological stresses. The tolerance to hypoxy depends on adaptive and physiological processes (Almeida Val et al. 1993). Several families of fishes are facultative air breathers while other families of fishes have special structures such as "lungs," diverticula in stomach and intestines that can contain reserve air (Val et al. 1996).

The concentrations of dissolved oxygen in the floodplains can vary from 120% saturation to as low as 0% saturation during 24 h. During anoxic periods, methane (CH_4) and hydrogen sulfide gas (H_2S) concentrations can rise very fast. Temperatures during daytime can be as high as 38°C at the surface of floodplain lakes. Such high temperatures occur mainly during dry periods at low water (Tundisi, personal observations).

7.1.4 Biogeochemical Cycles

The interactions of terrestrial and aquatic systems, the high transport of sediments and the high biomass and biodiversity play a qualitative and quantitative role in the floodplains. The high process rate due to high temperatures, the volume of detritus, and the transport by hydrofluvial mechanisms, have a significant influence on the cycles of carbon, phosphorus, and nitrogen. The hydrochemistry of the rivers depends on their origin. The classification in white waters, blackwaters, and clear waters originally proposed by Russel Wallace—Wallace (1953) and confirmed by Sioli (1951) determines the concentration of dissolved ions in the water, as shown in the Table 7.4.

TABLE 7.4

Concentration of the Main Elements and Nutrients in the Amazonian Waters

	Fundamental Elements (mg/L)				Nutrients			
	Na	Ca	Mg	Cl	SO_4	$PO_4\text{-}P$	$NO_2\text{-}N$	SiO_2
Andines rivers	2.0–3.0	1–2(23)	1.0–2.0	?	4.0–6.0	?	3.0–4.0	35
White waters	1.5–4.2	7.2–8.3	1.2–8.3	?	1–6.4	15	4.0–15.0	<9
Clear waters	1.0–2.0	<2	<1, 0.1–2.1	0–3	0	<1	<7.0–0.5	3.0–9.0
Blackwaters	0.55	<0.46	?	?	?	5.8	0.036	2.4

Source: Fittkau, E.J., *Verh. Int. Verein. Theor. Angew. Limnol.*, 15, 1092, 1964; Greisler e Schneide (1976); Oltmann (1966); Schmidt, G.W., *Amazoniana*, 2, 393, 1970; Schmidt, G.W., *Verh. Int. Verein. Limnol.*, 18, 613, 1972; Schmidt, G.W., *Amazoniana*, 4, 139, 1973a; Schmidt, G.W., *Amazoniana*, 4, 379, 1973b; Schmidt, G.W., *Amazoniana*, 5, 817, 1976; Schmidt, G.W., *Amazoniana*, 7, 335, 1982; Ungemach, H., *Amazoniana.*, 3, 175, 1972; Turcotte, P.R. and Harper, P.P., *Hydrobiologia*, 89, 141, 1982.

The floodplain lakes ("varzea lakes") are capacitors of biomass for the rivers. They have special mechanisms of influencing the biogeochemical cycles, due to their connection with the river during the inundation period and their isolation at low water (Mitamura et al. 1995). Pulses of phosphorus, nitrogen, and carbon with diurnal length time occur in these lakes (Mitamura et al. 1995). These diurnal pulses are common during the inundation period. Dissolved inorganic phosphorus, ammonia, and nitrate are assimilated by phytoplankton algae, macrophytes, and periphyton. Phosphorus, nitrogen, and carbon are released in the decomposition process of the high biomass in the lakes. The interaction of sediment and water with release and complex formation of phosphorus in the sediment depends on the oxidation–reduction conditions near the sediment–water interface. It may provide another "phosphorus pump" to the water (Tundisi et al. 1984). The biogeochemical cycles are regulated by the activity of the heterotrophic bacteria. The detritus and suspended sediments mineralized by these bacteria release high concentrations of phosphorus, nitrogen, and carbon at a fast rate due to the high temperature (Rai and Hill 1984a,b). The autochthonous organic matter produced during low water period in the lakes and mineralizing rapidly under the action of the heterotrophic bacteria is substituted by allochthonous organic matter originated from the inundation waters from the rivers. Thus, the carbon, nitrogen, and phosphorus transferred to higher trophic levels throughout the microbial loop have an origin in the lakes or in the inundation water during the seasonal cycle of inundation water in the floodplains (Junk and Nunes da Cunha 2005).

Fixation of carbon, nitrogen, and phosphorus by living organisms also play an important quantitative role in the biogeochemical cycles. The interaction of this living biomass with the biogeochemical cycles and the pools of carbon, phosphorus, and nitrogen is another fundamental process in the regulation of the cycles since the excreta are readily processed in the floodplain lakes (Araújo Lima et al. 1986). According to Richey et al. (1990), heterotrophic processes predominate in the metabolism of Amazonian waters where high heterotrophic activity was measured in blackwaters and clear waters. Besides high interaction of air and water,

TABLE 7.5

Describes the Processes of Nutrient Input, Nutrient Sinks, and Recycling
in the Floodplain Lakes and Wetlands

Sources	Recycling	Sinks
Input by advection (from rivers)	Decomposition by bacteria	Sedimentation loss due to water flushing
Desorption on the lakes	Zooplankton excretion	Denitrification
Input of phosphorus and nitrogen by rainfall	Interactions water–sediment	
N_2 fixation in the lakes	Macrophytes: assimilation of phosphorus and nitrogen and decomposition	
Surface drainage to the lakes	Phytoplankton: assimilation of phosphorus and nitrogen and decomposition	

Source: Tundisi, J.G. et al., *Hydrobiologia*, 108, 3, 1984; Junk, W.J., Aquatic plants of the Amazon system, in: Davies, B.R. and Walker, F.F. (Eds.), *The Ecology of River Systems*, W.J. Junk Publishers, Dordrecht, the Netherlands, pp. 319–337, 1986; Melack, J.M. and Fisher, T.R., *Archiv fur Hydrobiologie, Stuttgart*, 98(4), 422, 1983; Forsberg, B.R., *Ver. Int. Verein Limnol.*, 22, 1294, 1984; Zaret, T.M. et al., *Verh. Int. Verein. Limnol.*, 21, 721, 1981.

the concentration of dissolved oxygen depends of course on the decomposition of organic matter and photosynthesis by aquatic plants during the low water and inundation periods. Dynamics of biogeochemical cycles are shown in Table 7.5.

7.1.5 Impact of Human Activities on Neotropical Floodplains at the Amazon Basin

Several human activities affect the neotropical floodplains in the Amazon region. Agriculture; cattle and buffalo farming; civil construction such as roads, channels, and reservoirs; urbanization; solid waste deposits; and untreated sewage water (industrial and urban) are activities that impact the floodplain, reducing species diversity, changing the hydrological cycle and the pulse of inundation, interfering with migration of fishes, and changing the hydrosocial cycle (Junk 1993). One of the possible ways to develop the Amazonian region and the floodplains is to stimulate small-scale agriculture, low-density cattle ranching, and controlled fisheries. The combination of small-scale agriculture with ecotourism may provide a more adequate sustainable technology for exploitation of the floodplain and its biodiversity. The maintenance of spatial heterogeneity and habitats, the preservation of hydrological balance represented by the inundation period, and the maintenance of the hydrosocial cycles are fundamental priorities for further development of this region (Junk 2000).

7.1.6 High Paraná River Floodplain

The Paraná River basin is part of the large and complex La Plata basin, which is the second largest watershed in the South American continent.

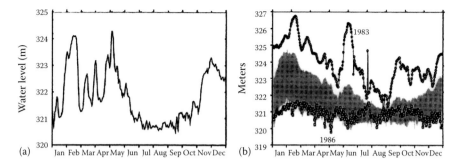

FIGURE 7.10

(a) Water level fluctuation in the Paraná floodplain during 1992 and (b) average water level from 1989 to 1997. (From Agostinho, A.A. et al., Biodiversity in the high Paraná floodplain, in: Gopal, B., Junk, W., and Davis, J.J. (Eds.), *Biodiversity in Wetlands: Assessment, Function and Conservation*, Vol. I, Backhuys Publishers, Leiden, the Netherlands, pp. 89–118, 353pp., 2000.)

The La Plata basin has an area of approximately 3,000,000 km² and a population of almost 150 million inhabitants. The Paraná River runs 4495 km in the south direction and discharges into the La Plata estuary after receiving water from the Paraguay and Uruguay rivers and other important tributaries. During the high water level in the inundation period fluctuates between 2 and 3 m; see (Figure 7.10). Considerable interannual variations occur in the water level in the Paraná River (Agostinho et al. 2000). It has a pattern of variations on the biodiversity, biogeochemical cycles, and species composition and niches. However, the Paraná Lima basin was very much affected by the dam construction in the river itself and in the tributaries such as Rio Grande, Rio Tietê, and Rio Paranapanema (Tundisi et al. 1981, 1988, Straskraba and Tundisi 1999, Tundisi and Matsumura-Tundisi 2008a,b, 2012).

The floodplain lakes during low water periods are subject to local processes such as turbulence induced by winds or inputs of suspended material, nutrients, and organic matter from the tributaries (Agostinho et al. 2000). High floods tend to homogenize river and floodplain habitats as it occurs for other floodplain systems.

Fluctuations from a stratified deeper lake system, during high water, to a homogenous nonstratified, shallow lake system, during low water, are a common feature. Low oxygen concentrations or even anoxic conditions occur when the water is at a high level and the lakes are deeper. The water from the Paraná River dilutes the water from the lakes and lagoons; therefore, the concentrations of phosphorus decrease when the flooding starts. This is partially explained by the phosphorus retention in the reservoirs upstream. Straskraba (1999) and Tundisi and Straskraba (1999) have shown that all the reservoirs in the cascade of reservoirs in the Tietê River retain phosphorus, turning this nutrient limiting for primary production. The contribution of autochthonous organic matter giving a peak in phosphorus and nutrients during low water levels is due to resuspension of sediments by wind action. This results in a higher primary productivity of phytoplankton in the lakes during low water (Thomaz et al. 1992, Agostinho et al. 1995).

Therefore, despite the similarities in the functioning of the floodplain ecosystem in the Amazon River and the Paraná River, two main differences should be pointed out: first, the spatial and temporal scale of events that is much larger in the Amazon floodplain and second, the presence of a great number of reservoirs in the Paraná River. These reservoirs regulate the biogeochemical cycles and the biodiversity patterns and interfere with the main forcing function that is the water level fluctuations.

7.1.7 Biodiversity in the Paraná Floodplain

The Paraná River and its tributaries are regulated by the construction of at least 50 large damns built up primarily for hydroelectricity productions. Today, the large reservoirs are used for multiple activities such as navigation, recreation, hydroelectricity, fisheries, and fish production in large scale (Tundisi and Matsumura-Tundisi 2008a,b). However, at each reservoir a region of floodplain is developed in the tail waters and at certain areas extensive floodplains are remained between reservoirs. The habitat heterogeneity and the water level fluctuations (despite the reservoir regulation, the water level fluctuations during dry and wet periods remain) are the main forcing functions that maintain a high diversity of phytoplankton, zooplankton, periphyton, and fishes. For phytoplankton a total of 300 taxa were recorded (Agostinho et al. 2000), with Chlorophyceae being the group of highest biodiversity. Changes in the algae diversity at different lagoons and channels were noted during periods of low water and high water (Train and Rodrigues 1998). Lowest density of phytoplankton occurred in the rivers Paraná, Ivinhema, and other tributaries (Thomaz et al. 1997). The zooplankton community of the Paraná floodplain is represented by 329 taxa, and the species richness varies with habitat. The densities of the zooplankton community are affected by water level fluctuations in the various floodplain habitats (Lansac-Tôha et al. 1997). The periphytic community is also affected by the water level fluctuations, with the greatest diversity occurring in the lagoons. The periphytic biomass is higher during water level (Rodrigues 1998).

For the zoobenthos, 80 invertebrate taxa were identified with microcrustacea, chaoboridae, and chironomidae, being the most important benthic invertebrate group (Takeda et al. 1997).

Macrophytes species were also determined in the studies of the high Paraná floodplain. The total number of species is 48, with 32 emergent, 3 floating leaved, 6 free floating, 5 rooted submerged, and 2 free submerged. *Panicum pionits* (Poaceae) and *Eichhornia azurea* are important species in the Paraná floodplain. *Salvinia auriculata*, *Pistia stratioides*, and *Utricularia* sp are commonly found in the stands of macrophytes (Souza et al. 1997).

As Agostinho et al. (2000) pointed out, the lagoons of the high Paraná floodplain are the richest environment in terms of species diversity of phytoplankton, zooplankton, benthic invertebrates, and periphyton. The highest abundance for phytoplankton, zooplankton, and aquatic macrophytes occurs also in these habitats. Changes in the biomass and diversity of species in the lagoons during low water and high water periods are due to better conditions for phytoplankton production and macrophyte growth at low water and increase in food resources during high water, especially detritus, enhancing the high diversity of rotifer, testate amoeba, and bacteria.

Therefore, the role of the lagoons and shallow lakes in the floodplains as capacitors of biomass has also been demonstrated by the studies on the high Paraná floodplain. The studies on fish community showed a fish fauna composed of 170 species. Six of these species were introduced from other basins. One hundred species were recorded from the main channel of the high Paraná River (Agostinho et al. 1994).

The floodplain offers a wide variety of refuges, spatial heterogeneity, and varied food supply. One hundred and three species were recorded in the lagoons and 101 in the channels that connect to the river. At least 13 species gained access to the upstream reaches of the floodplain after inundation of the Itaipu impoundment. This was due to the submersion of falls that were a barrier to fish migration. One hundred species were recorded in the main channel of the Paraná River (Agostinho et al. 1994, 1997). The upper tributaries of the Paraná River are nursery grounds and spawning areas for the migratory fishes of the basin. The fish species of the floodplain are subjected to high fluctuations of temperature, dissolved oxygen, especially in the lagoon environments. The main habitats of the fish fauna of the high Paraná floodplain are shown in Table 7.6.

Lowe-McConnell (1987) described the relative proportions of the different taxonomic orders in the ichthyofauna of the neotropical region. According to this author, 85% of the fish fauna belongs to the orders characiformes and siluriformes, with characiformes predominating slightly.

The fish fauna of the high Paraná floodplain as the fish fauna of the Amazonian floodplains is composed of many species of detritus feeders, herbivores, and piscivores that make use of the varied and abundant food resources resulting from the periodic inundation and isolation of the lakes and lagoons.

In the floodplain, species diversity and the density of each species are controlled by the periodic flood regime and its variations in duration, magnitude of flood, and variability of pulses. Therefore, the structure and functioning of the lakes, lagoons, channels, and small rivers and creeks in the floodplains are dependent upon the alternation of flood and drought periods (Agostinho et al. 2009).

Other components of the biological communities of the high Paraná floodplain depend upon this seasonal pulses. Amphibians, reptiles, birds, and mammals are adapted to the water level changes and the ecological diversity. Trophic dynamics, feeding, rhythm, physiological responses, and reproductive patterns are responses to the changes in the hydrological regime.

TABLE 7.6

Main Habitats of the Fish Fauna of the High Paraná Floodplain

Habitat	Characteristics	Conspicuous Species
Paraná River channel	Sandy or arenitic substrate, lotic	*Paulicea luetheni, Loricaria*, and adults of big fishes *Potamotrygon* sp, *Parodon tortuosus, Hemisorubim platyrhynchos, Leporinus elongatus, Schizodon altoparanae*
Larger tributaries of Paraná River	Meandering streams with sandy bottom (Ivinheima and Iguatemi), greater diversity, or rapid water and rocky bottom (Piquiri), lower diversity	Meandering streams: Doradidae, Ageneiosidae, *Schizodon, Hoplias, Rhaphiodon*. Auchenipteridae, *Pimelodus, Roeboides, Piaractus mesopotamicus, Pseudoplatystoma corruscans*
		Rapid water: *Leporinus amblirhynchus, Schizodon nasutus, Galeocharax knerii, Apareiodon, Myloplus*
Creeks	Variation in gradient, substrate, size, proportion riffles/pools, cover, and conservation of riparian vegetation	Small fishes: Cheirodontidae, Tetragonopterinae, small Pimelodidae, Loricaridae, Trichomycteridae
Marginal lagoons (floodplain)	Shallow; sand or mud bottom, abundant macrophytes, with diel stratification, low fish diversity	*Loricariichthys, Hoplosternum, Leporinus lacustris*, young of *Prochilodus* and other rheophilic species
Secondary channels (floodplain)	Semi-lenitic, drain the floodplain	*Lagoon* sp. plus *Trachydoras, Iheringichthys, Serrasalmus* sp.
Ephemeral pools	Form during water recession, drying up during different periods of the year, mud bottom, stressful environment	End of drying period: *Asfyanax bimaculatus, Cheirodon notomelas, Steindachnerina insculpta, Hoplias malabaricus*, and *Roeboides paranaensis*

Source: Agostinho, A.A. et al., Biodiversity in the high Paraná floodplain, in: Gopal, B., Junk, W., and Davis, J.J. (Eds.), *Biodiversity in Wetlands: Assessment, Function and Conservation*, Vol. I, Backhuys Publishers, Leiden, the Netherlands, pp. 89–118, 353pp., 2000.

7.1.8 Human Impacts on the Floodplain of the High Paraná River

Due to the large-scale reservoir construction in the upper reaches of the Paraná River and its tributaries, the first impact of the dam construction was the effect on the hydrological regime and the floodplain and river biodiversity. Each reservoir, however, develops connections with wetlands that are important for the maintenance of biodiversity and as a buffer system to reduce impacts on the reservoirs (Tundisi and Straskraba 1999, Tundisi et al. 2008). Nevertheless, the floodplain of the Paraná River is the only region of the floodplains of this large river still remaining. This remaining floodplain of approximately 200 km length is an important

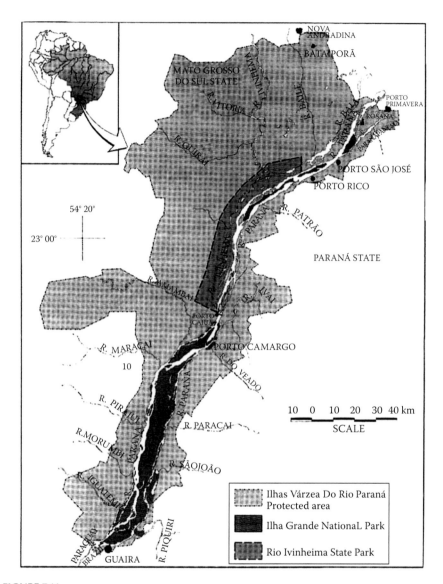

FIGURE 7.11
Remaining high Paraná floodplain. (From Agostinho, A.A. et al., Biodiversity in the high Paraná floodplain, in: Gopal, B., Junk, W., and Davis, J.J. (Eds.), *Biodiversity in Wetlands: Assessment, Function and Conservation*, Vol. I, Backhuys Publishers, Leiden, the Netherlands, pp. 89–118, 353pp., 2000.)

buffer for the entire high Paraná River (see Figure 7.11). It is a fish nursery ground and a general spawning region. The high biodiversity is still relatively untouched. In these regions, the ecohydrological processes are in full operation. The maintenance of this floodplain is thus fundamental. Cattle ranching, extensive agriculture, and fisheries are human activities that impact this floodplain (Agostinho et al. 2000). The area of environmental protection has approximately 2.500 km².

7.1.9 Colonization Patterns of the Reservoirs in the High Paraná River

The high Paraná floodplain and wetlands associated with the Paraná River and its tributaries were impacted by the construction of several large reservoirs, which changed the hydrological, the hydrosocial, and bioecological cycles (Tundisi et al. 1981, Tundisi and Straskraba 1999, Tundisi and Matsumura-Tundisi 2008a,b). These neotropical reservoirs were colonized immediately after the filling phase and transition from a lotic to a lentic environment took place. The filling phase that lasted from 10–100 days was relatively short (Agostinho et al. 1999). During the filling phase, rapid changes in water level, water temperature, and dissolved oxygen in the water occurred with several consequences on the dispersion of organisms and their distribution in the water column and in bays and embankments resulting from the new reservoir.

Tundisi (1994) compared these filling stages of the neotropical reservoirs to a chaotic state where several rearrangements of physical, chemical, and biological components changed almost daily. Straskraba (1999) considered this "trophic upsurge" during the filling stages of a reservoir as possibilities for growth of opportunistic species. The classical example of this colonization is the growth and development of a large biomass of *Cichla occelaris* (Tucunaré) in Amazonian reservoirs. After the trophic upsurge, a gradual colonization process occurred with changes in fish communities aquatic macrophytes, zooplankton, and benthos, as well as in phytoplankton (Matsumura-Tundisi et al. 1981, Tundisi and Straskraba 1999).

Agostinho et al. (1999) described that not only changes in biodiversity occurred but also changes in reproductive effort of the species, remaining in the reservoir. A general absence of fish species preadapted to colonize the pelagic zone of the reservoirs was observed (Agostinho et al. 2009). Fish diets changed from allochthonous resources such as detritus to autochthonous resources. Herbivores and zooplanktivorous fishes predominated after the stability has started. Changes in the primary producers also were conspicuous. Tundisi and Matsumura-Tundisi (2008a,b) described how the phytoplankton dominated by diatoms in the river was substituted by periphyton (due to the increase of the surfaces-indrowned vegetation) and a Chlorophyceae community.

The river zooplankton is mainly dominated by rotifers and this was changed toward a Cladocera–Copepoda dominance after the stability period started (Tundisi and Matsumura-Tundisi 2008a,b). Therefore, food chains changed completely when the system was stable. Changes in species diversity occurred after the transition period but Rocha et al. (2005) pointed out that with the reservoir aging the diversity can increase, too.

In general, the colonization of the reservoir depends on the existing biota of the pre-impoundment basin. This is especially observed for the fish fauna (Fernando and Holčík 1991), but it is also valid for other groups of organisms (Tundisi et al. 1993). One of the major impacts of the impoundment is the interference in the hydrological cycle reducing the water level fluctuation of the riverine environment and thereby introducing instability in the cycle that changes reproductive habitats of species, patterns of colonization and dispersion, interfering with biotic interactions and physical–biotic interactions (Agostinho et al. 1994). The changes in the environment caused by the impoundment have, as a consequence, several responses of the biota; see Figure 7.12. Some of these responses are very fast, others take more time, depending on the species and the capability to anticipate changes in their physiological and ecological behavior or in ecophysiological responses. Time and spatial scales, therefore, vary for different organisms.

This figure is representative of the large-scale changes induced by reservoir construction of the fish community, as an illustration of the overall changes in the aquatic biota of the floodplain. Variations in the diversity of species, food resources, abundance of organisms, trophic structure, size of organisms, and the organization of the communities are common features during the reservoir filling and after the stability process.

Besides the colonization patterns and the reorganization of the floodplain community after impoundment, the reservoir aging is another process that must be considered to understand the relationship watershed–reservoir and their continuous interactions.

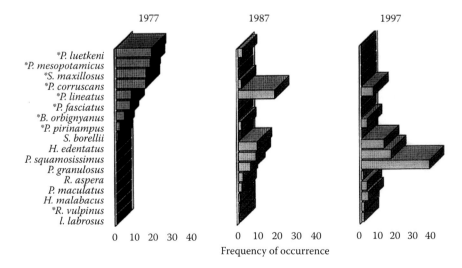

FIGURE 7.12
Changes in the fish species composition in the landings of commercial fishery at Itaipu reservoir before (1977) and after impoundment (1987 and 1997). (From Agostinho et al. 1999.)

The aging of lakes and reservoirs is a consequence of the interrelationships between the watershed and these freshwater ecosystems. The inputs from the watersheds depend upon the soil uses and the extension of activities. Reservoirs have a drainage area much larger than lakes (see Chapter 2). Therefore, the inputs from the watersheds and the loading of nutrients, suspended material, and toxic substances can be much larger. This accelerates the aging process and reservoirs age more rapidly than lakes (Kimmel and Groeger 1984).

However, the aging processes depend not only from the inputs from the watersheds. Retention time is a key variable as a controlling factor (Straskraba 1999), too. The retention times of the phosphorus, nitrogen, and suspended matter are considered. Suspended matter and siltation is one of the main inputs from the watersheds to the reservoirs. The deposition of silt reduces the volume of the reservoir, affecting storage capacity, and changing habitats. The accumulation of organic matter in the sediments, due to the load of suspended organic material from the watershed increases the oxygen demand of sediments and in some extreme cases increases the redox potential significantly and results in further phosphorus liberation from the sediment to the water (Tundisi et al. 1993). Large reservoirs in the Paraná River and its tributaries are losing their nutrient capacity of storage at a fast rate due to the agricultural activities in the watersheds. Estimates for these reservoirs range, however, from 150 to 250 years, so they have the capacity to last for a few centuries. Small reservoirs lose their capacity at a much faster rate and they can last for 25–70 years (Agostinho et al. 1999). The changes in nutrient concentration and in the biogeochemical cycles follow reservoirs aging. The most important changes in the high Paraná floodplain reservoirs are the shifts from an oligotrophic–mesotrophic state to eutrophic or hyper-entrophic one (Tundisi and Matsumura-Tundisi 1990).

Matsumura-Tundisi and Tundisi (2003) reported the changes in the nutrient cycles and water chemistry of reservoirs of the Tietê watershed in the upper reaches of the high Paraná River. Conductivity changes from 100 μS/cm to approximately 400 μS/cm in 20 years. This was followed by a change in the Calanoida in 1979 was substituted by *Notodiaptomus iheringi* in 1999. These changes were attributed to increase in conductivity, food availability, and ammonium content since species of the genus *Notodiaptoms* sp are more tolerant to higher ionic concentration or toxic substances.

The increase in conductivity and ionic concentration as well as the eutrophication process is a result of the intensive use of the watershed for sugarcane plantation and agricultural production in general, followed by a large-scale deforestation and loss of mosaic of vegetation including riparian forests and vegetation (Tundisi and Matsumura-Tundisi 2010).

The deterioration of habitats that is a consequence of the aging process, results in loss of nursery grounds for fishes, and macroinvertebrates. Moreover, littoral zones are vulnerable to eutrophication becoming the sites of fast growth of macrophytes such as *Eichhormia crassipes* or

Pistia stratioides. These aquatic plants can host larvae of insects or adults of mollusks that are vectors of important waterborne diseases (Tundisi and Matsumura-Tundisi 2008a,b).

7.1.10 Prognosis for Environmental Impact Assessment of New Reservoirs

Because of rapid and irregular environmental changes in reservoirs due to the impact of the impoundment on the rivers and the hydrographic basin, it is difficult to make a prognosis about the future development of a reservoir and the organization of the communities after the filling phase is terminated (Straskraba 1999). However, when knowing the conditions of the engineering project, the location of the reservoir and other information related to its future functioning, it is possible to develop scenarios and a prognosis of the future limnological and ecological characteristics (Table 7.7). It is furthermore possible to develop a model that can be used to make approximate prediction of the behavior of the reservoir in the years after the filling phase is terminated; see Chapter 19. It is often recommendable to combine the use of models anf model predictions with the application of ecological indicators; see Chapter 18.

The information listed in Table 7.7 is extremely useful in the anticipation of impacts and in the design of management guidelines that optimize multiple uses, and apply ecotechnological methods to reduce impacts and to control the water quality and environmental processes in a future reservoir.

Reservoirs offer the opportunity to develop ecological theories and experimentation in artificial ecosystems by following succession patterns, aging processes, and spatial and temporal organization of the communities. Reservoirs in the high Paraná River basin changed considerably

TABLE 7.7

Basic Information for Prognosis of Reservoir Impacts

Location of the reservoir in the river continuum
Size, volume, and morphometry of the future reservoir; possible compartmentalization
Retention time (planned for future operation)
Biodiversity in the impacted watershed
Soil uses and vegetation cover of the watershed
Human population, settlements, rural areas
Nutrient concentration in the main river and tributaries
The load of suspended organic and inorganic matter in the main river and its tributaries
The perspectives for multiple uses of the future reservoir
The projected future population in the watershed after dam construction
Future spatial heterogeneity patterns

Source: Straskraba, M. and Tundisi, J.G., *Theoretical Reservoir Ecology and Its Applications*, Brazilian Academy of Sciences/International Institute of Ecology, Backhuys Publishers, Leiden, the Netherlands, 592pp., 1999.

the floodplain ecosystems, but their study contributed considerably to the design of new reservoirs and to improve management alternatives with a good, technical, and scientific support.

7.1.11 Pantanal Wetlands

One of the important freshwater aquatic systems in the high Paraná watershed is the Pantanal wetland. This large wetland with an area of approximately 140,000 km^2 covers the Alto Paraguay River depression between the old crystalline shield of central Brazil and the geological uplifting of Andes Mountains, which are much younger (Junk and da Silva 1995). Since the Quaternary Period, the upper Paraguay River, which is a tributary of the Paraná River, deposited its sediments in this depression. There are considerable geomorphological subunits in the Pantanal. Adámoli (1986) distinguishes 10 different geomorphological subunits. The Pantanal as well as the Amazon floodplains and wetlands are subject to a predictable monomodal pulse as described by Junk and da Silva (1995). The altitude of the floodplain varies: in the Amazon River and its tributaries. Water level fluctuations between 5 and 15 m during the seasonal cycle are common. The variation in Pantanal is between 2 and 5 m. The total precipitation in the Amazon is around 2000–3000 mm and this is approximately twice as high as in the Pantanal (Salati and Marques 1984). The dry phase in the Amazon is much less pronounced than in the Pantanal. In this wetland, there are cyclical multiannual periods of severe floods and extreme droughts, which did not occur in Amazonia. The seasonality in the Pantanal is more pronounced than in the Amazon region. This can be observed with the cycles in temperature: for example, near Cuiabá, Mato Grosso (Lat 15° 35′ 46″, Long 56° 05′ 48′). The mean monthly temperature varies from 21.4°C in July up to 27.4°C in February. In Central Amazonia, air temperatures range from 25.9°C in February to 27.6°C in September. The Pantanal region is subject to the impact of cold polar air masses in winter, which lowers air temperatures (down to 0°C in some critical years) and causes overturns of shallow lakes with large fish hills. In the Amazon region, the classical distinction between white waters, blackwaters, and clear waters (Sioli 1984) shows the hydrochemical origin in the Andes and pre-Andean region. These are waters with conductivity of 60 μS/cm in average. The blackwaters drain watersheds with podzolic soils. They are waters with low dissolved minerals, conductivities in the range 10–15 μS/cm. Transparent waters have also low conductivity low dissolved solids and dissolved minerals. Conductivity is in the range μS/cm!

Several inputs from tributaries, channels, and connected rivers contribute hydrochemically to the chemical composition of Amazonian waters. The Pantanal receives water from several sources such as the Paraguay River and its tributaries. These tributaries drain water from watersheds with different geological formations. Therefore, the hydrochemical and mineralogical diversity of water reflects these different geological formations (conductivity of water, for example, can be as high as 350 μS/cm due to transport of alkaline waters with pH = 8.5). In other tributaries, transport of slightly acid water results in conductivity as low as 10 μS/cm and pH about 5.5 (Junk and da Silva 1995). The conductivity of the floodplain lakes in the Pantanal varies: can range from 50 to 900 μS/cm in low waters or isolated lakes.

The large Amazonian rivers exert a much stronger mechanical influence on the floodplain freshwater ecosystem, due to the magnitude of the pulse, greater discharge, and large sediment load (Junk et al. 2000). The Pantanal wetland is in a depression that traps sediments. Since the pulses are of low magnitude, the physical impact on the morphology and morphometry of channels and rivers is smaller. Sediment deposition has a strong impact on large parts of the wetland. The water flowing from north to south in the Paraguay River and tributaries takes 3–4 months to reach the outflow. Retention of sediments and impacts on the biodiversity of flora and fauna and on the biogeochemical cycles occur. The water level fluctuation is a driving force that promotes a spatial heterogeneity, and a complex mosaic of habitats with a diverse and abundant wildlife. The diversity of macroinvertebrates and crustaceans is a consequence of this pulse of inundation, but these also play an important role in ecological processes in the Pantanal since they are components of the food chain, such as herbivores, predators, decomposers, and prey (Alho 2011). As for the fish fauna, Britski (1992) listed 263 species for the Pantanal. Fish resources of the Pantanal are

1. An important biota component of the ecosystem
2. Food for the local human population
3. High interest for sport fishing and tourism
4. A genetic resource
5. An economical resource of food supply
6. An income by sale of ornamental fishes

The fauna of amphibians and reptiles is also an important link in the food chain, its diversity, physiology, and behavior depending on the variety of habitats, seasonal cycles of inundation, and dry periods.

Bird diversity is very high (Junk et al. 2006). There are 444 bird species recorded for the floodplain and the savannah (cerrado) biome in the Pantanal. Bird species in aquatic habitats are very diverse and abundant, playing an important role as predators and migrants. Large nesting grounds for these birds have an important quantitative and qualitative impact on the biogeochemical cycles.

The large floodplains of the Amazon watershed are covered with flood-adapted forests. Junk et al. (1989) estimates that about 1000 species of Amazonian trees are adapted to flooding. The Pantanal floodplain and wetlands are dominated by herbaceous vegetation (Junk and da Silva 1995). Human impacts on the Pantanal flood plains and wetlands are

1. Deforestation for cattle ranching and for agriculture production of large-scale commercial crops of soya bean and sugarcane for alcohol production
2. Navigation: oil spills, destruction of river banks, discharge of wastes
3. Overfishing for commercial purposes
4. Exploitation of the reptile fauna specially Caiman crocodiles
5. Unregulated tourism and damage to local habitats and to biodiversity
6. Introduction of exotic species: fish species from other watersheds, mollusks (such as *Limnoperna fortunei*), etc.

Conservation: the conservation of the Pantanal floodplains and wetlands should be directed for a better protection of the geomorphological sub-units, a control (by law enforcement) of economic activities, and a regulation of all navigation projects. Stimulus to rational exploitation of the fish and reptile biodiversity and protection of the headwaters of the high Paraguay River should be implemented (Junk et al. 2006, Wantzen et al. 2008).

The promotion of adequate agricultural practices to avoid erosion should be realized. A monitoring of water quality of rivers and shallow lakes is also needed. The conservation of wetlands within the framework of the RAMSAR conversion is another recommended action (Alho and Sabino 2011).

A recent publication describes the Pantanal biodiversity and the challenges for its conservation (Alho and Sabino 2011). Figure 7.13 shows the geographical map of the Pantanal in central Brazil.

7.1.12 Tropical Freshwater Ecosystems of South America and Their Ecological, Economical, and Social Significance

The flat relief of the landscape and the periodic cycle of a large annual rainfall, favored a dense drainage network in the Amazon, Paraná, and La Plata basins. The water availability and the large biodiversity gradients controlled by the water level fluctuation are the fundamental forces that drive the human occupation, the economical exploitation, and the hydro-social cycle (Ab'Saber 1988).

Roosevelt (1999) indicates that Amazonia was colonized by early hunters and gatherers as long as 12,000 years ago. Evidence of widespread colonization that exploited the "varzea" and the floodplain dates from 3000 years B.P. Various population densities developed at different periods upon arrival of Europeans on South America. The carrier capacity of the floodplains varied depending on the technologies applied for exploitation. The carrying capacity of the floodplains for human populations varies considerably between

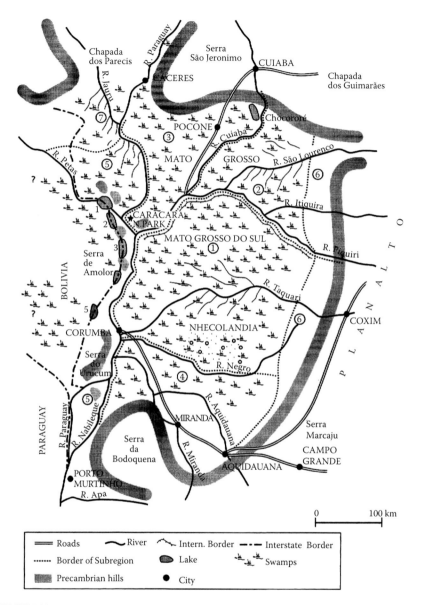

FIGURE 7.13
Pantanal wetlands. (From Moss, B., *Ecology of Fresh Waters: Man and Medium, Past to Future*, 3rd edn., Blackwell Publishing, Oxford, U.K., 557pp., 2007.)

the different systems. For example, Junk and da Silva (1995) pointed out that nutrient rich water river floodplains have a higher carrying capacity than nutrient poor blackwaters.

The present human activities in the floodplains are developed around the economic exploitation by agriculture, fisheries, timber extraction, and animal husbandry. Small-scale agriculture in the "varzea" or dry areas of wetlands during periods of low water seems to be very favorable to the hydrosocial cycle adapted to the water level fluctuations and water availability. Fisheries also sustain the supply of protein for the Amazonian population. According to Bayley and Petrere (1989), the fishery potential of the Amazonian inland waters is about 1 million tons a year. Only 20%–25% is actually used.

The multiple uses of the floodplains should, according to Ohly and Junk (1999), drive ecological conditions, environmental protection, and socioeconomic needs. Small-scale exploitation of the biodiversity, use of the water cycle as a guideline to develop agricultural, and fisheries activities seems to be one possible way to exploit the potential of the floodplains (Welcomme 1985, Padoch et al. 1999).

Projects such as the Mamirauá protection area and the use of natural resources in the settlements located in or near the "varzea" are examples of resource management activities and housing with complete interaction of the hydrosocial cycle with the social cycle, being this the hydrosocial cycle (Ayres et al. 1999). Large-scale developments such as reservoirs or agroindustrial projects should be carefully considered and their impact clearly examined before the interference in the floodplain and in the hydrological cycle. Disruption of the ecological connectivity, followed by loss of biodiversity and the hydrosocial cycle often occurs in these sites interfering with the ecological integrity of the floodplain and causing losses in ecosystems services difficult to recover (Tundisi et al. 2012, in preparation). Recent advances in the management of floodplains in the high Paraná River basin and in the conceptualization of sciences in the Pantanal wetlands were described by Wantzer et al. (2008), Agostinho et al. (2007) and Agostinho et al. (2007, 2008).

7.2 Continental Waters of Tropical Africa

Beadle (1981) has published an extensive revision of the limnological research work carried out in inland ecosystems of tropical Africa. The African great lakes originated by tectonic activity are located in the Rift Valley. The valley has its origin during the Miocene–Pliocene period. The uplifting and subsidence gave origin to a pattern of large depressions separated by ridges and this has determined the outlines of the African continent hydrology Figure 7.14 shows the courses of the two Great Rift Valleys and their drainage systems.

Lakes Malawi, Tanganyka, Kivu, Edward, George, and Albert are in a sequence south–north at the western Rift Valley and Lakes Marijara, Natron, Magadi, Naivaska, Nakuru, Bogoria, Baringo, Turkana, and Chew Bohr are located at the eastern Rift Valley. The uplift of the western edge blocked the west wind flowing of the rivers flowing to Lake Victoria. All these lakes are much older than most of the northern temperate lakes which are postglacial.

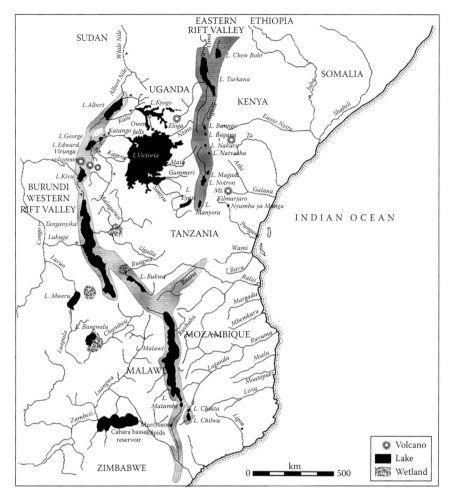

FIGURE 7.14
Rift Valley and other lakes of the African Continent. (Beadle, L.C., *The Inland Waters of Tropical Africa: An Introduction to Tropical Limnology*, 2nd edn., Longman, London, U.K., 475pp., 1981.)

The studies of the Great Rift Valley lakes were an important contribution to the knowledge of tropical freshwaters ecosystems. These studies contributed to the world knowledge in limnology related to tropical ecosystems in four main fields:

1. The relationship between the seasonal cycle, the nictemeral cycle, and the influence of biomass on these cycles, especially in shallow lakes such as Lake George (Viner and Smith 1973, Ganf and Horne 1975, Talling 1992)

2. Mechanisms of adaptations to desiccation and high salinity (Melack 1984, 1988)

3. The interactions of primary producers, invertebrates, and vertebrates in the food chain (McLachlan 1969, 1975)
4. The evolutionary processes in isolated systems and the cichlid fish species flocks Meyer et al. (1994, 1996) and Lowe-McConnell (1987)

Comparative studies of the Great Valley lakes, other African great lakes, and temperate lakes were carried out by Talling and Talling (1965), Talling (1986), and Lemoalle (1981, 1983).

These studies concluded that despite the higher photosynthetic rate and production of organic matter by these tropical lakes in comparison with temperate lakes, the respiration rates of the biomass and the decomposition of organic matter dissipates the accumulated carbon in the food chain (Talling and Talling 1965). The studies on shallow African lakes such as Lake George improved very much the knowledge about the cycles of organic matter and nutrients in tropical shallow freshwater ecosystems. There is a high rate of primary production (gross primary production) a low net primary production and a very rapid exchange of nutrients between the sediment and the water promoting a fast recycling of nutrients (Ganf 1975, Ganf and Horne 1975). The seasonal variation is small. Of high importance are the short cuts in the food chain; the consumption of phytoplankton by large populations of the cichlid fishes is a characteristic of this lake. Lake George has small fluctuations in biomass and a high rate of production with little variation throughout the seasonal cycle (Beadle 1981).

Intensive limnological studies carried out in the deep lakes of the Rift Valley such as Lake Taganyika (Coulter 1991), Lake Malawi (Beadle 1981), and other deep lakes including Lake Victoria (Talling 1966) describe the pattern of stratification and circulation, the nutrient cycles, and the diversity of invertebrates and fish fauna. These were fundamental research works that detailed the physical interactions (diurnal and annual levels of temperature and density stratification), physical–chemical interactions, for example, wind stress and circulation, and nutrient cycles, and the coupling between environmental–biological factors, such as the effects of water level fluctuations on the response of the biota to desiccation, seasonal cycles, and lunar–biological cycles (Talling and Lemonalle 1998).

Biological interactions such as patterns of biomass abundance and competition, predator–prey relationships, and food chain relationships were described by Burgis (1978).

Other studies in Africa lakes include Lake Chilwa, a shallow saline lake located to the east of the main Rift Valley, at an altitude of 622 m above sea level and with an area of 678 km². The open area of the lake is surrounded by almost 600 km² of swamps and marshes (Kalk et al. 1979). The basin of Lake Chilwa covers an area of 7500 km². The water level of the lake fluctuates annually by 0.8–1.0 m but larger fluctuations may occur in periods of 6 and 12 years (2–3 m). The water level can drop drastically during some years, and the lake can even dry up. The main cause of changes on lake level is variation of rainfall over the catchment confirming the hydrological factor as one fundamental forcing function for this and other African lakes including Lake Malawi, Lake Naivasha, and Lake Chad. The water level changes have, as a consequence, changes in the chemical composition of the water, conductivity (McLachlan 1979), and proportions of the major ions. Changes in the zooplankton composition during periods of low water level and during the drying and filling of the lake were observed and also of the benthic invertebrates (McLachlan 1979).

Fisheries also changed during the several periods of drying, dryness, filling, and normal water level of the Lake Chilwa (Furse et al. 1979). Thus, the fishery in this lake offers an economically unstable environment due to the water level fluctuations and the changes in the ecological conditions. An overview of the limnological functioning of the lake was produced by Moss (1979). Focus on social problems was summarized by Kalk et al. (1979).

Of the great lakes of the African Rift Valley, Lake Tanganyika was studied in detail; Lake Tanganyika is a meromitic lake with a well-marked stratification. The lake has an area of 650 km², a mean width of 50 km, and average depths of 570 m and maximum depth of 1470 m. Physical limnology, water chemistry, and primary productivity was studied. The studies on the pelagic fish fauna showed a simple grazing community (Hecky 1984). The recycling of fixed carbon from the sediments or the upper strata of the metalimnion could be a carbon and nutrient source that maintain a higher productivity. Recycling of the nutrients in the epilimnion by excretion of zooplankton and pelagic fishes could be another source of reduced inorganic substances. Hecky et al. (1991) discusses also that the high productivity of Lake Targanyika is due to the marine origin of the food chain—copepods and clupeidae. This trophic structure accordingly to Hecky et al. (1991) is similar to marine systems. This author considers that the pelagic community has emerged in Lake Targanyika due to the special hydrography and circulation and the internal cycling of nutrients. The tropical seasonality enforces competition in the pelagic compartment and colonization of clupeids and centropomids; *Stolothrissa* and *Limnothrissa* are clupeid species that evolved probably from a common ancestor (Coulter 1991). They are important components of the food chain of the lake and major source of protein for the human population.

Another well studied African lake is Lake Chad (Figure 7.15).

Carmouze et al. (1983) described in detail the lacustrine environment and its evolution, the communities of the lake, the adaptations to water level fluctuations, and the trophic relationships. The lacustrine basin has a temporary region of 25,000 km² and the mean depth of the lake is 4 m. Oscillations in the water level produce variation in the water area with profound changes in the lacustrine zones in space and time. Fluctuations caused a change of the area from 25,000 km² in 1964 to 9,000 km² in 1974 (Carmouze et al. 1983a).

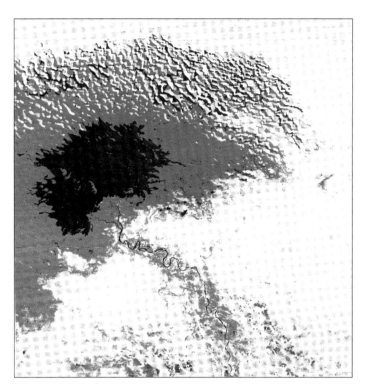

FIGURE 7.15
Satellite image of Lake Chad. (From Carmouze, J.P. et al., Physical and chemical characteristics of the waters, in: Carmouze, J.P., Durand, J.R., and Lévêque, C. (Eds.), *Lake Chad. Ecology and Productivity of a Shallow Tropical Ecosystem*, Monographiae Biologicae, Vol. 53, W. Junk Publishers, The Hague, the Netherlands, pp. 65–94, 1983; NASA.)

The water level fluctuation resulted in hydrochemical modifications. Also the influence of the biological activity in the nutrient and ionic composition of the water was demonstrated. For example, the rate of Ca liberation was indicated by the percentage of dead shells/live shells with the water level fluctuation (Lévêque 1973a,b).

Macrophytes also had a strong influence on the biogeochemical cycles of K^+, Ca^{++}, Mg^{++}, Na^+, and SiO_4H_4. When considering its endorheic nature and the arid environment, Lake Chad should be a lake of high salinity. However, as pointed out by Carmouze et al. (1983b), this is not the case due to

- Inflowing river's salinity, which is low
- The mollusks and macrophyte contribution to the sedimentation and fixation of Ca, Mg, and HCO_3/Co_3, which is high
- The deposition/redissolution of salt deposits in marginal regions during water level fluctuations that regulates the salinity, which is relatively stable

Changes in primary production of phytoplankton and macrophytes with the water level fluctuations, modifications in the lacustrine hydrology that caused disturbances in the fish communities, and fish stocks.

The mains conclusions about Lake Chad is that it is an ecosystem where most of the trophic levels (phytoplankton, macrophytes, benthos, and fish) are well represented and the decomposition of organic matter with a fast recycling of nutrients is one of the main explanation of the richness of the ecosystem. The bird fauna has also an impact in the lacustrine ecosystem (Dejoux 1983). Despite the strong changes in the hydrological cycle and on the hydrochemical cycle there is a high plasticity of the flora and fauna allowing for a high productivity.

The fish fauna of the African lakes and rivers was studied in detail by several specialists (Lévêque et al. 1988). The studies of the evolution, speciation, and distribution of freshwater fishes were carried out on the Great African lakes and in other aquatic systems. Africa has 2000 known species of freshwater fishes. African rivers and swamps are ecosystems with a large group of archaic and phyllogenetically isolated fish groups, mostly endemic. The most evident examples of adaptive radiation especially among the cichlid fishes (Lowe-McConnell 1988) can be found in the African lakes. There is a high degree of endemism in Africa and this occurs among the less advanced fishes. Africa is the only continental area where cypriniformes, characiformes, and siluriformes occur. The importance of Ciclids in African lakes is high as well as the characoids and cyprimids in rivers (Lowe McConnell 1969).

Lowe-McConnell (1988) summarizes the fish fauna of African rivers and lakes as

1. Endemic families that evolved in the African continent only
2. Components shared with South America—an evidence of a common origin (Gondwanaland)
3. Components shared with the Oriental region
4. Marine immigrants from which freshwater species have evolved
5. Remnants of archaic components with wide distribution in Africa and with relatives in South America, Australia, and Southeast Asia

The responses of the African freshwater fish faunas to environmental conditions are related to

1. Water level fluctuation
2. Mains sources of floods
3. Deoxygenation

4. Interactions between types of fishes and their relationship with the environments

5. Ionic concentration (higher in lakes than in rivers), patterns of stratification and deoxygenation in lakes, and turbidity (higher in rivers than in lakes)

Other studies in tropical lakes (Rzoska 1976) of interest studies on lakes and swamps related to the River Nile and watersheds were carried out Lake Tana (Source of the Nile); nilotic lakes of the western rift, Lake Turkana; these studies are covered in Dumont (2009), who also made an extensive review of River Nile, its environment, limnology and human use (see also Chapter 4).

The impacts on African lakes can be described as (Crul 1997)

1. Water pollution and contamination
2. Excessive water uses
3. Deforestation
4. Introduction of exotic species
5. Water conflicts on international water basins
6. Reservoir construction
7. Impacts of climatic change
8. Competitive uses of water

7.3 Floodplains and Wetlands in Asia

The floodplains and wetlands associated with large rivers in Asia play an important ecological and social role. The fluctuations in discharge related to the seasonal monsoon, produce oscillations of abundant water availability and water scarcity. The Indus, Ganges, and Brahmaputra rivers have floodplains of high importance for human life since they are used for several purposes as fisheries, water for irrigation, and extensive exploitation of the biodiversity of these areas (Dudgeon 2000).

The Indochinese peninsula has more than 930 species of fishes. The Mekong basin supports over 500 species (Welcomme 1979).

The floodplains and wetlands of Asia can be easily converted to agricultural land after draining. The riverine habitats host a diversity of mammal species especially in China, India, and Indonesia. Arboreal animals such as primates, inhabit riparian forests and swamp forests (Dudgeon 2000).

Of special importance in Asia are the coastal wetlands with a large fringe of mangrove trees and a high biodiversity (Welcomme 1979). These mangrove

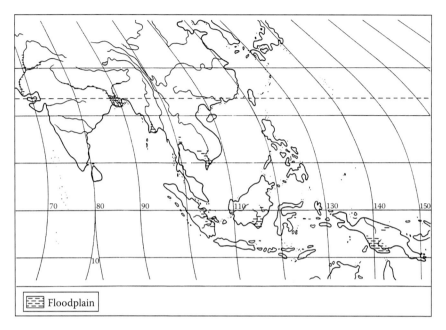

FIGURE 7.16
Floodplains of Asia. (From Welcomme, R.L., *Fisheries Ecology of Floodplain Rivers*, Longman, London, U.K., 317pp., 1979.)

regions are important sites for fish and crustaceans biodiversity and also macroinvertebrates.

Figure 7.16 shows the location of major floodplains in Asia.

The riverine wetlands and floodplains of Asia are threatened by deforestation, timber extraction, drainage and agricultural conversion, and hydroelectricity production by dam construction. In Indonesia and Malaysia, only 20% of the original wetlands remained. One of the most important sites protected since 1995 is Tasek–Bera a relict of a once well-developed extensive riparian inundation system in southeast Asia (Furtado and Mori 1982).

7.4 Tropical Freshwater Environments

Tropical freshwater ecosystems are as pointed out by Talling and Lemoalle (1998) influenced by the strong solar radiation input that is fundamental for the temperature regime, and the photoperiod that is limited in its seasonal range. The hydrological control of tropical seasonality influences discharges and fluctuations in water level which are important forcing functions. This hydrological control is a result of air masses circulation with a seasonal pattern of wind and rainfall. The same authors, Talling and Lemoalle (1998) emphasize that controlling processes for the tropical freshwater environment are climatic, geological, and biological. The higher water temperature of tropical lakes renders high stability of the standing water with a longer stratification period for well sheltered lakes with little influence of wind action (Tundisi and Saijo 1997). Advection water due to rainfall can contribute also to stabilization of water columns due to differences in density of the entrained water (Tundisi 1997).

Temperature depth profiles show stratification patterns followed by dissolved oxygen and chemical periods of thermal continuities. This chemical stratification is also followed by biological organization patterns as shown by Reynolds et al. (1983). Daily changes or day to day changes in the patterns of thermal structure were shown for Lewis (1973), MacIntyre and Melack (1988, 1995) and for reservoirs (Barbosa and Tundisi 1989, Tundisi 1992).

The *geologic* influence is a consequence of the contribution of rock and soil composition. Many tropical regions are subjected to chemical weathering and water leaching of rock masses; this results in waters of low conductivity with the ion Na^+ being dominant in many freshwaters of the tropics. However, the mineral products of weathering are influenced by the geological composition, rock types, and water discharge (Barbosa et al. 1989).

Weathering of siliceous rock results in soluble silicon and high sodium ion concentrations. Carbonated rocks result in more alkaline waters (Gibbs 1970, Forsberg 1984, Junk 1986, Junk and Welcomme 1990, Furch and Junk 1997).

The biological component of tropical freshwater ecosystem has a very high qualitative and quantitative influence in the water chemistry composition of lakes, reservoirs, shallow wetlands. The high photosynthetic rates of primary producers and the high respiration and decomposition rates of the aquatic biota in tropical freshwater influence the cycle and biogeochemistry of *carbon, nitrogen, phosphorus,* and ions such as *sodium, potassium, calcium,* and *magnesium* (Burgis 1972, 1986, Ganf 1972, 1974, Ganf and Viner 1973, Ganf and Horne 1975, Junk 1986, Talling and Lemoalle 1998).

The tropical freshwater ecosystems have three main characteristics (Talling andLemoalle 1998): absolute magnitudes of environmental factors, time variability of environmental factors, and the responses of the biota (Lewis 1990, Lewis et al. 1995, Loffler 1964, Mercante et al. 2011, Scheffer 1998, Schimdt 1970, Straskraba et al. 1993, Talling 1956, 1977, Tundisi 1983, Tundisi et al. 1994, 1995, 1997, 2008).

Evidence from the patterns of regularity of the factors with some short time irregularities arise from several studies in tropical and subtropical lakes and reservoirs; see (Talling 1956, 1963, 1969, 1986, 1987, 1976, Loffler 1964, Schimdt 1970, Burgis and Walker 1972, Burgis 1974, Melack 1982, 1988, 1996, Carmouze et al. 1983, Tundisi 1983, Lewis 1984, 1987, 1995, Sioli 1984, Forsberg et al. 1988, Lewis 1990, Melack and MacIntyre 1992, Straskraba et al. 1993, Tundisi et al. 1994, 1995, 1997, 2008, Lewis et al. 1995, Tundisi and Saijo 1997 , Scheffer 1998, Mercante et al. 2011).

According to Dumont (1992), the factors that limit and control the distribution and structure of communities in the shallow African lakes are geology, water level fluctuation, climate (rainfall/evaporation/wind/air–water temperature), water chemistry and influence of sediments, osmotic regulation, and balance competition/predation. One of the mains attributes of aquatic fauna and flora of shallow African lakes is the adaptation to desiccation and high conductivity/salinity. The "Floodplain Declaration" prepared during the Ecohydrological meeting at the Center of UNESCO, in Poland (Lodz), was described in Zaleswki (2008).

References

Ab'Saber, A.N. 1988. O Pantanal Mato Grossense e a Teoria dos Refúgios. *Revista Brasileira de Geografia* 50(1/2):9–57.

Adámoli, J. 1986. A dinâmica das inundações no Pantanal. In: *Simpósio Sobre Recursos Naturais e Sócio-Econômicos do Pantanal*. Embrapa, Anais, Brasília, pp. 71–76.

Adis, J. 1997. Terrestrial invertebrates: Survival strategies, group spectrum, dominance and activity patterns. In: Junk, W.J. (Ed.), *The Central Amazon Floodplain: Ecology of a Pulsing System*. Ecological Studies, Vol. 126. Springer Verlag, Berlin, Germany, pp. 299–318.

Agostinho, A.A.; Bonecker, C.C.; Rodrigues, L.; Gomes, L.C.; and Thomas, S.M. (Eds.). 2009. Biodiversity and conservation. *Braz. J. Biol.* 69(Suppl. 2):755.

Agostinho, A.A.; Gomes, C.L.; and Pelicice, M.F. 2007. *Ecologia e Manejo de Recursos Pesqueiros em reservatórios do Brasil*. EDUEM, Maringá, Brazil, 501pp.

Agostinho, A.A.; Gomes, L.C.; Pelicice, M.F.; Souza Filho, E.C.; and Tomanik, E. 2008. Application of the ecohydrological concept for sustainable development of floodplains: The care of upper Paraná River basin. *Ecohydrol. Hydrobiol.* 8(2–4):205–223.

Agostinho, A.A.; Julio, H.F. Jr.; Gomes, L.C.; Bini, L.M.; and Agostinho, C.S. 1997. Composição, abundância e distribuição espaço-temporal da ictiofauna. In: Vazzoler, A.E.A. de M.; Agostinho, A.A.; and Hahn, N.S. (Eds.), *A Planície de Inundação do Alto Rio Paraná: Aspectos Físicos, Biológicos e Socioeconômicos*. Editora Universidade Estadual de Maringá, Maringa State University Publisher, Maringá, Brazil, pp. 179–208.

Agostinho, A.A.; Julio, H.F. Jr.; and Petrere, M. Jr. 1994. Itaipu reservoir (Brasil): Impacts of the impoundment on the fish fauna and fisheries. In: Cowx, I.G. (Ed.), *Rehabilitation of Inland Fisheries*. Fishing New Books, Osney Mead, Oxford, U.K., pp. 171–184.

Agostinho, A.A.; Miranda, L.E.; Bini, L.M.; Gomes, L.C.; Thomaz, S.M.; and Suzuki, H.I. 1999. Patterns of colonization in neotropical reservoirs, and prognoses on aging. In: Tundisi, J.G. and Straskraba, M. (Eds.), *Theoretical Reservoir Ecology and Its Applications*. International Institute of Ecology, Brazilian Academy of Sciences and Backhuys Publishers, São Carlos, Brazil, pp. 227–265.

Agostinho, A.A.; Thomaz, S.M.; Minte-Vera, C.V.; and Winnemiller, K.O. 2000. Biodiversity in the high Paraná floodplain. In: Gopal, B.; Junk, W.; and Davis, J.J. (Eds.), *Biodiversity in Wetlands: Assessment, Function and Conservation*. Backhuys Publishers, Leiden, the Netherlands, Vol. I, pp. 89–118, 353pp.

Agostinho, A.A.; Vazzoler, A.E.A.de M.; and Thomaz, S.M. 1995. The high Paraná river basin: Limnological and ichthyological aspects. In: Tundisi, J.G.; Bicudo, C.E.M.; and Matsumura-Tundisi, T. (Eds.), *Limnology in Brazil*. Brazilian Academy of Science/Brazilian Limnological Society, Rio de Janeiro, Brazil, pp. 59–104.

Alho, C.J.V. 2011. Biodiversity of the Pantanal: Its magnitude, human occupation, environmental threats and challenges for conservation. *Braz. J. Biol.* 71(Suppl. 1):327–335.

Alho, C.J.V. and Sabino, J. 2011. A conservation agenda for the Pantanal's biodiversity. *Braz. J. Biol.* 71(Suppl. 1):229–232.

Almeida Val, V.M.F.; Val, A.L.; and Hochaschka, P.W. 1993. Hypoxia tolerance in Amazon fishes: Status of an under-explored biological "goldmine." In: Hochachka, P.W.; Van den Thillart, G.; and Luts, P. (Eds.), *Surviving Hypoxia: Mechanisms of Control Versus Adaptation*. CRC Press, Boca Raton, FL.

Araújo Lima, C.A.R.M.; Fersberg, B.; Victoria, R.L.; and Martinelli, L. 1986. Energy sources for detritivores fishes in the Amazon. *Science* 234:1256–1258.

Ayres, M. 1986. Some aspect of social problems facing conservation in Brazil. *Trends Ecol. Evol.* 1(2):48–49.

Ayres, J.M. et al. 1999. Mamirauá: The conservation of biodiversity in an Amazonian flooded forest. In: Padoch, C.; Ayres, J.M.; Pinedo-Vasquez, M.; and Henderson, A. (Eds.), *Varzea Diversity, Development and Conservation of Amazonian's Whitewater Floodplains*. Botanical Garden, New York, pp. 203–216.

Barbosa, F.A.R. and Tundisi, J.G. 1989. Diel variations in a shallow tropical Brazilian lake. I. The influence of temperature variation on the distribution of dissolved oxygen and nutrients. *Arch. Hydrobiol.* 116:333–349.

Barbosa, F.A.R.; Tundisi, J.G.; and Henry, R. 1989. Diel variations in a shallow tropical Brazilian lake. II. Primary production, photosynthetic efficiency and chlorophyll-*a* content. *Arch. Hydrobiol.* 116:435–448.

Barthem, R.C. and Goulding, M. 2007. *Um ecossistema inesperado. Amazônia revelada pela pesca*. ACA SCM, Lima, Peru, p. 241.

Bayley, P. and Petrere, M. 1989. Amazon fisheries: Assessment methods, current status and management options. In: Dodge, O.P. (Ed.), *Proceedings of the International Large Rivers Symposium*, Vol. 106. *Can Spec. Publ. Fish Aquatic Sci.*, Otawa, ON, Canada, pp. 385–398.

Beadle, L.C. 1981. *The Inland Waters of Tropical Africa: An Introduction to Tropical Limnology*, 2nd edn. Longman, London, U.K., 475pp.

Bigarella, J.J. 1973. Geology of the Amazon and Paranaiba basins. In: Naim, A.E.M. and Stehli, F.G. (Eds.), *Ocean, Basins and Margins*. Plenum Press, New York, pp. 25–86.

Britski, H.A. 1992. Conhecimento atual das relações filogenéticas d peixes neotropicais. In: Agostinho, A.A. and Benedito, C.E. (Eds.), *Situação Atual e Perspectivas da Lctiologia do Brasil (IX Encontro Brasileiro de Lctiologia, Maringá)*. Ed. Da UEM Cap 6, Brazil, pp. 42–57.

Burgis, M.J. 1972. A preliminary comparison of the zooplankton in a tropical and temperate lake (Lake George, Uganda and Loch Leven, Scotland). *Verh. Int. Verein. Limnol.* 18:647–655.

Burgis, M.J. 1974. Revised estimates for the biomass and production of zooplankton in Lake George, Uganda. *Freshwater Biol.* 4:535–541.

Burgis, M.J. 1978. Case studies of lake ecosystems at different latitudes: The tropics. The Lake George ecosystem. *Verh. Int. Verein. Limnol.* 20:1139–1152.

Burgis, M.J. 1986. Food chain efficiency in the open water of Lake Tanganyika. *Bulletin des séances. Académie r. des sciences d'outre mer* 30:282–298.

Burgis, M.J. and Walker, A.F. 1972. A preliminary comparison of the zooplankton in a tropical and a temperate lake (Lake George, Uganda and Loch Leven Scotland). *Ver. Int. Verein. Limnol.* 18:647–655.

Carmouze, J.P.; Chantraine, J.M.; and Lamoalle, J. 1983a. Physical and chemical characteristics of the waters. In: Carmouze, J.P.; Durand, J.R.; and Lévêque, C. (Eds.), *Lake Chad. Ecology and Productivity of a Shallow Tropical Ecosystem.* Monographiae Biologicae, Vol. 53. W. Junk Publishers, The Hague, the Netherlands, pp. 65–94.

Carmouze, J.P.; Durand, J.R.; and Lévêque, C. (Eds.), 1983b. *Lake Chad. Ecology and Productivity of a Shallow Tropical Ecosystems.* Monographiae Biologicae, Vol. 53. W. Junk Publishers, The Hague, the Netherlands, 575pp.

Coulter, G.W. (Ed.). 1991. *Lake Tanganyika and Its Life.* Natural History Museum Publications, London, U.K., 354pp.

Crul, R.C.M. 1997. *Limnology and Hydrology of Lakes Tanganyika and Malawi.* Studies and Reports in Hydrology 54, UNESCO, Paris, France, 111pp.

Dejoux, C. 1983. The fauna associated with aquatic vegetation. In: Carmouze, J.R.; Durand, L.; and Leveque, C. (Eds.), *Lake Chad, Ecology and Productivity of a Shallow Tropical Ecosystem,* Monographiae Biologicae, Vol. 53. W. Junk Publishers, The Hague, the Netherlands, pp. 279–292.

Depetris, P.J. and Paolini, J.E. 1991. Biogeochemical aspects of South American Rivers: The Parana and the Orinoco. In: Degens, E.T.; Kempe, S.; and Richey, J.E. (Eds.), *Biogeochemistry of Major World Rivers.* SCOPE 42, John Wiley & Sons, New York, pp. 105–122, 355pp.

Dudgeon, D. 2000. Riverine wetlands and biodiversity conservation in tropical Asia, pp. 35–60. In: Gopal, B.; Junk, W.; and Davis, J.J. (Eds.), *Biodiversity in Wetlands: Assessment, Function and Conservation,* Vol. I. Backhuys Publishers, Leiden, the Netherlands, 353pp.

Dumont, H.J. 1992. The regulation of plant and animal species and communities in African shallow lakes and wetlands. *Rev. Hydrobiol. Trop.* 25:303–346.

Dumont, H.J. (Ed.). 2009. *The Nile: Origin Environments Limnology and Human Use.* Monogaphiae Biologica 89, Springer, Berlin, Germany.

Fernando, H.C. and Holčík, J. 1991. Fish in reservoirs. *Int. Revue Ges. Hydrobiol.* 76:149–167.

Fittkau, E.J. 1964. Remarks on the limnology of central-Amazon rain-forest streams. *Verh. Int. Verein. Theor. Angew. Limnol.* 15:1092–1096.

Forsberg, B.R. 1984. Nutrient processing in Amazon floodplain lakes. *Verh. Int. Verein. Limnol.* 22:1294–1298.

Forsberg, B.R.; Devol, A.H.; Richey, J.E.; Martinelli, L.A.; and Dos Santos, H. 1988. Factors controlling nutrient concentrations in Amazon floodplain lakes. *Limnol. Oceanogr.* 33:41–56.

Furch, K. and Junk, W.J. 1997. Physico-chemical conditions in floodplain lakes. In: Junk, W.J. (Ed.), *The Central Amazon Floodplain: Ecology of a Pulsing System.* Ecological Studies, Vol. 126. Springer Verlag, Berlin, Germany, pp. 69–108.

Furse, M.T.; Morgan, P.R.; and Kalk, M. 1979. The fisheries of Lake Chilwa. In: Kalk, M.; Machachlan, A.J.; and Howard Williams, C. (Eds.), *Lake Chilwa. Studies of Change in a Tropical Ecosystem.* W. Junk Publishers, Dordrecht, the Netherlands, pp. 211–229, 462pp.

Furtado, J.I. and Mori, S. 1982. *Tasek Bera: The Ecology of a Freshwater Swamp.* Monographiae Biologicae, Vol. 47. W. Junk, The Hague, the Netherland.

Ganf, G.G. 1972. The regulation of net primary production in Lake George, Uganda, East Africa. In: Kajak, Z. and Hillbricht-Ilkowska, A. (Eds.), *Productivity Problems of Freshwaters.* Polish Scientific Publishers, Kraków, Poland, pp. 693–708.

Ganf, G.G. 1974. Diurnal mixing and the vertical distribution of phytoplankton in a shallow equatorial lake (Lake George, Uganda). *J. Ecol.* 62:611–629.

Ganf, G.G. 1975. Photosynthetic production and irradiance—Photosynthesis relationships of the phytoplankton from a shallow equatorial lake (Lake George Uganda). *Oecologia. Berlin.* 18:165–183.

Ganf, G.G. and Horne, A.J. 1975. Diurnal stratification, photosynthesis and nitrogen-fixation in a shallow equatorial lake (Lake George, Uganda). *Freshwater Biol.* 5:13–39.

Ganf, G.G. and Viner, A.B. 1973. Ecological stability in a shallow equatorial lake (Lake George, Uganda). *Proc. R. Soc. B* 184:321–346.

Gibbs, R.J. 1970. Mechanisms controlling world water chemistry. *Science* 170(3962):1088–1090, New York.

Goulding, M. 1980. *The Fishes and the Forest.* University of California Press, Berkeley, CA, 280pp.

Goulding, M.; Carvalho, M.L.; and Ferreira, E.G. 1988. *Rio Negro: Rich Life in Poor Water.* SPB Academic Publishing, The Hague, the Netherlands, 200pp.

Haffer, J. 2008. Hypothesis to explain the origin of species in Amazonia. *Braz. J. Biol.* (São Carlos, SP) 68(4):917–948.

Hecky, R.E. 1984. Afeican Lakes and trophic efficiencies: A temporal perspective. In: Meyers, D.G. and Strickler, J.R. (Eds.), *Trophic Interations within Aquatic Ecosystems*, pp. 405–408.

Hecky, R.E.; Spigel, R.H.; and Coulter, G.W. 1991. The nutrient regime. In: Coulter, G.W. (Ed.), *Lake Tanganyika and Its Life.* Oxford University Press, Oxford, U.K., pp. 76–89.

Howard Williams, C. 1979. The distribution of aquatic macrophytes in Lake Chilwa: Annual and long term environmental fluctuations. In: Falk, M.; Mchaclan, A.J.; and Howard Williams, C. (Eds.), *Lake Chilwa: Studies of Change in a Tropical Ecosystem.* W. Junk Publishers, Dordrecht, the Netherlands, pp. 107–121, 462pp.

Junk, W.J. 1984. Ecology of the varzea, floodplain of Amazonian white rivers. In: Sioli, H. (Ed.), *The Amazon: Limnology and Landscape Ecology of a Mighty Tropical River and Its Basin.* Dr. W. Junk Publishers, Dordrecht, the Netherlands.

Junk, W.J. 1986. Aquatic plants of the Amazon system. In: Davies, B.R. and Walker, F.F. (Eds.), *The Ecology of River Systems.* W.J. Junk Publishers, Dordrecht, the Netherlands, pp. 319–337.

Junk, W.J. 1993. Wetlands of tropical South America. In: Whigham, D.F. (Ed.), *Wetlands of the World.* Kluwer Academic Publishers, Dordrecht, the Netherlands, pp. 679–739.

Junk, W.J. 1997. *The Central Amazon Floodplain: Ecology of a Pulsing System.* Ecological Studies, Vol. 126. Springer Verlag, Berlin, Germany, p. 526.

Junk, W.J. 2006. Flood pulsing and the linkages between terrestrial, aquatic and wetland systems. *Verh. Internat. Verein. Limnol. Stuttgart* 29:11–38.

Junk, W.J. and Bayley, P.B. 2004. The scope of the flood pulse concept regarding riverine fish and fisheries given geographic and manmade differences among systems. In: *World Fisheries Congress*, Vol. 4. American Fisheries Society, London, U.K.

Junk, W.J.; Bayley, P.B.; and Sparks, R.E. 1989. The flood pulse concept in river-floodplain-systems. *Spec. Publ. Can. Soc. Fish Aquat. Sci.* 106:110–127.

Junk, W.J. and da Silva, C.J. 1995. Neotropical floodplains: A comparison between the Pantanal of Mato Grosso and the large Amazonian River floodplains. In: Tundisi, J.G.; Bicudo, C.E.M.; and Matsumura-Tundisi, T. (Eds.), *Limnology in Brazil.* Brazilian Academy of Sciences, Brazilin Limnological Society, Rio de Janeiro, Brazil, pp. 195–217.

Junk, W.J. and Furch, K. 1985. The physical and chemical properties of Amazonian Waters and their relationships with the biota. In: Prance, G.T. and Lovejoy, T.E. (Eds.), *Amazonia*. Pergamon Press, Oxford, U.K., pp. 3–17.

Junk, W.J. and Nunes da Cunha, C. 2005. Pantanal: A large South American wetland at a crossroads. *Ecol. Eng.* 24:391–401.

Junk, W.J.; Nunes da Cunha, C.; Wantzen, K.M.; Petermann, P.; Strüssmann, C.; Marques, M.; and Adis, J. 2006. Biodiversity and its conservation in the Pantanal of Mato Grosso, Brazil. *Aquatic Sci.* 68(3):278–309.

Junk, W.J.; Ohly, J.J.; Piedade, M.T.F.; and Soares, M.G.M. (Eds.). 2000. *The Central Amazon Floodplain: Actual Use and Options for a Sustainable Management*. Backhuys Publishers, Leiden, the Netherlands, 584pp.

Junk, W.J. and Piedade, M.T.F. 1997. Plan life in the floodplain with special reference to herbaceous plants. In: Junk, W.J. (Ed.), *The Central Amazon Floodplain: Ecology of a Pulsing System*. Ecological Studies, Vol. 126. Springer Verlag, Berlin, Germany, pp. 147–186.

Junk, W.J. and Weber, G.E. 1997. Amazonian floodplains: A limnological perspective. *Verh. Int. Verein. Limnol*. 26:526.

Junk, W.J. and Welcomme, R.L. 1990. Floodplains. In: Patten, B.C. et al. (Eds.), *Wetlands and Shallow Continental Water Bodies*. SPB Academic Publishing, The Hague, the Netherlands, pp. 491–524.

Kalk, M.; McLachlan, A.J.; and Howard Williams, C. (Eds.). 1979. *Lake Chilwa: Studies of Change in a Tropical Ecosystem*. W. Junk Publishers, Dordrecht, the Netherlands, 462pp.

Kimmel, B.L. and Groeger, A.W. 1984. Factors controlling primary production in lakes and reservoirs: A perspective. Lake and reservoir Management. In: *Proceedings of the Third Annual Conference October* 18–20. Knoxville, TN, USEPA Washington, DC, pp. 277–281.

Lansac-Tôha, F.A.; Bonecker, C.C.; Velho, L.F.; and Lima, A.F. 1997. Composição, distribuição e abundância da comunidade zooplanctônica. In: Vazzoler, A.E.A.de M.; Agostinho, A.A.; and Hahn, N.S. (Eds.), *A Planície de Inundação do Alto Paraná: Aspectos Físicos, Biológicos e Socioeconômicos*. Editora Universidade Estadual de Maringa. Maringá University Publisher, Maringá, Brazil, pp. 117–156.

Lemoalle, J. 1981. Phytosynthetic production and phytoplankton in the euphotic zone of some African and temperate lakes. *Rev. Hydrobiol. Trop*. II:31–37.

Lemoalle, J. 1983. Phytoplankton production. In: Carmmouze, J.P.; Durand, J.R.; and Levecque, C. (Eds.), *Lake Chad: Ecology and Productivity of a Shallow Tropical Ecosystem*. Monographic Biologique, W. Junk Publisher, The Hague, the Netherlands, pp. 357–384, 575pp.

Lévêque, C. 1973a. Dynamique des peuplements, biologie, et estimation de la production des mollusques benthiques du lac Tchad. *Cah. ORSTOM sér. Hydrobiol*. 7:117–147.

Lévêque, C. 1973b. Bilans énergétiques des populations naturelles de molluesques benthiques du lac Tchad. *Cah. ORSTOM sér. Hydrobiol*. 7:151–165.

Lévêque, C.; Brutom, M.N.; and Setongo, W.W. (Eds.). 1988. *Biology and Ecology of African Freshwater Fishes*. Orstom Collection Tranaux et Documents 216, Paris, France, 508pp.

Lewis, W.M. 1973. The thermal regime of Lake Lanao (Philippines) and its theoretical implications for tropical lakes. *Limnol. Oceanogr*. 18:200–217.

Lewis, W.M. 1984. A five-year record of temperature, mixing and stability for a tropical lake (Lake Valencia, Venezuela). *Arch. Hydrobiol*. 99:340–346.

Lewis, W.M. 1987. Tropical limnology. *Ann. Rev. Ecol. Syst*. 18:158–184.

Lewis, W.M. 1990. Comparisons of phytoplankton biomass in temperate and tropical lakes. *Limnol. Oceanogr*. 35:1838–1845.

Lewis, W.M. 1995. Tropical lakes: How latitude makes a difference. In: Timotius, K.H. and Göltenboth, F. (Eds.), *Tropical Limnology*, Vol. 1. Satya Wacana Christian University, Salatiga, Indonesia, pp. 29–44. Reprinted 1996 in Schiemer, F. and Boland, K.T. (Eds.), *Perspectives in Tropical Limnology*. SPB Academic Publishing, Amsterdam, the Netherlands, pp. 43–64.

Lewis, W.M.; Hamilton, S.K.; and Saunders, J.F. III. 1995. Rivers of Northern South America. In: Cushing, C.E.; Cummins, K.W.; and Minshall, G.W. (Eds.), *River and Stream Ecosystems*. Ecosystems of the World 22, Elsevier, Amsterdam, the Netherlands, pp. 219–256.

Löffler, H. 1964. The limnology of tropical high-mountain-lakes. *Verh. Int. Verein. Limnol.* 15:176–193.

Lowe-McConnell, R.H. 1969. Speciation in tropical freshwater fishes. In: Lowe-McConnell, R.H. (Ed.), *Speciation in Tropical Environments*. Academic Press, London, U.K., pp. 51–75.

Lowe-McConnell, R.H. 1984. The status of studies on Southern America freshwater food fish. In: Zaret, T.M. (Ed.), *Evolutionary Ecology of Neotropical Freshwater Fishes*. Dr. W. Junk, Haia, the Netherlands.

Lowe-McConell, R.H. 1987. *Ecological Studies in Tropical Fish Communities*. Cambridge University Press, Cambridge, U.K., 382pp.

Lowe-McConell, R.H. 1988. Broad characteristics of the Ichthyofaune. In: Levecque, C.; Bruton, M.N.; and Ssetongo, G.W. (Eds.), *Biology and Ecology of African Freshwater Fishes*. Orstom Collections Taraux et Documents 216, Paris, France, pp. 93–110, 508pp.

MacIntyre, S.R. and Melack, J.M. 1988. Frequency and depth of vertical mixing in an Amazon floodplain lake (L. Calado, Brazil). *Verh.int. Verein. Limnol.* 23:80–85.

MacIntyre, S.R. and Melack, J.M. 1995. Vertical and horizontal transport in lakes: Linking littoral, benthic and pelagic habitats. *J. North America Benthol. Soc.* 14:599–615.

Margalef, R. 1983. *Limnologia*. Ediciones Omega, Barcelona, Spain, 1025pp.

Margalef, R. 1997. *Our Biosphere*. In: Kinnee, O. (Series Ed.), Ecology Institute Publication, Olderndorf/Luke, Germany, 215pp.

Matsumura-Tundisi, T.; Hino, K.; and Claro, S.M. 1981. Limnological studies at 23 reservoirs of Southern part of Brazil. *Verh. Int. Verein. Limnol.* 21:1046–1056.

Matsumura-Tundisi, T. and Tundisi, J.G. 2003. Calanoida (Copepoda) species composition changes in the reservoirs of S. Paulo State (Brazil) in the last twenty years. *Hydrobiologia* 504:215–222.

McLachlan, A.J. 1969. Some effects of water level fluctuation of the benthic fauna of two Central African Lakes *Limnol. Soc. South Africa. Newsletter* 13:58–63.

McLachlan, A.J. 1975. The role of aquatic macrophytes in the recovery of the benthic fauna of a tropical lake after a dry phase. *Limnol. Oceanogr.* 20:54–63.

McLachlan, A.J. 1979. Decline and recovery of the benthic invertebrate communities. In: Kalk, M.; McLachlan, A.J.; and Howard-Williams, C. (Eds.), *Lake Chilva: Studies of Change in a Tropical Ecosystem*, Monographie Biologique (J. Illies-Edition) Vol. 35. W. Junk Publishers, The Hague, the Netherlands, pp. 147–159, 462pp.

Melack, J.M. 1979. Temporal variability of phytoplankton in tropical lakes. *Oceanologica* 44:1–7.

Melack, J.M. 1982. Photosynthetic activity and respiration in an equatorial African soda lake. *Freshwater Biol.* 12:381–400.

Melack, J.M. 1984. Amazon floodplain lakes: Shape, fetch, and stratification. *Verh. Int. Verein. Limnol. Stuttgart* 22:1278–1282.

Melack, J.M. 1988. Primary producer dynamics associated with evaporative concentration in a shallow, equatorial soda lake (Lake Elmeteita, Kenya). *Hydrobiologia* 158:1–14.

Melack, J.M. 1996. Recent developments in tropical limnology. *Verh. Int. Verein. Limnol.* 26:211–217.

Melack, J.M. and Fisher, T.R. 1983. Diel oxygen variations and their ecological implications in Amazon floodplain lakes. *Archiv fur Hydrobiologie, Stuttgart*, 98(4):422–442.

Melack, J.M. and MacIntyre, S. 1992. Phosphorus concentrations, supply and limitation in tropical African lakes and rivers. In: Tiessen, H. and Frossard, E. (Eds.), *Phosphorus Cycles in Terrestrial and Aquatic Ecosystems*. Africa SCOPE, Saskatchewan Institute of Pedology, Saskatoon, Canada, pp. 1–18.

Menezes, N.A. 1970. Distribuição e origem da fauna de peixes de água doce das grandes bacias fluviais do Brasil. In: Comissão Interestadual da Bacia Paraná-Uruguai. Poluição e piscicultura; notas sobre poluição, ictiologia e piscicultura. São Paulo, Faculdade de Saúde Pública da USP, Secretaria da Agricultura, Instituto de Pesca, 216pp.

Mercante, M.A.; Rodrigues, S.C.; and Ross, J.L.S. 2011. Geomorphology and habitat diversity in the Pantanal. *Braz. J. Biol.* 71(Suppl. 1):233–240.

Meyer, A.; Montero, C.; and Spreinat, A. 1994. Evolutionary history of the cichlid fish species flocks of the East African great lakes inferred rom molecular phylogenetic data. *Arch. Hydrobiol. Beih. Ergebn. Limnol.* 44:407–423.

Meyer, A.; Montero, C.; and Spreinat, A. 1996. Molecular phylogenetic inferences about the evolutionary history of East African cichlid fish radiations. In: Johson, T.C. and Odada, E.O. (Eds.), *The Limnology, Climatology, and Paleoclimatology of the East African Lakes*. Gordon & Breach, Amsterdam, the Netherlands, pp. 303–323.

Mitamura, O.; Saijo, Y.; Hino, K.; and Barbosa, F.A.R. 1995. The significance of regenerated nitrogen for phytoplankton productivity in the Rio Doce Valley Lakes, Brazil. *Arch. Hydrobiol.* 134:179–194.

Moss, B. 1979. The Lake Chilawa—A limnological overview. In: Falk, M.; Mchaclan, A.J.; and Howard Williams, C. (Eds.), *Lake Chilwa: Studies of Change in a Tropical Ecosystem*. W. Junk Publishers, Dordrecht, the Netherlands, pp. 401–413, 462pp.

Moss, B. 2007. *Ecology of Fresh Waters: Man and Medium, Past to Future*, 3rd edn. Blackwell Publishing, Oxford, U.K., 557pp.

Ohly, J.J. and Junk, W.J. 1999. Multiple use of Central Amazonian floodplains: Combining ecological conditions requeriments for environmental protection and socio economic needs. In: Padoch, C.; Ayres, J.M.; Pinedo-Vasquez, M.; and Henderson, A. (Eds.), *Várzea: Diversity, Development, and Conservation of Amazonia's Whitewater Floodplains*. Botanical Garden Press, Bronx, NY, pp. 283–300.

Oltman, R.E. 1970. Reconnaissance investigations of the discharge and water quality of the Amazon. In: *Atas do Simpósio sobre a Biota Amazônica*. Rio de Janeiro, Brazil, CNPq, Vol. 3, pp. 163–185.

Padoch, C.; Ayres, J.M.; Pinedo-Vasquez, M.; and Henderson, A. (Eds.). 1999. *Varzea Diversity, Development and Conservation of Amazonian's Whitewater Floodplains*. Botanical Garden, New York, 407pp.

Rai, H. and Hill, G. 1984a. Microbiology of Amazonian waters. In: Sioli, H. (Ed.), *The Amazon*. Monographiae Biologicae, Vol. 56. W. Junk Publishers, Dordrecht, the Netherlands, pp. 413–41.

Rai, H. and Hill, G. 1984b. Primary production in the Amazonian aquatic systems. In: Sioli, H. (Ed.), *The Anaton: Limnology and Landscape Ecology of a Mighty Tropical River and Its Basin.* Dr. W. Junk Publishers, Dordrecht, the Netherlands, pp. 311–335, p. 763.

Reynolds, C.S.; Tundisi, J.G.; and Hino, K. 1983. Observations on a metalimnetic Lyngbya population in a stably stratified tropical lake (Lagoa Carioca, Eastern Brazil). *Arch. Hydrobiol.* 97:7–17.

Richey, J.E.; Hedges, J.I.; Devol, A.H.; Quay, P.D.; Victoria, R.; Martinelli, L.; and Forsberg, B.R. 1990. Biogeochemistry of carbon in the Amazon River. *Limnol. Oceanogr.* 35:352–371.

Rocha, O.; Gaeta Espindola, E.L.; Fenerich-Verani, N.; Verani, J.R.; and Rietzler, A. 2005. *Espécies invasoras em águas doces. Estudos de caso e propostas de manejo.* UFSCar, São Paulo, Brazil, 416pp.

Rodrigues, L. 1998. Sucessão do Perifíton na Planície de Inundação do Alto Rio Paraná: Interação entre Nível Hidrológico e Regime Hidrodinâmico. PhD dissertation, Universidade Estadual de Maringá, Maringá, Brazil, 208pp.

Roosevelt, A.C. 1999. Twelve thousand years of human-environment interaction in the Amazon floodplain. In: Padoch, C.; Ayres, J.M.; Pinedo-Vasquez, M.; and Henderson, A. (Eds.), *Varzea Diversity, Development and Conservation of Amazonian's Whitewater Floodplains.* Botanical Garden, New York, pp. 371–392.

Rzóska, J. (Ed.). 1976. *The Nile, Biology of an Ancient River.* Monographiae Biologicae, Vol. 29. W. Junk Publishers, The Hague, the Netherlands.

Salati, E. and Marques, J. 1984. Climatology of Amazon region. In: Sioli, H. (Ed.), *The Limnology and Landscape-Ecology of a Mighty River and Its Basin.* W. Junk Publishers, Dordrecht, the Netherlands, pp. 85–126.

Salo, J.; Kalliola, R.; Hakkinen, J.; Makonen, Y.; Niemela, P.; Puhakka, M.; and Coley, P.D. 1986. River dynamics and the diversity of Amazon lowland forest. *Nature* 322:254–258.

Scheffer, M. 1998. *Ecology of Shallow Lakes.* Chapman & Hall, Chichester, U.K., 357pp.

Schmidt, G.W. 1970. Numbers of bacteria and algae and their interrelations in some Amazonian waters. *Amazoniana* 2:393–400.

Schmidt, G.W. 1972. Seasonal changes in water chemistry of a tropical lake (Lago do Castanho, Amazonia, South America). *Verh. Int. Verein. Limnol.* 18:613–621.

Schmidt, G.W. 1973a. Primary production of phytoplankton in three types of Amazonian waters II. The limnology of a tropical floodplain in central Amazonia (Lago do Castanho). *Amazoniana* 4:139–203.

Schmidt, G.W. 1973b. Primary production of phytoplankton in three types of Amazonian waters. III. Primary productivity of phytoplankton in a tropical flood-plain lake of central Amazonia, Lago do Castanho, Amazonas, Brazil. *Amazoniana* 4:379–404.

Schmidt, G.W. 1976. Primary production of phytoplankton in three types of Amazonian waters. IV. On the primary productivity of phytoplankton in a bay of lower Rio Negro (Amazonas, Brazil). *Amazoniana* 5:817–828.

Schmidt, G.W. 1982. Primary production of phytoplankton in three types of Amazonian waters. V. Some investigations on the phytoplankton and its primary productivity of phytoplankton in the clearwater of the lower Rio Tapajóz (Pará, Brazil). *Amazoniana* 7:335–348.

Sioli, H. 1951. Zum Alterungsprozess von Flüsen, und Flusstypen im Amazonasgebiet. *Naturwiss* 41:456–457.

Sioli, H. 1984. *The Amazon: Limnology and Landscape Ecology of a Mighty Tropical River and Its Basin.* Dr. W. Junk Publishers, Dordrecht, the Netherlands, p. 763.

Souza, M.C.; Cislinski, J.; and Romagnolo, M.B. 1997. Levantamento florístico. In: Vazzoler, A.E.A.de M.; Agostinho, A.A.; and Hahn, N.S. (Eds.), A Planície de Inundação do Alto Rio Paraná: Aspectos Físicos, Biológcos e Socioeconômicos. Editora Universidade Estadual de Maringa. Maringá State University Publisher. Maringá, Brazil, pp. 343–368.

Straskraba, M. 1999. Retention time as a key variable of reservoir limnology. In: Tundisi, J.G. and Straskraba, M. (Eds.), *Theoretical Reservoir Ecology and Its Application*. Backhuys Publishers, Brazilian Academy of Sciences, International Institute of Ecology, Leiden, the Netherlands, pp. 385–410, 585pp.

Straskraba, M. and Tundisi, J.G. 1999. *Theoretical Reservoir Ecology and Its Applications*. Brazilian Academy of Sciences/International Institute of Ecology, Backhuys Publishers, Leiden, the Netherlands, 592pp.

Straskraba, M.; Tundisi, J.G.; and Duncan, A. (Eds.). 1993. *Comparative Reservoir Limnology and Water Quality Management*. Kluwer Academic Publishers, Dordrecht, the Netherlands, 291pp.

Talling, J.F. 1956. Comparative problems of phytoplankton production and photosynthetic activity in a tropical and temperate lake. *Mem. Ist. Ital. Hydrobiol.* 18(Suppl.):399–424.

Talling, J.F. 1963. Origin of stratification in a African rift lake. *Limnol. Oceanogr.* 18:68–78.

Talling, J.F. 1966. The annual cycle of stratification and phytoplankton growth in Lake Victoria (East Africa). *Int. Rev. Ges. Hydrobiol.* 51:545–621.

Talling, J.F. 1969. The incidence of vertical mixing, and some biological and chemical consequences, in tropical African lakes. *Verh. Int. Verein. Limnol.* 17:998–1012.

Talling, J.F. 1976. Water characteristics. In: Rzóska, J. (Ed.), *The Nile, Biology of an Ancient River*. Monographiae Biologicae, Vol. 29. W. Junk Publishers, The Hague, the Netherlands, pp. 357–84.

Talling, J.F. 1986. The seasonality of phytoplankton in African lakes. *Hydrobiologia* 138:139–160.

Talling, J.F. 1987. The phytoplankton of Lake Victoria (East Africa). *Arch. Hydrobiol. Beih. Ergebn. Limnol.* 25:229–256.

Talling, J.F. 1992. Environmental regulation in African shallow lakes and wetlands. *Ver. Hydrob. Trop.* 25:87–144.

Talling, J.F. and Lemoalle, J. 1998. *Ecological Dynamics of Tropical Inland Waters*. Cambridge University Press, Cambridge, U.K., 441pp.

Talling, J.F. and Talling, I.B. 1965. The chemical composition of African lake waters. *Int. Rev. Ges. Hydrobiol.* 50:421–463.

Thomaz, S.M.; Bini, L.M.; and Alberte, S.M. 1997. Limnologia do reservatório de Segredo: Padrões de variação espacial e temporal. In: Agostinho, A.A. and Gomes, L.C. (Eds.), *Reservatorio de Segredo: Bases Ecológicas Para o Manejo Maringá*. Eduem, Maringá-Paraná, Brazil, pp. 19–37.

Thomaz, S.M.; Lansac Toha, F.A.; Roberto, M.C.; Esteves, F.A.; and Lima, A. 1992. Seasonal variation of some limnological factors of lagoa do Guaraná a várzea lake of The High Rio Paraná State of. Mato Grosso do Sul. *Revue D. Hydrobiologie Tropicale*. Paris. 25(4):269–276.

Train, S. and Rodrigues, L.C. 1998. Temporal fluctuations of the phytoplankton community of the Bacia River in the upper Paraná floodplain. Mato Grosso do Sul, Brazil. *Hydrobiologia* 361(1–3):125–134.

Tundisi, J.G. 1983. A review of basic ecological processes interacting with production and standing crop of phytoplankton in lakes and reservoirs in Brazil. *Hydrobiologia* 100:223–243.

Tundisi, J.G. 1992. Comparative studies of the mechanisms of ecological functioning of Barra Bonita reservoir (Tiete river S. Paulo State) and Jurumirim reservoir (Paranapanema river S. Paulo State) under the impact of their watershed uses. Report to Fapesp (S. Paulo State Foudation.) Thematic Project 0922/91-5.

Tundisi, J.G. 1994. Tropical South America: Present and perspectives. In: Margalef, R. (Ed.), *Limnology Now: A Paradigm of Planetary Problems*. Elsevier, Amsterdam, the Netherlands, pp. 353–424.

Tundisi, J.G. 1997. Climate. In: Tundisi, J.G. and Saijo, Y. (Ed.), *Limnological Studies on the Rio Doce Valley Lakes*. USP, Brazilian Academy of Sciences, S. Paulo, Brazil, pp. 7–13.

Tundisi, J.G.; Bicudo, C.E.M; and Matsumura-Tundisi, T. (Eds.). 1995. *Limnology in Brazil*. Brazilian Academy of Sciences/Brazilian Limnological Society, São Carlos, Brazil, 376pp.

Tundisi, J.G.; Forsberg, B.R.; Devol, A.H.; Zaret, T.M.; Matsumura-Tundisi, T.; Dos Santos, A.; Ribeiro, J.; and Hardy, E.R. 1984. Mixing patterns in Amazon lakes. *Hydrobiologia* 108:3–15.

Tundisi, J.G. and Matsumura-Tundisi, T. 1990. Limnology and eutrophication of Barra Bonita reservoir, S. Paulo State, Southern Brazil. *Arch. Hydrobiol. Beih. Ergebn. Limnol*. 33:661–676.

Tundisi, J.G. and Matsumura-Tundisi, T. 1994. Plankton diversity in a warm monomicticlake (Dom Helvécio, MG) and polymictic reservoir (Barra Bonita, SP): A comparative analysis of the intermediate disturbance hypothesis. *An Acad. Bras. Cienc*. 66(Suppl. 1):16–28.

Tundisi, J.G. and Matsumura-Tundisi, T. 2008a. Biodiversity in the neotropics: Ecological, economical and social values. *Braz. J. Biol*. 68(4):913–915.

Tundisi, J.G. and Matsumura-Tundisi, T. 2008b. *Limnologia*. Oficina de Textos, São Paulo, Brazil, 632pp.

Tundisi, J.G. and Matsumura-Tundisi, T. 2010. Impactos potenciais das alterações do Código Florestal nos recursos hídricos. *Biota Neotrop*. 10(4):67–76.

Tundisi, J.G. and Matsumura-Tundisi, T. 2012. *Limnology*. CRC Press Taylor & Francis, Boca Raton, FL, 853pp.

Tundisi, J.G.; Matsumura-Tundisi, T.; and Abe, D.S. 2008. The ecological dynamics of Barra Bonita reservoir: Implications for its biodiversity. *Braz. J. Biol*. 68(Suppl. 4):1079–1098.

Tundisi, J.G.; Matsumura-Tundisi, T.; and Calijuri, M.C. 1993. Limnology and management of reservoirs in Brazil. In: Straskraba, M.; Tundisi, J.G.; and Duncan, A. (Eds.), *Comparative Reservoir Limnology and Water Quality Management*. Kluwer Academic Publishers, Dordrecht, the Netherlands, pp. 25–55, 291pp.

Tundisi, J.G.; Matsumura-Tundisi, T.; Henry, R.; Rocha, O.; and Hino, K. 1988. Comparação do estado trófico de 23 reservatórios do Estado de São Paulo: eutrofização e manejo. In: Tundisi, J.G. (Ed.), *Limnologia e Manejo de Represas*. Série: Monografias em Limnologia 1(1). Academia de Ciências do Estado de São Paulo, São Paulo, Brazil, pp. 165–204.

Tundisi, J.G.; Matsumura-Tundisi, M.; Pontes, M.C.F.; and Gentil, J.G. 1981. Limnological studies at Quaternary Lakes in Eastern Brazil. I. Primary production of phytoplankton and ecological factors at Lake D. Helvecio. *Rev. Braz. Bot*. 4:5–14.

Tundisi, J.G.; Matsumura-Tundisi, T.; Saraiva, A.; and Campagnoli, F. 2012. How many more dams in the Amazon? (in preparation).

Tundisi, J.G. and Saijo, Y. (Eds.). 1997. *Limnological Studies on the Rio Doce Valley Lakes.* Brazilian Academy of Sciences, São Carlos, Brazil.

Tundisi, J.G.; Saijo, Y.; Henry, R.; and Nakamoto, N. 1997. Primary productivity, phytoplankton biomass and light photosynthesis responses in four lakes. In: Tundisi, J.G. and Saijo, Y. (Eds.), *Limnological Studies on the Rio Doce Valley, Brazil.* Brazilian Academy of Science, São Carlos, Brazil, pp. 199–225.

Tundisi, J.G. and Straškraba, M. (Eds.). 1999. *Theoretical Reservoir Ecology and Its Applications.* Instituto Internacional de Ecologia, São Carlos, Brazil, 858pp.

Turcotte, P.R. and Harper, P.P. 1982. Drift patterns in a high Andean Stream. *Hydrobiologia* 89:141–151.

Ungemach, H. 1972. Die Ionenfracht dês Rio Negro, Staat Amazonas, Brasiliem Nach Untersuchungem. *Amazoniana.* 3:175–185.

Val, A.L. and Almeida Val, V.M.F. 1995. *Fishes of the Amazon and Their Environmental: Physiological and Biochemical Aspects (Zoophysiology).* Springer, Berlin, Germany, 224pp.

Val, A.L.; Almeida Val, V.M.A; and Randall, D.J. 1996. *Physiology and Biochemistry of the Fishes of the Amazon.* INPA, Manaus, Brazil, 402pp.

Viner, A.B. and Smith, I.R. 1973. Geographical, historical and physical aspects of Lake George. *Proc. R. Soc. B* 184:235–270.

Walker, I. 1995. Amazonian streams and small rivers. In: Tundisi, J.G.; Bicudo, C.E.M.; and Matsumura-Tundisi, T. (Eds.), *Limnology in Brazil.* Brazilian Academy of Sciences, Rio de Janeiro, Brazil, pp. 167–194.

Walker, I. and Henderson, P.A. 1998. Ecophysiological aspects of Amazonian blackwater litterbank fish communities. In: Val, A.L.; Almeida-Val, V.M.F.; and Randall, D.J. (Eds.), *Physiology and Biochemistry of the Fishes of the Amazon.* Instituto Nacional de Pesquisas da Amazônia (INPA), Manaus, Brazil, pp. 7–22.

Walker, I.; Henderson, P.A.; and Sterry, P. 1991. On the patterns of biomass transfer in the benthic fauna of the Amazonian black-water river, as evidenced by ^{32}P label experiment. *Hydrobiologia* 215:153–162.

Walker, T. and Tyler, P. 1984. Tropical Australia, a dynamic limnological environment. *Verh. Int. Verein. Limonol.* 22:1727–1734.

Wallace, A.R. 1853. *A Narrative of Travels on the Amazon and Rio Negro with an Account of Native Tribes.* Ward, Lock & Co., London, U.K.

Wantzen, K.M. et al. 2008. Towards a sustainable management concept for ecosystem services of the Pantanal Wetland. *Ecohydrol. Hydrobiol.* 8(2, 4):115–138.

Wantzen, K.M. and Junk, W.J. 2006. Aquatic-terrestrial likages from strems to Rivers: Biotic hot spots and hot momments. *Archiv für Hydrobiologie Suppl.* 158:595–611.

Ward, J.V.; Tockner, K.; and Schiemer, F. 1999. Biodiversity of river ecosystems: Ecotones and connectivity. *Regul. Rivers: Res. Manage.* 15:321–328.

Welcomme, R.L. 1979. *Fisheries Ecology of Floodplain Rivers.* Longman, London, U.K., 317pp.

Welcomme, R.L. 1985. Rivers fisheries. FAO, Rome, Italy, Fisheries Technical Paper, p. 262.

Wilson, E.O. 1998. *Biodiversity.* National Academic Press, Washington, DC.

Zaleswki, M. 2008. Rationale for the "Floodplain Declaration" for, environmental conservation toward sustainability science. *Ecohydrol. Hydrobiol.* 8(2, 4):107–113.

Zaret, T.M.; Devol, A.D.; and Dos Santos, A. 1981. Nutrient addition experiments in Lago Jacaretinga, Central Amazon Basin, Brazil. *Verh. Int. Verein. Limnol.* 21:721–724.

8

Freshwater Temperate Lakes and Reservoirs

8.1 Introduction

The development of limnology in the last 100 years is due undoubtedly to extensive studies of lakes, wetlands, estuaries, and reservoirs in temperate zones. Only in the second half of the twentieth century, tropical limnology contributed significantly to improve the conceptual basis of limnology. Many lake districts of temperate zones have been formed and modified by glacial action. Comparative studies in lake districts in temperate regions have developed many principles and mechanisms of the lake ecosystem functioning (Horne and Goldman 1994, Tundisi and Matsumura Tundisi 2011). Comparative studies in different lakes within the same geographical region are an important conceptual route to advance limnological knowledge. If the lakes are situated in regions with similar climate, soil, vegetation, geomorphological setup, comparison can be made in relation to the biogeochemical cycles, the composition of the aquatic biota, the introduction of a fish species, or the impact of pollution in a particular lake of a lake district. Therefore, by developing regional limnology in temperate lakes, several new concepts were proposed by different authors. A sequence of some contributions is shown in Table 8.1.

Scientists were associated with each lake district in temperate regions. In the lake district of the English lakes, Macan, Lund, Reynolds, and Talling are well-known names. In the German and Austrian lakes, Ruttner and Ohle were the names associated with the lake comparative studies. Birge and Juday studied Wisconsin lakes, Welch and Eggleton, Michigan lakes. Goldman dedicated intensive studies to Lake Tahoe. Beadle and Talling studied east African lakes; Rawson studied Canadian lakes; and Schindler created the Experimental Lakes Area in Canada (Macan 1970, Schelske and Roth 1973, Vollenweider 1976, Likens 1992, Schindler et al. 2008).

In Japan, several studies were developed by Yoshimura (1938), Mori and Yamamoto (1975), and Matsuyama (1978).

TABLE 8.1

Main Developments and Conceptual Advances in Limnology Promoted by Studies in Freshwater Temperate Lakes

Forel (1901)—Physical classification based on thermal structure of lakes
Birge and Juday (1911)—Chemical classification based on thermal structure and dissolved oxygen
Chemical and zoological classification based on the sediment biota. Classification based on organic matter in the sediment
Yoshimura (1938)—Comparative analysis of dissolved oxygen and thermal structure in lakes of Japan
Mortimer (1941)—Lake circulation and sediment—water interactions
Lindeman (1942)—Trophic dynamic theory applied to lakes
Odum (1956)—Development of techniques to measure river metabolism
Margalef (1958)—Introduction of the information theory in the phytoplankton succession
Vollenweider (1965)—Concept of load to lakes and reservoirs. Advances in eutrophication studies
Likens and Borman (1974)—The watershed as a unit of study
Imberger (1994)—The hydrodynamics of lakes
Reynolds (1997)—Phytoplankton succession temporal scales
Jørgensen and Svirezhev (2004)—A thermodynamic theory for ecological systems. The concept of exergy applied as ecological indicator

8.2 English Lake District

The 18 lakes of the English Lake District were intensively studied and a synthesis of the research was described in Macan (1970), Lund (1971, 1979), Walsby and Reynolds (1980), and many other authors. Most of the lakes are monomictic; some lakes are polymictic and few lakes are dimictic. The lakes have a wide variety of physical, chemical, and biological conditions that are summarized in Table 8.2.

One important feature described in the studies in the English Lake District is the stratification and the circulation process and the phytoplankton that develops with these processes. The comparative studies ranked the lakes by an eutrophication classification as oligotrophic, mesotrophic, and eutrophic. The biological studies focused on the phytoplankton composition and succession (Reynolds 1997), the bottom fauna and studies of groups of benthic organisms (Macan 1970), the indicator species of bottom organisms, and the fish fauna and fisheries. The mentioned topics are all examples of approaches on how to do research in a lake district and how to contribute significantly to comparative studies. One of these studies by Talling (1965a,b, 1966) compared the productivity of phytoplankton and the seasonal cycle between Lake Victoria in Africa (of tectonic origin) and Lake Windermere (of glacial origin) in the English Lake District. Victoria has higher primary production (950 g C/m^2 year) as compared to Lake Windermere (20 g C/m^2 year). This is probably due to an extensive solar radiation in Lake Victoria, higher temperature

TABLE 8.2

Main Physical, Chemical Conditions of the English Lake District

Characteristics of Selected Lakes in the English Lake District

All are monomictic except for Blelham Tarn and Esthwalte Water, which are occasionally dimictic, and Bassenthwaite Water, which is polymictic (P)

Lake	L (km) B (km) A (km²)	Depth Z_{max} (Z) (m)	Drainage Basin Area (km²)	Thermocline Depth (m)	Typical Max (min) Surface Temp. (°C)	Approximate Period of Thermal Stratification	Winter (Summer) Nutrients (µg/L)		
							NO_3—N	PO_4—P	SiO_2
Windermere	17 1.5 15	67 (24)	231	5–20(N) 5–20(S)	17 (3–5)	May–end Nov.	300 (100)	3.0 (0.5)	600 (200)
Wastewater	4.8 0.8 2.9	79 (41)	49	15–20	15 (<4)	June–Dec.	>140 (−100)		
Ennerdale	3.8 0.9 2.9	45 (19)	44	15–25	19 (<4)	May–Oct.	>140 (50)		
Esthwait	2.5 0.6 1.0	16 (6.4)	14	9–12	20	Apr.–mid-Sept.	400 (100)	2–4 (0.5)	1000 (100)
Blelham	0.8 0.3 0.11	15 (6.6)	2.3	8–10	20 (3.5)	Apr.–mid-Sept	600 (200)	4 (0.5)	1000 (200)
Bassenthwaite	6.2 1.2 5.4	21 (5.5)	238	P	21 (<4)	Irregular	~250		

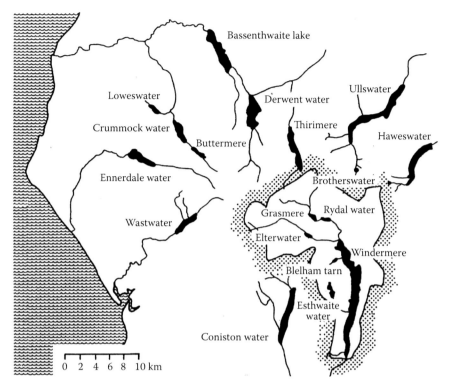

FIGURE 8.1
Lakes of the English lake District. (From Lund, J.W.G., *Investigations on Phytoplankton with Special Reference to Water Usage*, Freshwater Biological Association, Ambleside, U.K., Occasional Publication 13, pp. 5–56, 1981.)

(10°C–12°C higher) in Lake Victoria than in Lake Windermere, and a deeper euphotic zone in the African lake as compared with Windermere.

Figure 8.1 shows the geography of the lakes in the English Lake District.

8.3 Laurentian Great Lakes in North America

The five lakes—Lake Superior, Lake Huron, Michigan, Erie, and Ontario, together represent the greatest volume of freshwater on Planet Earth, slightly more than the volume of Lake Baikal.

Their age is approximately 10,000 years and the retention time of Lakes Superior and Michigan is around 200 and 100 years, respectively. The chemical composition of the water of the Great Lakes is shown in Table 8.3. The Great Lakes are monomictic and their seasonal cycle is similar to smaller lakes in temperate regions: a summer circulation with growth of phytoplankton

TABLE 8.3

Physical and Chemical Characteristics of the Great North American Lakes

Physical and Chemical Characteristics of Pelagic Waters of the Laurentian Great Lakes

Note that, as in the English Lake District, the trophic classification based on transparency (Secchi disk) or chlorophyll a is also supported by the distribution of phytoplankton, zooplankton, and zoobenthos

Lake	L (km) B (km) A (km²)	Z_{max} (Z) (m)	Drainage Basin Area (km² × 10³)	Thermocline Depth (m)	Max/Min Summer (Winter) Temp (°C)	Hydraulic Residence Time, Year	Approximate Period of Thermal Stratification	Winter (Summer) Surface Nutrients (µg/L) NO_3–N$^+$ NH_4–N	PO_4–P	SiO_2
Superior	560 256 82,000	406 (149)	125	10–13	14 (0.5)	184	Aug.–Dec	280 (220)	0.5 (0.5)	2200 (2000)
Michigan	490 188 58,000	281 (85)	118	10–15	18–20 (<4)	104	July–Dec.	300 (130)	6 (5)	1300 (700)
Huron	330 292 60,000	228 (59)	128	15–30	18.5 (<4)	21	End June– Oct. or Nov.	260 (180)	0.5 (0.5)	1400 (800)
Erie[a]	385 91 26,000	w: 13(7,3) c: 24(18) e: 70(24)	59	W: p c: 14–20 e: 30	24 (<4)	w: 0.13 c: 1.7 e: 0.85 alt: 3	Mid-June–Nov.	w: 640 (80) c: 140 (20) e: 180 (20)	23 (2) 7 (1) 7 (1)	1300 (60) 350 (30) 300 (30)
Ontario	309 85 20,000	244 (86)	70	15–20	20.5 (<4)	8	End June– Nov.	280 (40)	14 (1)	400 (100)

Sources: Schelske, C. and Roth, J.C., Limnological survey of the lakes Michigan, Superior, Huron and Erie, Vol. 17, Great Lakes Research Division, University of Michigan, Ann Arbor, MI, 108pp, 1973; Dobson, H.F.H. et al., *J. Fish. Res. Board Can.,* 13, 731, 1974; Ragotzkie, R.A., *Am. Sci.* 62, 454, 1974; Bennett, E.B., *J. Great Lake Res.,* 4, 310, 1978.

[a] For Lake Erie, w = western; c = central, e = eastern, p = polymictic.

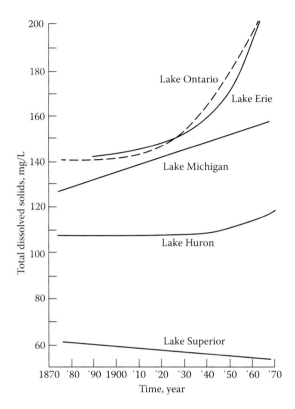

FIGURE 8.2
The evolution of total dissolved solids in the Great Lakes for a period of 100 years.

limited by phosphorus and silicon, dominance of phytoplankton by diatoms, such as *Fragilaria, Tabellaria, Aulacoseira, Asterionella*; in some areas subjected to eutrophication such as in Lake Erie phytoplankton blooms of cyanobacteria such as *Anabaena, Microcystis, Aphanizomenon* are common.

The lakes are influenced by their watersheds; therefore, they may undergo a slow process of eutrophication. Figure 8.2 shows the changes and progression of total dissolved solids in 100 years (Beeton 1969).

8.4 Japanese Lakes

The classical work of Yoshimura (1938) the contribution of Japan to IBP (Mori and Yamamoto 1975), and the studies in Lake Biwa (Nakamura and Nakagima 2002) are other fundamental studies in lakes of the temperate regions. In Lake Biwa, the control of eutrophication as compared with land use (Figure 8.3) is an important contribution to understand the process of eutrophication in a large lake surrounded by urban area. It also is

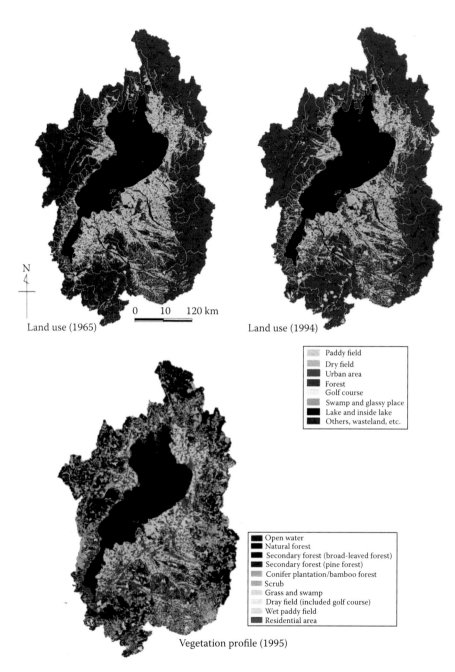

FIGURE 8.3
(See color insert.) Changes in vegetation cover and land use in the Lake Biwa Watershed. (From LBRI, *Lake Biwa Study Monographs*, Lake Biwa Research Institute, Otsu, Japan, 118pp, 1984.)

an example of eutrophication control and abatement due to application of technology and comprehensive and profound limnological and ecological studies. The eutrophication of Lake Biwa has been at almost the same level from about 1980 to about 2005 due to massive investments in wastewater treatment. The eutrophication has not been reduced because the abatement of nonpoint sources was not considered sufficiently. In the last 5 years or more, the abatement of the agricultural nutrient pollution has grown considerably and at least a minor reduction in the eutrophication is expected in the coming years due to construction and restoration of wetlands for removal of nonpoint pollution nutrients.

Long-term changes in river water quality were monitored (Azuma and Okubo 2002), nutrient influx from ground water was determined and the loads of phosphorus and nitrogen from point and nonpoint sources were studied. Furthermore, bioassay tests in river water were carried out. The bioassay tests in river water were performed with both phytoplankton and zooplankton (Okubo et al. 1998). One important study was the large-scale surface vortices determination that showed how the water masses were distributed in the lake and influenced the water quality. The overall mass balance for Lake Biwa for phosphorus and nitrogen is shown in Figure 8.4 (Somiya 2000).

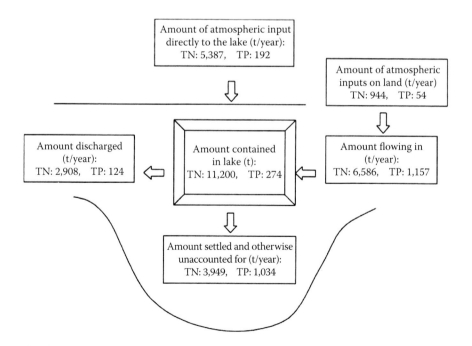

FIGURE 8.4
Total nitrogen (TN) and total phosphorus (TP) mass balance of Lake Biwa (simplified version of a diagram presented by Fujii and Somiya. (From Somiya, I., *Lake Biwa, Its Environmental and Water Quality Formation*, Gihodo Co., Tokyo, Japan (in Japanese), p. 165, 2000.)

The work of Yoshimura (1938) is a classical contribution to the study of freshwater lakes of Japan, in relation to biological productivities and thermal condition. Approximately 200 freshwater and brackish water lakes were studied. Vertical profiles of dissolved oxygen and water temperature were obtained considering oligotrophic, mesotrophic, and eutrophic lakes. Lakes sampled ranged from temperate to subtropical regions. Some meromictic lakes were studied in seasonal cycle. The work of Yoshimura (1938) can be considered a fundamental study of temperate lakes with a comparative approach and by utilizing only two state variables: water temperature and dissolved oxygen. The vertical profiles in each lake enabled the author to compare the freshwater ecosystems and their characteristic.

8.5 Reservoirs in Temperate Regions

Several studies in reservoirs of temperate regions were developed in order to understand these artificial ecosystems (Straskraba et al. 1993, Jørgensen et al. 2005). The contributions promoted conceptual advances and provided many alternatives for a proper integrated management. The important role of retention time to control water quality of the reservoir, the design of the level of outlets to remove water of undesirable quality downstream and the knowledge of patterns of circulation provided a basis for interventions and control of the water quality of the reservoirs. Equally important was the integration of watershed management with the reservoir management (Kennedy 1999, Kennedy et al. 2003).

The intensive studies on Rimov reservoir (Czech Republic) with several years of monitoring provided detailed observations on the spring overturn of the reservoir. The conclusion is that aging is not only a consequence of the increasing load from the watershed but is also due to the changes in biotic interactions, development of fresh populations (Straskraba et al. 1993), and changes in the organization of tropic relationships. In addition, the microbial food chain was intensively studied (Straskraba 1994, Masin et al. 2003). Moreover, studies of biomanipulation of the reservoir as a possible control mechanism were also intensively carried out (Hejzlar 1989). Other studies in temperate reservoirs were developed in Spain (Margalef et al. 1976). A comparative study of 100 reservoirs was developed and provided an important conceptual basis for management (Armengol et al. 1999). Important studies were also performed in the London drinking water reservoirs (Duncan 1975, Straskraba et al. 1993).

The studies of freshwater lakes and reservoirs in temperate regions have been fundamental to the development of limnological knowledge but have also contributed intensively to improved management strategies of the natural and artificial freshwater ecosystems in general (Vollenweider 1965, 1975, 1976).

This work stimulated again comparative studies of lakes and reservoirs of tropical, subtropical, and arctic regions, which enhanced the theoretical knowledge. The results were major conceptual advances in the understanding of the functioning of lakes and reservoirs (Walsby and Reynolds 1980). Improvement of management technology benefitted from these advances too. The studies of temperate freshwater ecosystems were furthermore important as major contributions to theoretical ecology. Problems such as succession of primary producers; food chain interactions; and climatological, hydrological, and biological responses of ecosystems were solved by these studies (Talling 1957, 1971, Lund et al. 1971, Lund 1972a,b, 1979). The long-term research developed intensively in some watersheds (Likens 1992) or lakes (Horne and Goldman 1994) was also fundamental to promote observational references that are and will be extremely useful to face and understand climatic changes in the future (Kumagai and Vincent 2003).

References

Armengol, J.; Garcia, J.C.; Comerma, M.; Romero, M.; Dolz, J.; Roura, M.; Han, B.H.; Vidal, A.; and Simek, K. 1999. Longitudinal processes in Canyon type reservoirs: The case of Sal (N. E. Spain). In: Tundisi, J.G. and Straskraba, M. (Eds.), *Theoretical Reservoir Ecology and Its Applications*. Backhuys Publishers, Leiden, the Netherlands, pp. 313–345.

Azuma, Y. and Okubo, T. 2002. Characteristic of nutrient loads from Rivers into Lake Biwa. In: Nakamura, N. and Nakajima, T. (Eds.), *Lake Biwa and Its Watershed: A Review of LBRI Research Notes*. LBRI, Otsu, Japan, 145pp.

Bennett, E.B. 1978. Characteristics of the thermal regime of Lake Superior. *J. Great Lake Res.* 4:310–319.

Birge, E.A. and Juday, C. 1911. The inland waters of Wisconsin: The dissolved gases in the water and their biological significance. *Wisconsin Geological and Natural History Survey Bulletin*, Vol. 22, No. 2.

Dobson, H.F.H.; Gilbertson, M.; and Sly, P.G. 1974. A summary and comparison of nutrients and related water quality in lakes Erie, Ontario, Huron and Superior. *J. Fish. Res. Board Can.* 13:731–738.

Duncan, A. 1975. The importance of zooplankton in the ecology of reservoirs. In: *The Effects of Storage on Water Quality*. Water Research Centre, Medmenham, U.K., pp. 247–272.

Forel, F.A. 1901. The study of lakes. In: *Handbook der Seenkunde*. Verlag Von Engelhorn, Stuttgart, Germany.

Hejzlar, J. 1989. Dissolved amino sugars in the Rinov. Reservoir (Czechoslovaquia) Arch. *Hydrobiol. Beih. Erbegn. Limnol.* 33:291–302.

Horne, A. and Goldman, C. 1994. *Limnology*. Mc Graw Hill International Editions, New York, 576pp.

Imberger, J. 1994. Transport processes in lakes: A review. In: Margalef, R. (Ed.), *Limnology Now: A Paadigm of Planetary Problems*. Elsevier Science, New York, pp. 99–194.

Jørgensen, S.E. et al. 2005. *Lake and Reservoir Management*. Elsevier, Amsterdam, the Netherlands, 502pp.

Jørgensen, S.E. and Svirezhev, Y.M. 2004. *Towards a Thermodynamic Theory for Ecological Systems*. Elsevier, Amsterdam, the Netherlands, 366pp.

Kennedy, R.H. 1999. Reservoir design and operation: Limnological implications and management opportunities. In: Tundisi, J.G. and Straskraba, M. (Eds.), *Theoretical Reservoir Ecology and Its Applications*. Backhuys Publishers, Leiden, the Netherlands, pp. 1–28.

Kennedy, R.H.; Tundisi, J.G.; Straskraba, V.; Lind, O.D.; and Hejzlar, I. 2003. Reservoir and the limnologists growing role in sustainable water resources. *Hydrobiology* 504:XI–XII. In: Straskrabova, V. et al. (Eds.), *Reservoir Limnology and Water Quality*. Kluwer Academic Publishers, Dordrecht, the Netherlands, 325pp.

Kumagai, M. and Vincent, W.F. 2003. *Freshwater Management: Global versus Local Perspectives*. Springer, Berlin, Germany, 33pp.

LBRI. 1984. *Lake Biwa Study Monographs*. Lake Biwa Research Institute, Otsu, Japan, 118pp.

Likens, G. 1992. *The Ecosystem Approach: Its Use and Abuse*. Ecology Institute, Oldendorf/Luhe, Germany, 166pp.

Likens, G.E. and Borman, F.H. 1974. Linkages between terrestrial and aquatic ecosystems. *Bio-Sci.* 24980:447–456.

Lindemman, R.L. 1942. The trophic dynamic aspect of ecology. *Ecology* 23:399–418.

Lund, J.W.G. 1971. The seasonal periodicity of three planktonic desmids in Lake Windermere. *Mitt. Int. Verein. Theor. Angew. Limnol.* 19:3–25.

Lund, J.W.G. 1972a. Changes in the biomass of blue-green and other algae in an English lake from 1945–1969. In: Desikachary, T.V. (Eds.), *Taxonomy and Biology of Blue-Green Algae*. University of Madras, Madras, India, pp. 305–327.

Lund, J.W.G. 1972b. Preliminary observations on the use of large experimental tubes in lakes. *Verh. Int. Verein. Theor. Angew. Limnol.* 18:71–77.

Lund, J.W.G. 1979. The uses of large experimental tubes in lakes. In: *The Effects of Storage on Water Quality*. Water Research Centre, Medmenham, U.K., pp. 291–311.

Lund, J.W.G. 1981. *Investigations on Phytoplankton with Special Reference to Water Usage*. Freshwater Biological Association, Ambleside, U.K., Occasional Publication 13, pp. 5–56.

Lund, J.W.G.; Jaworski, G.H.M.; and Bucka, H. 1971. A technique for bioassay of fresh water, with special reference to algal ecology. *Acta Hydrobiol.* (Krakow) 13:235–249.

Macan, T. 1970. *Biological Studies of English Lakes*. Longman, London, U.K., 276pp.

Margalef, R. 1958. Temporal succession and spatial heterogeneity in phytoplankton. In: Buzzati-Traverso, A.A. (Ed.), *Perspectives in Marine Biology*. University of Clifornia Press, Berkeley, CA, pp. 323–349.

Margalef, R. et al. 1976. *Limnologia de los embalses españoles*. Direccion delas obras hídricas, Madrid, Spain.

Masin, M. et al. 2003. Changes in bacterial community composition and microbial activities along the longitudinal axis of two canyon-shaped reservoirs with different inflow loading. *Hydrobiologia* 504:99–113. In: Straskrabova, V.; Kennedy, R.H.; Lind, O.T.; Tundisi, J.G.; and Hezlar, J. (Eds.), *Reservoir Limnology and Water Quality*. Kluwer Academic Publishers, Dordrecht, the Netherlands.

Matsuyama, M. 1978. Limnological aspects ofmeromictic Lake Suigetsu: Its environmental conditions and biological metabolism. *Bulletin of the Faculty of Fisheries.* Nagasaki University, Nagasaki, Japan.

Mori, S. and Yamamoto, G. (Eds.). 1975. Productivity of communities in Japanese Island waters. In: *JIBP Synthesis*, Vol. 10, Tokyo University, Tokyo, Japan, 436pp.

Mortimer, C.H. 1941. The exchange of dissolved substances between mud and water in lakes (Parts I and II). *J. Ecol.* 29:280–329.

Nakamura, N. and Nakagima, T. 2002. *Lake Biwa and Its Watershed: A Review of LBRI Research Notes.* LBRI, Otsu, Japan, 145pp.

Odum, H.T. 1956. Primary production in flowing Waters. *Limnol. Ocean.* 1:102–117.

Okubo, T.; Sudo, M.; Okamura, H.; Hamada, J.; Imaizumi, M.; Ishiki, A.; and Fushiwak, Y. 1998. Seasonal changes in water quality of Farmland Rivers and assessment of the effect of the changes on lake ecosystems. *Lake Biwa Res. Inst. Bull.* 16:28–37 (In Japanese).

Ragotzkie, R.A. 1974. The Great Lakes rediscovered. *Am. Sci.* 62:454–464.

Reynolds, C.S. 1997. *Vegetation Processes in the Pelagic: A Model for Ecosystem Theory*, Excellence in Ecology, Vol. 9. Ecology Insitute, Oldendorf, Germany.

Schelske, C. and Roth, J.C. 1973. Limnological survey of the lakes Michigan, Superior, Huron and Erie, Vol. 17. Great Lakes Research Division, University of Michigan, Ann Arbor, MI, 108pp.

Schindler, D.W. et al. 2008. Evolution of phosphorus limitation in lakes. *Science* 195:260–262.

Somiya, I. 2000. Total nitrogen and total phosphorus balance of Lake Biwa: Lake Biwa its environmental and water quality formation. Gihodo Co., Tokyo, Japan (in Japanese).

Straskraba, M. 1994. Ecotechnological models of reservoir water quality management. *Ecol. Mod.* 74:1–38.

Straskraba, M.; Blazka, P.; Brand, Z.; Hejlan, P. et al. 1993. Framework for investigation and evaluation of water quality in Czchoslovaquia. In: Straskraba, M.; Tundisi, J.G.; and Duncan, A. (Eds.), *Comparative Reservoir Limnology and Water Quality Management.* Kluwer Academic Publishers, Dordrecht, the Netherlands, pp. 169–212.

Talling, J.F. 1957. The phytoplankton population as a compound photosynthetic system. *New Phytol.* 56:133–149.

Talling, J.F. 1965a. Comparative problems of phytoplankton production and photosynthetic activity in a tropical and a temperate lake. *Mem. Ist. Ital. Hydrobiol.* 18(Suppl.):399–424.

Talling, J.F. 1965b. Comparative problems of phytoplankton in East African lakes. *Int. Rev. Ges. Hydrobiol.* 50:1–32.

Talling, J.F. 1966. The annual cycle of stratification and phytoplankton growth in Lake Victoria (East Africa). *Int. Revue Hydrob.* Weinheins, Germany 51(4):545–621.

Talling, J.F. 1971. The underwater light climate as a controlling factor in the production ecology of freshwater phytoplankton. *Mitt. Int. Verein. Theor. Angew. Limnol.* 19:214–243.

Tundisi, J.G. and Matsumura-Tundisi, T. 2012. *Limnology.* CRS Press, Taylor & Francis, Baca Raton, FL, 854pp.

Vollenweider, R.A. 1965. Water management research. Scientific fundamentals of the eutrophication of lakes and flowing waters, with particular reference to nitrogen and phosphorus as factors in eutrophication. O.E.C.D. DAS/SCI/68.27.

Vollenweider, R.A. 1975. Input-output models with special reference to the phospho-
rus loading concept in limnology. *Schweiz. Z. Hydrol.* 37:53–84.
Vollenweider, R.A. 1976. Advances for defining critical loading levels for phosphorus
in lake eutrophication. *Mem. Ist. Ital. Idrobiol.* 33:53–83.
Walsby, A.E. and Reynolds, C.S. 1980. Sinking and floating. In: Morris, I.G. (Ed.),
The Physiological Ecology of Phytoplankton. Blackwell, Oxford, U.K., pp. 371–412.
Yoshimura, S. 1938. Dissolved oxygen of the lake waters of Japan. *Sci. Rep.* Tokyo
Bunrika Daigaku: Sect. C 2(8):64–215.

9

Application of the Conservation Principles

9.1 Mass Conservation: An Important Basic Principle

According to the law of mass conservation, mass can neither be created nor destroyed but only transformed from one form to another. Thus everything must go somewhere. The notion of cleaning up the environment or pollution-free products is a scientific absurdity. We can never avoid pollution effects. Nobody—neither man nor nature—consumes anything; we only borrow some of the earth's resources for a while, extract them from the earth, transport them to another part of the planet, process them, use them, and discard, reuse, or reformulate them (Cloud 1971). These principles are applied again and again in environmental management of aquatic ecosystems.

The law of mass conservation assumes that no transformation of mass into energy takes place, which, however, is possible in accordance with Einstein as follows:

$$E = mc^2 \tag{9.1}$$

$$c = 3 \times 10^8 \, m/s$$

The transformation from energy to mass or from mass to energy, however, does not take place in the environment but only by nuclear processes, which are very minor in nature.

If we consider a system that exchanges mass with the environment, then the following equation is valid for an element or a chemical compound, c:

$$\frac{dm_c}{dt} = \text{import} - \text{export} \pm \text{result of chemical reactions} \tag{9.2}$$

where m_c is the mass of c in the system. It is possible to compute concentrations, $c = m_c/V$, where V is the volume, in ecosystems by use of Equation 9.2.

Mass is conserved; it can be neither created nor destroyed, but only transformed from one form to another.

TABLE 9.1

Cu-Concentrations (Characteristic)

Item	Sphere Represented	Concentration
Atmospheric particulates (unpolluted area)	Atmosphere	$2\,mg/m^3$
Seawater (unpolluted)	Hydrosphere	$2\,\mu g/L$
River water	Hydrosphere	$10\,\mu g/L$
Soil	Lithosphere	$20\,mg/kg$
Freshwater sediment	Lithosphere	$40\,mg/kg$
Algae	Biosphere	$20–200\,\mu g/L$

The form and location is of great importance for the effect of pollutants. We should always attempt to discharge our waste in such a way that the change in concentration of the most harmful forms becomes as low as possible. It is therefore noticeable that the four spheres have a completely different composition, as demonstrated for copper in Table 9.1. Waste should generally be discharged in the sphere where the concentration would be changed least by the discharge. It is also possible to get a first rough idea about the impact by a comparison of the actual concentration in an ecosystem with the background concentration. If the difference between the two concentrations is minor, there is a high probability that the discharge is harmless, while a major difference could indicate that the discharge is harmful.

Let us take a concrete example to illustrate these considerations: where should waste containing copper be deposited? To answer this question we need more information than that available in Table 9.1, although this table indicates that the highest concentration of copper is in the lithosphere.

Therefore, we can assume that the discharge of copper to the lithosphere will produce the smallest change in copper concentration of the four spheres, but we need to know something about the effect of copper in its different forms, because the different forms have different availability. For instance, the availability is very low of copper ions adsorbed to various forms of sediment and soil. Copper complexes are also generally less available and toxic to most organisms. The toxicity is in this context expressed, for instance, by the LC-value. Free copper ions are, however, extremely toxic to some aquatic plants and animals; see Table 9.2. Furthermore, free copper ions are bound to soil and sediments, which means that the most toxic form, free copper ions, often are present in the environment in low concentrations. Copper complexes are also less toxic to, for instance, phytoplankton than the free copper ions. It has been found that the lehtal concentration killing 50% of the test animals (LC_{50}) value for phytoplankton toward copper is 5–10 times higher at a water hardness of about $200\,mg/L$ than at a water hardness of $15\,mg/L$, which can be explained by the higher extent of complex formation at higher hardness (Jørgensen 2000). A waste deposition site is, in the first instance, therefore selected in the sphere where the relative change in concentration is smallest. Furthermore, it is necessary to compare the form

TABLE 9.2

Lethal Concentrations of Copper Ions (LC_{50}-Values)
and Lethal Doses of Copper (LD_{50}-Values)

Species	Values
Asellus meridianus	$LC_{50}{}^a = 1.7–1.9\,mg/L$
Daphnia magna	$LC_{50}{}^a = 9.8\,\mu g/L$
Salmo gairdneri	$LC_{50}{}^a = 0.1–0.3\,mg/L$
Rats	$LC_{50}{}^a = 300\,mg/(kg\ body\ weight)$ (as sulfate)

[a] Dependent on pH, temperature, water hardness, and other experimental conditions.

and the processes of the waste in the four spheres and consider this information for the deposition site specifically. The conclusion is that it is necessary to make the necessary chemical calculation of the concentration of the various forms of copper compounds; see Chapters 10 and 11, where the methods are presented.

Cadmium forms complexes with one, two, three, or four chlorine ions. These complexes are less toxic than the free cadmium ions due to differences in availability, and it is therefore of importance to determine the concentrations of the free cadmium ions and the cadmium chloro -complexes when cadmium is discharged, for instance, to a marine environment with a certain chlorine ion concentration. The relevant aquatic chemical calculations are shown in Chapters 10 and 11.

We can collect garbage and remove solid waste from sewage, but they must be burned, which causes air pollution; dumped into rivers, lakes, and oceans, which causes water pollution; or deposited on land, which will cause soil pollution. The management problem is not solved before the final deposit site for the waste is selected. Furthermore, environmental management requires that *all* consequences of environmental technology be considered. Eliminating one form of pollution can create a new form, as described previously.

Finally, as the production of machinery and chemicals for environmental technology may also cause pollution, the entire mass balance must be considered in environmental management, including deposit of waste products and pollution from service industries. This problem can be illustrated by considering, as an example, the Lake Tahoe wastewater plant, where municipal wastewater is treated through several steps to produce a very high water quality. The conclusion is that it hardly pays, from an environmental point of view, to make such comprehensive wastewater treatment. *A total solution to an environmental problem implies that all environmental consequences are considered by use of a total mass balance, including all wastes produced and the service industries.*

It is important to recognize that complete elimination of pollution is an unachievable goal. The task is to balance the cost of environmental degradation with the cost required to control that degradation. It is, however, generally much more difficult to identify and assess the costs associated with uncontrolled pollution or environmental degradation. Costs of pollution

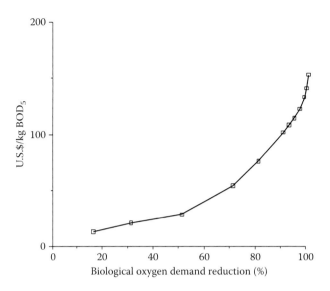

FIGURE 9.1

Relation between removal of biological oxygen demand measured during a period of 5 days (BOD$_5$) (%) and the cost per kg BOD$_5$ removed from typical industrial water with a high concentration of biodegradable material. BOD$_5$ expresses the concentration of organic matter measured by the oxygen demand. (From Jørgensen, S.E., *Principles of Pollution Abatement*, Elsevier, Amsterdam, the Netherlands, 520pp., 2000.)

should include increased medical costs for sensitive people and loss of resources, but reduction in the quality of life and long-term environmental effects are difficult to estimate.

Typically, most of the pollution from a particular source may be controlled relatively inexpensively up to a certain efficiency. Figure 9.1 shows the relation between the degree of purification (in this case removal of organic matter, measured by its consumption of oxygen over a period of 5 days) (in %) and the cost of treatment. Increasing the efficiency of the treatment toward 100% may produce exponential growth of the treatment costs. As the number of possible forms of by-product pollution often follows the trend of the cost, the same relation might exist between the environmental side effects and the percent reduction of pollution. Consequently, our problem is not the elimination of pollution, but its control. There exists, at least in principle, a minimum total cost for any activity as illustrated in Figure 9.2. A desirable goal for an environmental manager is to define that minimum cost.

Technology is essential in helping us to reduce pollution levels below a dangerous level, but in the long term pollution control must also include population control and control of the technology including its pattern of production and consumption. Wise use of existing technology can buy us some time to develop new methods, but the time we can buy is limited. The so-called energy crises are the best demonstration of the need for new and far more advanced technology: to provide energy by alternative and new methods based on renewable sources. We do

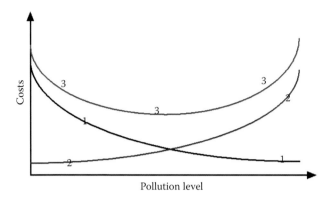

FIGURE 9.2
Cost of pollution versus pollution level. Curve 1 shows the cost of pollution control, Curve 2 presents the cost associated with pollution and environmental degradation, and Curve 3 is the sum of the two costs.

not know how much time we have—probably 30–60 years—so we had better get started now to make sure, we have the solution to man's many serious problems in time. The increasing cost of treatment with efficiency (Figure 9.1) must be taken into account in urban planning. If it is decided to maintain environmental quality, increased urbanization (which means increased amounts of wastes from an increasing urban population) will require higher treatment efficiency, which leads to a higher cost for waste treatment per inhabitant or per kg of waste. This fact renders the solution of environmental problems of metropolitan areas in many developing countries economically almost prohibitive. *The more waste that is required to treat in a given area, the higher the treatment efficiency that is needed and the higher the costs per kg of waste will be to maintain an acceptable environmental quality.*

9.2 Threshold Levels

It is important to recognize that there are both natural and man-generated pollutants. Of course, the fact that nature is polluting does not justify the extra addition of such pollutants by man, as this might result in the threshold level being reached.

In general, we can classify pollutants into two groups (Jørgensen 2000):

1. Nonthreshold or gradual agents, which are potentially harmful in almost any amount
2. Threshold agents, which have a harmful effect only above or below some concentration or threshold level

This classification is illustrated in Figure 9.3.

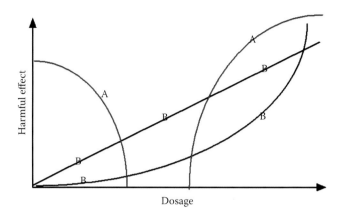

FIGURE 9.3
A—Threshold agent and B—Nonthreshold or gradual agent. To have a threshold agent, it is sufficient that one of the two A-plots is valid. The two B-plots represent two different dose-response curves.

For the latter class, we come closer to the limit of tolerance for each increase or decrease in concentration, until, finally, like the last straw that broke the camel's back, the threshold is crossed. For non-threshold agents, which include several types of radiation, many man-made organic chemicals, which do not exist in nature, and some heavy metals such as mercury, lead, and cadmium, for which there is no theoretically safe level.

Threshold agents include various nutrients, such as phosphorus, nitrogen, silica, carbon, vitamins, and minerals (calcium, iron, zinc, etc.). When they are added or taken in excess, the organism or the ecosystem can be over stimulated, and the ecological balance is damaged. Examples are the eutrophication of lakes, streams, and estuaries from fertilizer runoff or municipal wastewater. The threshold level and the type and extent of damage vary widely with different organisms and stresses. The thresholds for some pollutants may be quite high, while for others they may be as low as 1 part per million (1 ppm—aquatic chemistry ppm is most often used to cover mg/L) or even 1 part per billion (1 ppb, in aquatic chemistry covering $1\,\mu g/L$). *The threshold level is closely related to the concentration found in nature under normal environmental conditions.*

9.3 Steady State and Equilibrium

When the mass conservation principles are applied to find unknown concentrations or amounts in various environmental compartments, it is convenient to distinguish between (thermodynamic equilibrium) and steady state. The latter means "unchanging with time," that is, all time derivates are zero. Thermodynamic equilibrium on the other hand means that the system has reached the concentrations corresponding to an equilibrium expression, for instance, that the concentrations of carbon dioxide in air and water correspond to Henry's law that expresses the equilibrium between air and water. Thermodynamic equilibrium implies also that the system has no Gibb's free energy available to do work on the environment, because this is the thermodynamic definition of thermodynamic equilibrium (see any textbook of thermodynamics or physical–chemistry).

There are, in principle, five possibilities combining flows with thermodynamic equilibrium and steady state (see Mackay 1991):

1. No flows, which implies that the system will be in steady state and thermodynamic equilibrium, provided (of course) that the system has had sufficient time to reach the thermodynamic equilibrium.

2. Flows, thermodynamic equilibrium and steady state. Figure 9.4 gives an example.

3. Flows, steady state and no equilibrium. Figure 9.5 illustrates this example.

4. Flows, not steady state and thermodynamic equilibrium. Figure 9.6 gives an example.

5. Flows, not steady state and no equilibrium, see Figure 9.7.

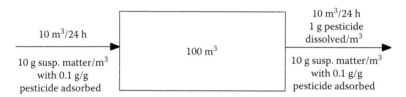

FIGURE 9.4
Flow to a pond of $100\,m^3$ is constant and the equilibrium between the adsorbed and dissolved pesticide has been established: by $1\,mg/L$ pesticide in solution, $100\,mg$ will be adsorbed per g of suspended matter. There are no changes in volumes or concentrations, and the equilibrium is maintained. Therefore, steady state, equilibrium.

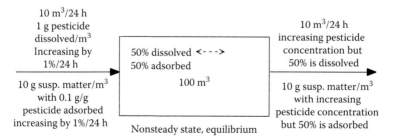

FIGURE 9.5
As Figure 9.4 but the concentration of pesticide in the incoming water increases by 1%/24 h but the equilibrium presumed to be a 50% allocation between adsorbed and dissolved is maintained. Therefore, nonsteady state, but equilibrium.

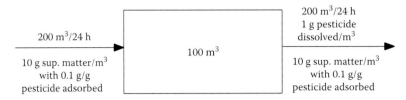

FIGURE 9.6
Inflow is constant in concentrations (pesticide and suspended matter) and volume, but the 50% allocation between dissolved and adsorbed pesticide is not established. It is presumed that all the pesticide is still dissolved in the pond to maintain the steady state for all concentrations and volumes. The retention is too short to yield any adsorption.

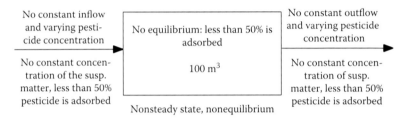

FIGURE 9.7
Flows and the concentrations vary and the allocation between the adsorbed and dissolved pesticide is varying too, that is, it is not at the 50% allocation corresponding to equilibrium.

If the thermodynamic equilibrium is established, it will often be beneficial to set up mass balances for the total amount of species that are in equilibrium. After the total concentrations have been determined the concentrations of the species are found applying the equilibrium equation. If the reaction is

$$aA + bB = cC + dD \tag{9.3}$$

then the following equation is valid at equilibrium

$$\frac{[C]^c [D]^d}{[A]^a [B]^b} = K \tag{9.4}$$

where [ ] indicates concentration in M (moles/liter).

9.4 Basic Concepts of Mass Balances

The simplest case is an isolated system, where no processes take place, and therefore the concentration of all components is constant. An ecosystem is never an isolated system, but may be either an open system or a closed system. The former system exchanges mass as well as energy with the environment. The input of energy to an ecosystem will cause cyclic processes (Morowitz 1968), in which the important elements will play a part. Very few pollutants are completely chemically inert, most are converted to other components or degraded. The common degradation process including the degradation of persistent chemicals can often be described by a first-order reaction scheme:

$$\frac{dC}{dt} = -kC \tag{9.5}$$

where
  C is the concentration of the considered compound
  t is the time
  k is the rate constant

k varies widely from the very easily biodegradable compounds, such as carbohydrates and proteins, to pesticides, such as DDT.

The so-called biological half-life time, $t_{1/2}$, is often used to express the degradability. $t_{1/2}$ is the time required to reduce the concentration to half the initial value. The relation between k and $t_{1/2}$ can easily be found:

$$\ln \frac{C_o}{C(t)} = kt$$

$$\ln 2 = kt_{1/2} \tag{9.6}$$

Here C_o is the initial concentration.

k and $t_{1/2}$ depend on the reaction conditions, pH, temperature, ionic strength, etc. Indication of half-life time or k should therefore always be accompanied by indication of the reaeration conditions.

The continuous mixed flow reactor (abbreviated CMF) closely approximates the behavior of many components in ecosystems. The CMF reactor is illustrated schematically in Figure 9.8. The input concentration of component i is C_{i0} and the flow rate is Q. The tank (ecosystem) has the constant volume V and, in the tank, the concentration of i, denoted C_i, is uniform. The effluent stream also has a flow rate Q, and because the tank is considered to be perfectly mixed the concentration in the effluent will be C_i.

The principles of mass conservation can be used to set up the following simple differential equation (no reactions take place in the tank):

$$\frac{VdC_i}{dt} = QC_{i0} - QC_i \tag{9.7}$$

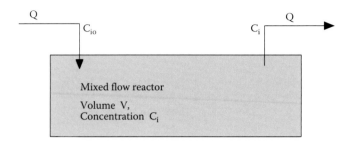

FIGURE 9.8
Principle of a mixed flow reactor.

In a steady state situation, $dC_i = 0$, this means we have

$$QC_{i0} = QC_i \tag{9.8}$$

If a first-order reaction has taken place in the tank (ecosystem), the equation will be changed to

$$\frac{VdC_i}{dt} = QC_{i0} - QC_i - kC_iV \tag{9.9}$$

And, in the steady state situation,

$$QC_{i0} - QC_i = VkC_i \tag{9.10}$$

By dividing this equation with QC_i, the following equation is obtained:

$$\frac{C_{i0}}{C_i} - 1 = tr\,k \tag{9.11}$$

where $tr = V/Q$ is the retention time or the mean residence time in the tank (ecosystem). Rearrangement yields

$$\frac{C_i}{C_{i0}} = \frac{1}{(1 + tr\,k)} \tag{9.12}$$

Figure 9.9 shows this equation on a graph: $C_i = f(t)$ with $C = 2$ at $t = 0$. C_i will approach asymptotically the steady state concentration, which can be found from Equations 9.11 and 9.12. For a number of reaction tanks, m, in series, each with volume V, a similar set of equations can be set up:

$$\frac{C_i}{C_{i0}} = \frac{1}{(1 + tr\,k)^m} \tag{9.13}$$

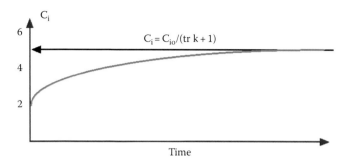

FIGURE 9.9
$C_i = f(t)$ with $C_i = 2$ at $t = 0$. C_i will approach asymptotically the steady state concentration.

If the set of equations set up to describe the concentration is far more compli-cated, more processes must be taken into account, and it is necessary to use an ecological model (see Jørgensen and Bendoricchio [2001] and Jørgensen and Fath [2011]). Space does not permit a detailed examination of more com-plicated models, but some processes of interest in an environmental con-text can be mentioned (in addition to *hydrophysical* and *meteorological* ones) (Jørgensen 2000):

1. Leaching of ions and organic compounds in soil
2. Evaporation of organic chemicals from soil and surface water
3. Atmospheric washout of organic chemicals
4. Sedimentation of heavy metals and organic chemicals in aquatic ecosystems
5. Hydrolysis of organic chemicals
6. Dry deposition from the atmosphere
7. Chemical oxidation
8. Photochemical processes

All these processes may be described as first-order reactions at least in some situations. Table 9.3 gives some typical examples of the eight aforemen-tioned processes and an idea of where these processes are of importance. The list has not included *biotic* processes, which would further complicate the picture. These processes will be mentioned several times later. The mass conservation principle and the corresponding equations presented in this book can often be applied directly on lakes, because lakes with a sufficient high water retention time can be considered as chemostate. In addition, many of the processes, biological decomposition, photolysis, hydrolysis, and sedimentation, can with good approximations be described as first-order reactions. Example 9.1 illustrates these applications of the mass conserva-tion principles.

TABLE 9.3

Some Chemical–Physical Processes of Environmental Interest

Process	Examples
Leaching of ions and organic compounds in soil	Nutrient runoff from agricultural areas to lake ecosystems
Evaporation of organic chemicals from soil and surface water	Evaporation of pesticides
Sedimentation of heavy metals	Most heavy metals have a low and solubility in seawater and will therefore precipitate and settle
Hydrolysis of organic chemicals	Hydrolytic degradation of pesticides in aquatic ecosystems
Dry deposition from the atmosphere	Dry deposition of heavy metals on land and surface water
Chemical oxidation	Sulfides are oxidized to sulfates, sulfur dioxide to sulfur trioxide, which forms sulfuric acid with water
Photochemical processes	Many pesticides are degraded photo-chemically

Example 9.1

Consider a lake with a volume of $500{,}000\,m^3$. $2000\,m^3/24\,h$ of wastewater containing $0.5\,mg/L$ methoxychlor is discharged into the lake. The natural flows to the lake correspond to 4 months' retention time. Methoxychlor follows a first-order decomposition rate, and the approximate half-life time is $58\,h$. Find the equilibrium concentration in the lake water. Assume that precipitation and evaporation are balanced.

Solution

Half-life $58\,h$ or 2.42 days
  $2.42\,k = 0.7$
  $k = 0.289\ 1/24\,h$
  Average conc. for all water to the lake: $2{,}000 \times 0.5/(2{,}000 + 500{,}000/120)$
    $= 0.162\,g/m^3$
  Mean retention time $= 500{,}000/(2{,}000 + 500{,}000/120) = 81$ days
  $c_i/0.162 = 1/(1 + 81.0.289)$
  $c_i = 0.00664\,g/m^3$ or mg/L

9.5 Mass Conservations in a Food Chain

The food taken in by one level in the food chain, I, is used in respiration, R; nonutilized (wasted) food, NUF; undigested food (faces), F; excretion (urine), E; and growth and reproduction, G. If the growth and reproduction are considered as the net production, NP, we can state that (Jørgensen 2012)

$$NP = G = I - (NUF + F + R + E) \qquad (9.14)$$

and we can call the ratio of the net production to the intake of food as the efficiency. The efficiency is dependent on several factors, but may be as low as 10% or even below 10%. The efficiency of toxic matter according to Equation 9.14 is often higher than for normal food components, and as a result several chemicals, such as chlorinated hydrocarbons, including DDT and some heavy metals, can be magnified at each level in the food chain.

The assimilated food, A, is the food used for (growth + reproduction) + respiration + excretion

$$A = I - (NUF + F) \tag{9.15}$$

Many organic toxic compounds are taken up (assimilated) by a high efficiency (more than 90%), that is, the loss by feces is low; see Equation 9.15.

Heavy metals have fortunately a low assimilation efficiency. Approximately only 5%–10% of their content in food is assimilated, but as they are excreted slowly and not removed by respiration, they still yield a relatively high biomagnification, defined as the magnification of the concentration through the food chain.

Many organic compounds, including chlorinated hydrocarbons, have a particularly high biomagnification, *because they have*

1. *A high assimilation efficiency, sometimes more than 90%*
2. *A very low biodegradability*
3. *Are only excreted from the body very slowly, because they are dissolved in fatty tissue*

This is illustrated for DDT in Table 9.4. As man is the last level of the food chain, relatively high DDT concentrations have unfortunately been observed in human body fat.

TABLE 9.4

Biological Magnification

Trophic Level	Concentration of DDT (mg/kg Dry Matter)	Magnification
Water	0.000003	1
Phytoplankton	0.0005	160
Zooplankton	0.04	~13,000
Small fish	0.5	~167,000
Large fish	2	~667,000
Fish-eating birds	25	~8,500,000

Source: Data after Woodwell, G.M. et al., *Science*, 156, 821, 1967.

The amount, Tox, of a toxic component in an organism can be followed approximately by use of a simplified differential equation

$$\frac{dTox}{dt} = \text{daily uptake via respiration and food} - kTox \qquad (9.16)$$

where the total daily uptake is found from the concentration in the ambient air or water of the toxic component times the efficiency of uptake via respiration plus the concentration in the food times the assimilation efficiency. It is assumed that the excretion follows a first-order reaction, which is approximately correct. k is the first-order excretion coefficient expressed, for instance, in the unit 1/24 h. Equation 9.16 explains why the concentration of many toxic substances increases with increasing weight and age of the organism. This is exemplified in Figure 9.10 for fish with increasing weight. A steady state concentration, dTox/dt = 0, can be found from Equation 9.16 as Tox = daily uptake via respiration and food/k. k is low for many toxic compounds including most toxic heavy metals such as mercury, lead, and cadmium. This implies that the concentration of Tox becomes high and that it takes many years to reach a concentration close to the steady state situation.

Some toxic substances reach considerably faster an equilibrium concentration in various organisms, that is, the concentration that corresponds to an equilibrium between the environment and the organism. This is particularly the case for toxic lipophilic substances that are taken up from water mainly by the gills of aquatic organism. In this case, it is possible to determine a so-called biological concentration factor (BCF), which expresses the ratio between the concentration in the organisms and in the environment (in casu

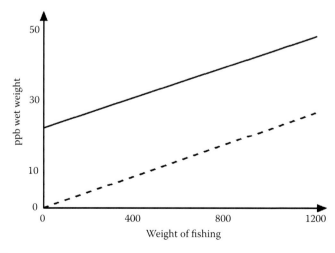

FIGURE 9.10
Increase in pesticide residues in fish as weight of the fish increases. Top line (full) = total residues; bottom line (dotted) = dichlorodiphenyltrichloroethane (DDE) only, Cox (1970). Notice that the plot is semilogarithmic.

water). BCF is a very important factor for evaluation of the effects of toxic substances in aquatic ecosystems, particularly freshwater ecosystems, because they receive a high discharge of toxic substances from municipal wastewater, from agriculture and from industries. As a solubility equilibrium occurs, the BCF value is dependent on the lipid fraction, f_{lipid}, in the considered organism. BCF may be estimated by the following equation (Jørgensen 2000):

$$\text{Log BCF} = \log f_{lipid} + b \log K_{ow} \qquad (9.17)$$

where
  b is a constant close to 1
  K_{ow} is the octanol-water distribution coefficient

As see if $f_{lipid} = 1.0$, BCF $\approx K_{ow}$ in accordance to what it is expected for a solubility equilibrium. In this context, it should be emphasized that BCF of course does not take into consideration the biomagnification through the food chain, which requires mass balance calculations trophic level by trophic level.

Many toxic substances are widely dispersed and a global increase in the concentration of heavy metals and pesticides has been recorded. The relationship between a global and a regional pollution problem and the role of dilution for this relationship are illustrated in Table 9.5, where the ratios of heavy metal concentrations in the River Rhine and in the North Sea are shown. Independent of the level, the usual first step in a solution procedure for a toxic substance problem is to set up a list of pollution sources with information about the quantities emitted to the environment, the form in which the toxic substance is emitted, and where in the environment (which spheres) the emission takes place. Based upon this information it is possible to assess the concentration of the toxic substance in various compartments of the environment. It is an essential step in an environmental risk assessment (ERA). It is often appropriate to start the solution of a pollution problem by setting up a mass balance for the considered component or element to clarify the sources of pollution and to state the most effective means of solving the problem.

TABLE 9.5

Heavy Metal Pollution in the River Rhine

Metal	The River Rhine (t/year)	Ratio: Conc. in the Rhine/Conc. in the North Sea
Cr	1,000	20
Ni	2,000	10
Zn	20,000	40
Cu	200	40
Hg	100	20
Pb	2,000	700

Source: Jørgensen, S.E., *Principles of Pollution Abatement*, Elsevier, Amsterdam, the Netherlands, 520pp., 2000.

9.6 Hydrological Cycle

Water is the most abundant chemical compound on earth (see Table 9.6), and has some unique properties. Its importance for all life on earth can be demonstrated as follows:

1. *Our body consists of 50%–65% water* and we need at least 1.5 L/day to survive. We can survive without food for perhaps 80 days, but only a few days without water.
2. *Water serves as a basic transport medium for life-giving nutrients.*
3. *Water removes and dilutes many natural and man-made wastes.*
4. *Water has a great ability to store heat energy and to conduct heat and has an extremely high vaporization temperature compared with its molecular weight.* These thermal properties are major factors influencing the climatic pattern of the world and in minimizing sharp changes in temperature on the earth.
5. *Water has its maximum density at 4°C above its freezing point,* so solid water, ice, is less dense than liquid water. This is the reason why a water body freezes only on the top. If ice was denser than liquid water, lakes, rivers, and oceans would freeze from the bottom up, killing most higher forms of aquatic life.

It is of course possible to quantify the global cycling of all compounds. It is because of the importance of water (see the five points mentioned earlier), of particular interest to obtain an overview of the global water

TABLE 9.6

Water Resources and Annual Water Balance of the Continents of the World

Component	Europe	Asia	Africa	North America	South America	Australia	Total
Area (1E6 km²)	9.8	45	30.3	20.7	17.8	8.7	132.3
Precipitation (10³ km³)	7.2	32.7	20.8	13.9	29.4	6.4	110.4
Total river runoff (km³)	3,110	13,190	4,225	5,960	10,380	1,965	38,830
Underground runoff (km³)	1,065	3,410	1,465	1,740	3,740	465	11,885
Infiltration (km³)	5,120	22,910	18,020	9,690	22,715	4,905	83,360
Evaporation (km³)	4,055	19,500	16,555	7,950	18,975	4,440	71,475
Percent underground runoff of total	34	26	35	32	36	24	31

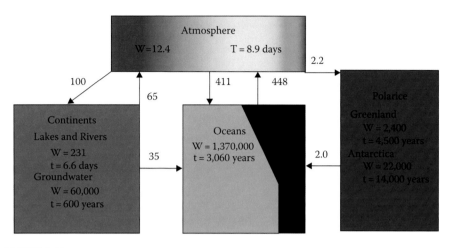

FIGURE 9.11
(See color insert.) Water cycle is shown. Water in the compartment is indicated in $1000\,km^3$ and the retention time is indicated in years. Number of fluxes represents $1000\,km^3/year$. The estimation of the ground water volume is to a depth of $5\,km$ of the earth's crust; much of this water is not actively exchanged.

cycle. *Water shows a physical cycling* (compare with the chemical cycling of the elements), *as demonstrated in Figure 9.11. In this vast cycle, driven by solar energy, our supply of water is recycled again and again.* Water evaporates from the oceans, rivers, lakes, and continents, and gravity pulls it back down as rain. Some of the water falls on the land sinks or percolates into the soil and ground to form ground water. The soil can, like a sponge, hold a certain amount of water, but if it rains faster than the rate at which the water percolates, water begins to collect in puddles and ditches and runs off into nearby streams, rivers, and lakes. This runoff may cause erosion. The water runs eventually into the ocean, which is the largest water storage tank. Because of this cycle, water is continually replaced, as indicated in Table 9.7.

TABLE 9.7

Water Cycle

Water In	Is Replaced Every
Human body	Month
The air	12 days
A tree	1 week
Rivers	A few days
Lakes	0.1–100 years
Oceans	3,600 years
Polar ice	15,000 years

Long-term average water-budget equations for extensive hydrological systems can be expressed as

$$P = E \tag{9.18}$$

where
  P is the precipitation inflow
  E is the evaporation outflow

The storage change, S, is zero. Water-budget equations for entire land and water masses must also contain the total water discharge from land to ocean, Q

$$P + Q = E \quad \text{for oceans} \tag{9.19}$$

$$P = Q + E \quad \text{for land} \tag{9.20}$$

The numerical equality is illustrated in Table 9.8.

The short-term water-budget equation for a terrestrial ecosystem must include a storage term, S:

$$P = Q + E + S \tag{9.21}$$

If subsurface flows are included, we have

$$P + Q^i + L^i = E + Q^o + L^o + S \tag{9.22}$$

where
  Q^i and Q^o are surface inflow and outflow, respectively
  L^i and L^o are the corresponding subsurface flows

TABLE 9.8

Mean Annual Water Balance Components for the Earth

Item	Land	Ocean	Earth
Area (10^6 km^2)	148.9	361.1	510.0
Volume (10^3 km^3)			
Precipitation	111	385	496
Evaporation	−71	−425	−496
Discharge	−40	40	0
Mean depth (mm)			
Precipitation	745	1066	973
Evaporation	−477	−1177	−973
Discharge	−269	111	0

References

Cloud, P.E. 1971. Resources, population and quality of life. In: Singer, S.F. (Ed.), *Is There an Optimum Level of Population*. Mc Graw Hill, New York, 442pp.

Cox, J.L. 1970. Accumulation of DDT residues in *Triphoturus mexicanus* from the Gulf of California. *Nature* 227:192–193.

Jørgensen, S.E. 2000. *Principles of Pollution Abatement*. Elsevier, Amsterdam, the Netherlands, 520pp.

Jørgensen, S.E. 2012. *Fundamentals of Systems Ecology*. CRC Press, Boca Raton, FL, 320pp.

Jørgensen, S.E. and Bendoricchio, G. 2001. *Fundamentals of Ecological Modelling*, 3rd edn. Elsevier, Amsterdam, the Netherlands, 628pp.

Jørgensen, S.E. and Fath, B. 2011. *Fundamentals of Ecological Modelling*, 4th edn. Elsevier, Amsterdam, the Netherlands, 400pp.

Mackay, D. 1991. *Multimedia Environmental Models*. Lewis Publishing, Chelsea, MI, 258pp.

Morowitz, H.J. 1968. *Energy Flow in Biology. Biological Organization as a Problem in Thermal Physics*. Academic Press, New York, 179pp.

Woodwell, G.M. et al. 1967. DDT residues in an Eastern Coast estuary. *Science* 156:821–824.

10

Application of Aquatic Chemistry in Environmental Management I: Calculations of Equilibria

10.1 Equilibrium Constant

We know several million different compounds, and 100,000 of these chemicals have environmental interest, because they may threaten the environment (Jørgensen 2000). The number of possible reactions among these chemicals is enormous, and it is of course not possible to set up a table of the equilibrium constants for all these reactions. We can, however, apply the standard free energies of formation of chemical compounds, ΔG_o. The standard free energy of formation of a compound is the free energy of reaction by which it is formed from its elements when all the reactants and products are in the standard state; that is, the activities are all one or by ideal conditions, it means that the concentrations are all one. Free energy equations can be added and subtracted just as thermochemical equations. It implies that the free energy of any reaction can be calculated from the sum of the free energies of the products minus the sum of the free energies of the reactants:

$$\Delta G_o = \sum G_o(products) - \sum G_o(reactants) \tag{10.1}$$

Free energy describes the chemical affinity under conditions of constant temperature and pressure: $\Delta G = G$ (products) – G (reactants). When the free energy is zero, the system is in a state of thermodynamic equilibrium. When the chemical energy change is positive for a proposed process, network must be put into the system to effect the reaction; otherwise, it cannot take place. When the free energy change is negative, the reaction can proceed spontaneously by providing useful network. As it has been shown, the equilibrium constant K is related to $-\Delta G_o$ by the following equation (for further details see also Jørgensen 2012):

$$-\Delta G_o = RT \ln K \tag{10.2}$$

If we consider the process aA + bB = cC + dD, we get

$$K = \frac{\{C\}^c \{D\}^d}{\{A\}^a \{B\}^b} \tag{10.3}$$

where {} indicates the fugacity or activity (when partial pressure is considered applied as unit, the equilibrium constant is often indicated as K_p) or activity (concentration units are applied, the equilibrium constant is indicated as K_c) in contrast to [], which indicates concentrations.

The equilibrium expression is also denoted as the mass law. The equilibrium constant may have different names corresponding to application of the mass law on different reactions. For instance, the equilibrium constant may be called the formation constant when A + B form AB, or if AB is a complex, the constant may be called a complexity constant or stability constant. The equilibrium constant is called dissociation constant when AB is dissociated into A^+ and B^- and acidity constant when an acid HA is dissociated into a hydrogen ion and the corresponding base HA = H^+ + A^-. The equilibrium constant for the opposite process is called a base constant. For a dissolution process, we talk about the solubility product.

Example 10.1

A chemical plant discharges wastewater containing 26 g/L cyanide. The wastewater is treated by complete oxidation by which cyanide is oxidized to cyanate, using NaClO by pH > 10.5. After oxidation, cyanate is hydrolyzed by addition of acid whereby it is transformed to ammonium and carbon dioxide. The ΔG_o values at room temperature (25°C) are the following:

Cyanate (CNO^-)	−98.7 kJ/mol
Oxonium (H_3O^+)	−237.2 kJ/mol
Ammonium (NH_4^+)	−79.5 kJ/mol
Carbon acid (H_2CO_3)	−623.4 kJ/mol

a. Balance the two equations for the two reactions applied in the wastewater treatment.
b. Which pH value must be applied to ensure that the cyanate concentration after the hydrolysis is $\leq 0.43\,\mu g/L$?

The system is considered closed.

Solution

Oxidation:

$$CN^- + OCl^- \rightarrow CNO^- + Cl^- \tag{I}$$

Redox balance:

$$C: 2 \rightarrow 4 \quad 2\uparrow$$

$$Cl: 1 \rightarrow -1 \quad 2\downarrow$$

Hydrolysis:

$$CNO^- + 2H_3O^+ \rightarrow NH_4^+ + H_2CO_3 \tag{II}$$

K_{II} for (II):

$$\Delta G_{II^\circ} = \Sigma \Delta G_{\circ Produkt} - \Sigma G_{\circ Reaktanter}$$

$$= (-623.4 - 79.5 - (-98.7 + 2(-237.2))) \text{ kJ/mol}$$

$$= 129.8 \text{ kJ/mol}$$

$$\log(K_{II}) = \frac{-\Delta G_{II}\circ}{(RT\ln(10))} = \frac{129.8}{5.7} = 22.8$$

If it is assumed that pH is sufficiently low to ensure that the two resulting compounds (see the reaction) are not dissociated, then

$$[NH_4^+]_{slut} \approx [H_2CO_3]_{slut} \approx [CN^-]_{start} = \frac{26 \text{ g cyanid/L}}{26 \text{ g cyanid/mol}} = 1 \text{ M}$$

$$[CNO^-]_{slut} = 0.43 \text{ µg cyanate/L} = 1 \times 10^{-8} \text{ M}$$

$$K_{II} = 10^{22.8} = \frac{[NH_4^+][H_2CO_3]}{[CNO^-][H_3O^+]^2}$$

$$[H_3O^+] = 10^{-7.4} \text{ M, at pH} = 7.4$$

The assumption was in other words correct and it can be assumed that all C (IV) is as HCO_3^-, that is, $[HCO_3] \approx 1$ M. A combination of the equilibrium expression for process (II) and the equilibrium expression for carbon acid's proteolysis yields

$$K_{s1}K_{II} = \frac{[NH_4^+][HCO_3^-]}{[CNO^-][H_3O^+]}$$

By solution for the unknown, $[H_3O^+]$, we obtain pH = 8.46. HCO_3^- is dominating at this pH. Our assumption was now fully acceptable, as $pK_s = 9.25$ for NH_4^+.

10.2 Activities and Activity Coefficients

Any activity can be written as the product of concentration and activity coefficient: $\{A\} = q\,[A]$. The activity is basically defined in such a way that the activity coefficient $q = \{A\}/[A]$ approaches unity as the concentration of all solutes approaches zero. It means for a solution such as water the activity coefficient becomes unity as the solution approaches the pure ionic medium, that is, when all concentrations other than the medium ions approach zero.

The activity coefficient, q, can be found for individual ions by empirical expressions as given in Table 10.1, where I is the ionic strength $I = 0.5\Sigma C_i Z_i^2$ and Z_i = charge of the ion. A in the table is $= 1.82 \times 10^6 (\acute{e}T)^{2/3} \approx 0.5$ for water at room temperature. $\acute{e}$ is the dielectric constant. $B \approx 0.33$ for water at room temperature, and it is an adjustable parameter corresponding to the size of the ion (see Table 10.2). log q for ions is negative, which implies that q is less than one and decreases with increasing ionic strength and charge of the ion. The activity is less than the concentration because negative ions form a shield around a positive ion and positive ions form a shield around negative ions. The shield is stronger; the more ions the solution contains; it means that the higher the ionic strength is. The electrical force is furthermore proportional to the charge in second, which at least explains that the effect is increasing more than proportional to the charge of the ion. The equations in Table 10.1 can therefore be understood as a consequence of the electrical forces in solutions. They can however not be proved but are useful empirical correlations.

It is clear from Sections 10.1 that we by application of (10.1 through 10.3) can find the equilibrium constant that presumes activities or fugacities. Equilibrium constants taken from handbook tables are also based on activities and fugacities, while we most often are interested in concentrations. Introduction of the activity coefficient, q, makes it possible, however, to set up the following relationships:

$$K = \frac{\{C\}^c\{D\}^d}{\{A\}^a\{B\}^b} = \left(\frac{[C]^c[D]^d}{[A]^a[B]^b}\right)\left(\frac{q_C^c q_D^d}{q_A^a q_B^b}\right) \tag{10.4}$$

Concentrations can now be determined provided that the activity coefficients are known.

TABLE 10.1

Equations for Individual Activity Coefficients

The Name of the Approximation	Equation: log q =	Valid at I <
Debye–Hückel	$-AZ^2\sqrt{I}$	0.005 M
Extended Debye–Hückel	$-AZ^2\sqrt{I}/(1 + Ba\sqrt{I})$	0.1 M
Güntelberg	$-AZ^2\sqrt{I}/(1 + \sqrt{I})$	0.1 M
Davies	$-AZ^2(\sqrt{I}/(1+\sqrt{I}) - 0.2I)$	0.5 M

TABLE 10.2

Parameter a for Individual Ions

Ion Size Parameter a	For the Following Ions
9	H^+, Al^{3+}, Fe^{3+}, La^{3+}, Ce^{3+}
8	Mg^{2+}, Be^{2+}
6	Ca^{2+}, Zn^{2+}, Cu^{2+}, Sn^{2+}, Mn^{2+}, Fe^{2+}
5	Ba^{2+}, Sr^{2+}, Pb^{2+}, CO_3^{2-}
4	Na^+, HCO_3^-, $H_2PO_4^-$, acetate, SO_4^{2-}, HPO_4^{2-}, PO_4^{3-}
3	K^+, Ag^+, NH_4^+, OH^-, Cl^-, ClO_4^-, NO_3^-, I^-, HS^-

10.3 Mixed Equilibrium Constant

Usually, many equilibrium calculations are carried out for the same solution with a well-defined ionic strength. It would therefore be beneficial to find an equilibrium constant, K', valid for concentrations for the considered solution. From Equation 10.4, we obtain

$$K' = K \frac{q_A^a q_B^b}{q_C^c q_D^d} \tag{10.5}$$

In accordance with the IUPAC's convention for determination of pH, we should consider pH = $-\log \{H^+\}$. It is therefore suggested to use a so-called mixed acidity constant, K'_{am}, which can be found from K_a for the process $HA \Leftrightarrow A^- + H^+$

$$K_a = \frac{\{A^-\}\{H^+\}}{\{HA\}} = \frac{[A^-]q_{A^-}\{H^+\}}{[HA]q_{HA}} = K'_{am} \frac{q_{A^-}}{q_{HA}} = K'_a \frac{q_{A^-} \cdot q_{H^+}}{q_{HA}} \tag{10.6}$$

q_{HA} is of course 1.0, if HA has no charge, see the equation in Table 10.1. Calculations of pH presume application of K'_{am}. In environmental calculations, where the accuracy generally is lower than for laboratory calculations, it is recommended to apply activity coefficients for salinities above 0.1%, while it is normally not necessary to apply activity coefficients for aquatic ecosystems with a salinity below 0.1%.

Example 10.2

Find the mixed acidity constant and the concentration quotient (equilibrium constant for all dissolved carbon dioxide hydrated or not and for the ammonium ion in marine environment with the salinity of 2.6% (assume that it is sodium chloride), when logarithm to the acidity constant in distilled water for all dissolved carbon dioxide at the actual temperature is known to be 6.2 and for the ammonium ion at the actual temperature is 9.2.

Solution

$$I = 0.5((26/(23 + 35.5)) + (26/(23 + 35.5)) = 0.445 \text{ M}$$

Davies equation is applied

For the hydrogen carbonate ion, the ammonium ion, and the hydrogen ion

$$\log q = -0.5 \, (\sqrt{0.445}/(1 + \sqrt{0.445}) - 0.2 \times 0.445) = -0.156$$

$$q = 0.698 \approx 0.7$$

The acidity constants for all dissolved carbon dioxide

$$\log K'_a = \log K_a + 2 \log q = 6.2 - 0.312 \approx 5.9$$

$$\log K'_{am} = 6.2 - \log q = 6.2 - 0.156 \approx 6.0$$

The acidity constant for the ammonium ion (Figure 10.1)

$$\log K'_a = \log K_a + \log q - \log q = 9.2$$

$$\log K'_{am} = \log K_a - \log q = 9.2 - (-0.156) = 0.9356 \approx 9.4$$

10.4 Classification of Chemical Processes and Their Equilibrium Constants

The chemical processes may be divided into four classes, and all four classes occur very frequently in all aquatic ecosystems (Jørgensen and Bendoricchio 2001):

1. *Acid–base reactions* are processes characterized by a transfer of a proton. Acids are hydrogen ion donors, and bases are hydrogen ion acceptors:

$$HA \Leftrightarrow A^- + H^+ \tag{10.7}$$

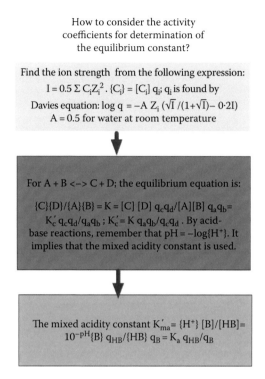

How to consider the activity
coefficients for determination of
the equilibrium constant?

Find the ion strength from the following expression:
$I = 0.5 \, \Sigma \, C_i Z_i^2 . \{C_i\} = [C_i] \, q_i$; q_i is found by
Davies equation: $\log q = -A \, Z_i \, (\sqrt{I} \, /(1+\sqrt{I}) - 0.2I)$
$A = 0.5$ for water at room temperature

For $A + B <-> C + D$; the equilibrium equation is:

$\{C\}\{D\}/\{A\}\{B\} = K = [C] \, [D] \, q_c q_d /[A][B] \, q_a q_b = K'_c \, q_c q_d /q_a q_b$; $K'_c = K \, q_a q_b /q_c q_d$. By acid-base reactions, remember that $pH = -\log\{H^+\}$. It implies that the mixed acidity constant is used.

The mixed acidity constant $K'_{ma} = \{H^+\} \, [B]/[HB] = 10^{-pH}\{B\} \, q_{HB}/\{HB\} \, q_B = K_a \, q_{HB}/q_B$

FIGURE 10.1
Step-wise calculation of the mixed equilibrium constant and the concentration quotient is shown. For more details see Weber and DiGiano (1996).

HA is therefore an acid, and A⁻ a base. HA-A⁻ is denoted an acid–base pair. This is called a half reaction, because the hydrogen ion cannot under normal chemical conditions exists alone and will inevitably be taken up by another component that therefore is a base:

$$HA + B^- \Leftrightarrow A^- + HB \qquad (10.8)$$

Water is both an acid and a base—it is an ampholyte—and can therefore react with both acids and bases:

$$HA + H_2O \Leftrightarrow A^- + H_3O^+ \qquad (10.9)$$

H_3O^+ is called the oxonium ion. The equilibrium constant for process (10.9) is called the acidity constant. The acidity constants listed in most handbooks of chemistry are however the acidity constants for process (10.9), but we use them as if they were the equilibrium constants for process (10.7); corresponding to that the concentration of water in water often is implicitly included in the equilibrium constant for aquatic solutions. Furthermore, we often do not distinguish between hydrogen ions and oxonium ions, because hydrogen ions in solution will anyhow "carry" a water molecule.

When the process is red that the base A^- takes up a hydrogen ion, the equilibrium constant is called a base constant, K_b.

$$\text{As seen } K_b = \frac{1}{K_a} \tag{10.10}$$

The equilibrium constant for process (10.8) can easily be found from the two acidity constants for HA, K_{aA} and HB, K_{aB}:

$$K \text{ for (10.8)} = \frac{\{A^-\}\{HB\}}{\{HA\}\{B^-\}} = \frac{K_{aA}}{K_{aB}} \tag{10.11}$$

The acidity constant for water is

$$\frac{\{H^+\}\{OH^-\}}{\{H_2O\}} = 10^{-15.74} \text{ (room temperature)} \tag{10.12}$$

$\{H_2O\} \approx [H_2O] = 1000/18 = 55.56$, because water has no charge. It implies that what is called water's ionic product

$$K_w = \{H^+\}\{OH^-\} = 10^{-14.00} \text{ (at room temperature)}$$

2. *Precipitation and dissolution* are reactions characterized by a change in solubility. Dissolution and precipitation processes are generally slower than reactions among dissolved species. Electrolytes may dissolve according to the following reaction:

$$A_m B_n(s) \Longleftrightarrow m\,A^{n+} + n B^{m-} \tag{10.13}$$

The opposite process is the corresponding precipitation process. If the equilibrium expression is used on process (5.13), we obtain the solubility product

$$K_s = \{A^{n+}\}^m \{B^{m-}\}^n \tag{10.14}$$

3. *Complex formation* is a reaction by which two or more components form a (more complex) compound. This reaction type is particularly known for a reaction between metal ions (named central atom) and various organic compounds (named ligands). Frequently, more than one ligand can be complex bound to the central atoms, for instance,

$$M + L = ML \quad K_1 = \frac{\{ML\}}{\{M\}\{L\}} \tag{10.15}$$

$$ML + L = ML_2 \quad K_2 = \frac{\{ML_2\}}{\{ML\}\{L\}} \tag{10.16}$$

$$ML_2 + L = ML_3 \quad K_3 = \frac{\{ML_3\}}{\{ML_2\}\{L\}} \tag{10.17}$$

$$ML_i + L = ML_{i+1} \quad K_i = \frac{\{ML_{i+1}\}}{\{ML_i\}\{L\}} \tag{10.18}$$

The process whereby two or more ligands react simultaneously with the central atom is of course also possible:

$$M + L_2 = ML_2 \quad \beta_2 = K_2 K_1 = \frac{\{ML_2\}}{\{M\}\{L\}^2} \tag{10.19}$$

As seen, the equilibrium constant, where i ligands are simultaneously reacting with the central atom, is denoted as $\beta_I = K_1 K_2 K_3 \cdots K_i$. Reactions can also take place by addition of protonated ligands:

$$M + HL = ML + H^+ \quad *K_1 = \frac{\{ML\}\{H^+\}}{\{M\}\{HL\}} \tag{10.20}$$

A parallel expression is used for reaction with the second HL, the third HL, and so on. By multiplication by {L} in the nominator and denominator of the expression in Equation 10.20, it is seen that $*K_1 = K_1 K_{aL}$.

Notice that we have here used the more generally expressed rule that the equilibrium constant for a process, K^*, that consists of i steps, is equal to the product of the i equilibrium constants of the steps, K_1, K_2, K_3, ..., K_i:

$$*K = K_1, K_2, K_3, \ldots, K_i \tag{10.21a}$$

or

$$\log *K = \log K_1 + \log K_2 + \log K_3 + \cdots \log K_i \tag{10.21b}$$

4. *Redox reactions* are processes characterized by a transfer of electrons. Reductants are electron donors, and oxidants are electron acceptors. The mass law may of course also be applied on redox processes, for instance, for (e symbolizes the electron with one negative charge)

$$Fe^{3+} + e = Fe^{2+} \tag{10.22}$$

$$K = \frac{\{Fe_{2+}\}}{\{Fe^{3+}\}}\{e\}$$

This process is, similar to what was mentioned under acid–base reactions, called a half reaction. Free electrons do not exist. They are inevitably taken up by another electron acceptor. For the process (10.22), it is possible to determine an equilibrium constant. log K can in handbooks be found to be 12.53, but the realization of the process requires a coupling to another half reaction, for instance, oxygen

$$O_2 + 4H^+ + 4e = 2H_2O \tag{10.23}$$

with the equilibrium expression

$$K_o = 1/p_{O2}\{H^+\}^4\{e\}^4 \tag{10.24}$$

$$\log K_o = 83.1$$

It is possible to couple the two processes to a feasible redox process:

$$O_2 + 4H^+ + 4Fe^{2+} = 2H_2O + 4Fe^{3+} \tag{10.25}$$

The equilibrium constant for this process, K_r, is found from the equilibrium constants of the two half reactions:

$$K_r = \frac{K_o}{K^4} \tag{10.26}$$

$$\log K_r = \log K_o - 4 \log K = 83.1 - 4 \times 13.0 = 31.1$$

In environmental chemistry, it is difficult to separate the many reactions that can take place simultaneously. We are therefore forced in aquatic environmental chemistry to make equilibrium calculations of several processes simultaneously. It will be shown in the next chapter how it is possible to overview many processes simultaneously, and particularly to assess which of the many simultaneous processes are of importance and which are negligible. In addition, it may often be advantageous to apply what is called a conditional equilibrium constant. It is an equilibrium constant that is only valid under given conditions, for instance, pH has a certain value or chloride has a certain concentration (of interest for a marine environment). It is, in other words, realized that certain components have a constant or almost constant concentration. Many aquatic ecosystems have, for instance, a stable pH at least for a short period. If water of pH = 7.0 is considered and as the partial pressure of oxygen is 0.21 atmosphere, we get for process (10.24) the following equilibrium expression, replacing activities by concentrations:

$$K_r = \frac{[Fe^{3+}]^4}{0.21 \times 10^{-28}[Fe^{2+}]^4} = 10^{31.1}$$

The ratio $[Fe^{3+}]/[Fe^{2+}]$ is therefore at pH = 7.0 and at equilibrium with oxygen in the atmosphere ≈ 4.0.

By incorporating 0.21×10^{-28} into the equilibrium constant, we get a conditional equilibrium constant:

$$\frac{[Fe^{3+}]}{[Fe^{2+}]} = 5.0$$

Which of course is valid only under the condition that pH = 7.0 and that the aquatic ecosystem is in equilibrium with the oxygen in the atmosphere.

10.5 Many Simultaneous Reactions

Many processes occur simultaneously in aquatic ecosystems. The next chapter will treat the four types of processes previously presented and include in this presentation how to provide an overview of many simultaneous processes of the *same* reaction type. The overview makes it possible to distinguish between processes of importance and processes that are negligible in the context. The double logarithmic presentation will be presented as an excellent tool to overview the processes that are very important in management of aquatic ecosystems as they are the chemical processes that determine the biological processes in the aquatic ecosystems.

10.6 Henry's Law

An increase or decrease in the concentration of components or elements in ecosystems are of vital interest, but *the observation of trends in global changes of concentrations might be even more important as they may cause changes in the life conditions on earth.*

The concentrations in the four spheres, the atmosphere, the lithosphere, the hydrosphere, and the biosphere, are of importance in this context. They are determined by the transfer processes and the equilibrium concentrations among the four spheres. *The solubility of a gas at a given concentration in the atmosphere can be expressed by means of Henry's, law, which determines the distribution between the atmosphere and the hydrosphere:*

$$p = H * x \qquad (10.27)$$

where
  p is the partial pressure
  H is the Henry's constant
  x is the molar fraction in solution

H is dependent on temperature (see Table 10.3) and H is expressed (usually) in atmospheres. It may be converted to Pascals, as 1 atm = 101,400 Pa.

TABLE 10.3

Henry's Constant (atm) for Gases as a Function of Temperature

Gas	Temperature (°C)						
	0	5	10	15	20	25	30
Acetylene	0.72	0.84	0.96	1.08	1.21	1.33	1.46
Air (atm)	0.43	0.49	0.55	0.61	0.66	0.72	0.77
Carbon dioxide	73	88	104	122	142	164	186
Carbon monoxide	0.35	0.40	0.44	0.49	0.54	0.58	0.62
Hydrogen	0.58	0.61	0.64	0.66	0.68	0.70	0.73
Ethane	0.13	0.16	0.19	0.23	0.26	0.30	0.34
Hydrogen sulfide	26.80	31.50	36.70	42.30	48.30	54.50	60.90
Methane	0.22	0.26	0.30	0.34	0.38	0.41	0.45
Nitrous oxide	0.17	0.19	0.22	0.24	0.26	0.29	0.30
Nitrogen	0.53	0.60	0.67	0.74	0.80	0.87	0.92
Nitric oxide	—	1.17	1.41	1.66	1.98	2.25	2.59
Oxygen	0.25	0.29	0.33	0.36	0.40	0.44	0.48

Source: Jørgensen, S.E. et al., *Handbook of Ecological Parameters and Ecotoxicology,* Elsevier, Amsterdam, the Netherlands, 1991.
The values in the table are Henry's constant ($\times 10^{-5}$).

A dimensionless Henry's constant may also be applied. As $p = RT$, $n/v = RTc_a$, and $x = c_h/(c_h + c_w)$, where c_a is the molar concentration in the atmosphere of component h expressed in mol/L, c_h is the concentration in the hydrosphere expressed also in mol/L, and c_w is the mol/L of water (and other possible components). If we consider only two components in the hydrosphere, h and water, and that $c_h \ll c_w$, we can replace $(c_h + c_w)$ with the concentration of water in water = 1000/18 = 55.56 mol/L. We obtain according to these approximations the following equation:

$$\frac{c_a}{c_h} = \frac{H}{(R \times T \times 55.56)} \tag{10.28}$$

where $H/(R \times T \times 55.56)$ is the dimensionless Henry's constant.

In aquatic environmental chemistry, we often know the partial pressure in atm, p_a, and want to calculate the concentration in water. In this case, we often use the following expression:

$$C_h = K_H p_a \tag{10.29}$$

K_H is a constant = 55.56/H. If we use the values in Table 10.3 for carbon dioxide, for instance, we get that K_H at 20°C is $55.56/1.42 \times 10^5 = 3.91 \times 10^{-2} = 10^{-1.41}$. In the year 2003, the partial pressure of carbon dioxide in the atmosphere was close to 0.0004 atm, which corresponds at 20°C to a concentration of $3.91 \times 10^{-2} \times 0.0004 = 0.0000156$ M in aquatic ecosystems in equilibrium with the atmosphere.

Example 10.3

The solubility of oxygen in freshwater is 11.3 mg/L at 10°C. Show that it corresponds to the Henry's constant found in Table 10.3 for oxygen.

Solution

Henry's law is applied to find the molar fraction in water, x (the partial pressure of oxygen is 0.21 atm, corresponding to 21% oxygen in the atmosphere):

$$x = 0.21/33,000 = 6.36 \times 10^{-6}$$

This is translated to mg/L

$$\text{oxygen dissolved (mg/L)} = 6.36 \times 10^{-6} \times 55.56 \times 32 \times 1000 = 11.31 \text{ mg/L}.$$

10.7 Adsorption

Water is in contact with suspended solid or sediment and the equilibrium between solid and water is therefore very important for the water quality. The soil–water distribution may be expressed by one of the following two adsorption isotherms:

$$a = kc^b \tag{10.30}$$

$$a = \frac{k'c}{c + b'} \tag{10.31}$$

where
a is the concentration in soil
c is the concentration in water
k, k', b, and b' are constants

Equation 10.30 corresponds to Freundlich adsorption isotherm and is a straight line with slope b in a log–log diagram, since log a = log k + b log c. This is shown in Figure 10.2.

Equation 10.31 represents the Langmuir adsorption isotherm, which is an expression similar to Michaelis–Menten's equation. If $1/a$ is plotted versus $1/c$, see Figure 10.3, we obtain a straight line, the so-called Lineweaver–Burk's plot, as $1/a = 1/k' + b'/k'c$. When $1/a = 0$, $1/c = -1/b'$ and when $1/c = 0$, $1/a = 1/k'$.

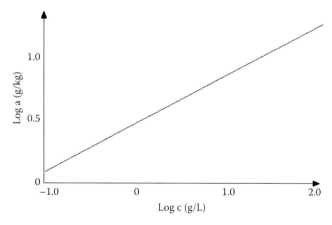

FIGURE 10.2
Log–log plot of the Freundlich adsorption isotherm is shown. The slope that is $1.15/3 = 0.383$ represents b in Equation 6.4 and log k = 0.48, which means that k = 3.1. The equation for the plot shown is therefore $a = 3.1\,c^{0.383}$.

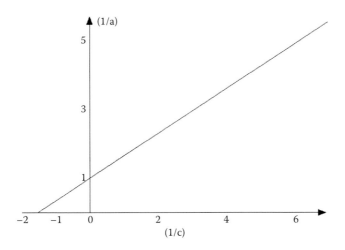

FIGURE 10.3
Lineweaver–Burk's plot where 1/a is plotted versus 1/c. From the plot, it is possible to read that $1/b' = -(-1.5)$ or $b' = 2/3$ and that $k' = 1$ ($1/k' = 1$). This means that it is a Langmuir adsorption isotherm $a = c/(c + 2/3)$.

b is often close to 1 and c is for most environmental problems small. This implies that the two adsorption isotherms get close to $a/c = k$, and k becomes a distribution coefficient. k for 100% organic carbon, usually denoted K_{oc}, may be *estimated* from K_{ow}, the octanol–water distribution coefficient, which is the solubility in octanol divided by the solubility in water. Several estimation equations have been published in the literature; see for instance Jørgensen et al. (1997). The following log–log relationships between K_{oc} (100% organic carbon presumed) and K_{ow} are typical examples:

$$\log K_{oc} = -0.006 + 0.937 \log K_{ow} \text{ (Jørgensen 2000)} \tag{10.32}$$

$$\log K_{oc} = -0.35 + 0.99 \log K_{ow} \text{ (van Leeuwen and Hermens 1995)} \tag{10.33}$$

In the case that the carbon fraction of organic carbon in soil is f, the distribution coefficient, K_D, for the ratio of the concentration in soil and in water can be found as $K_D = K_{oc} f$.

If the solid is activated sludge (from a biological treatment plant) instead of soil, K_D can be found as described earlier or the equation log DCS = 0.39 + 0.67 log K_{ow}, which was presented in Section 4.8, can be applied.

K_{ow} can be found for many compounds in the literature, but if the solubility in water is known it is possible to estimate the partition coefficient n-octanol-water at room temperature by use of a correlation between the water solubility in μmol/L and K_{ow}. A graph of this relationship is shown Figure 10.4.

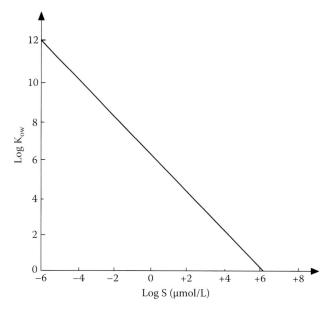

FIGURE 10.4
Linear regression between log K_{ow} and log S, the solubility in water expressed in μmol/L. The regression is based on a wide range of organic compounds.

Example 10.4

A mixture of polyaromatichydrocarbons (PAH) with log K_{ow} = 5.5 is discharged to an aquatic ecosystem with 1 mg/L suspended matter. Experiments have shown that PAH adsorbs to the suspended matter with the following results as basis:

PAH in Solution (μg/L)	PAH Adsorbed (g/kg) Suspended Matter
1	2.5
2	5.0
3	7.4
5	12.0
8	17.0
10	17.5
12	17.8

The discharge is 0.120 kg/day of the PAH-mixture. The aquatic ecosystem can be considered a mixed flow reactor; see Section 2.4. The hydrological retention time is 60 days. The volume is 1,000,000 m³. BCF (L/kg) of the PAH-mixture for fish can be found from Log BCF = 0.8 · Log(K_{ow}) – 0.25.
   Adsorption Isotherms

   a. Find an adsorption isotherm that can describe the adsorption of the PAH-mixture to the suspended matter.
   b. Find the concentration of dissolved PAH in the aquatic ecosystem.
   c. What is the expected concentration of PAH in fish?

Solution

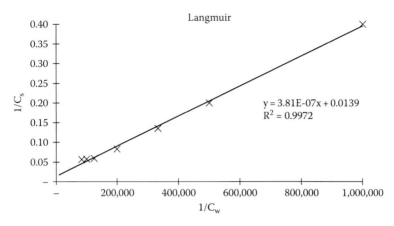

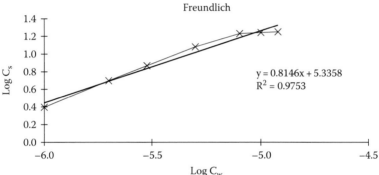

a. It is seen that R^2 is biggest for the Langmuir plot. It implies that the adsorption can be described by an adsorption isotherm with the following constants:

$$C_{s,max} = 1/\beta = 71.9 \text{ g/kg}$$

$$K_{ads.} = 1/(\alpha \cdot C_{s,max}) = 36.5 \times 10^3 \text{ L/g}.$$

b. The concentration totally in the aquatic ecosystem can be found from the discharge rate and the retention time:

$$C_{PAH} = \frac{120 \,(\text{g/day}) \cdot 60 \text{ days}}{10^6 \text{ m}^3 \times 10^3 \text{ m}^{-3}} = 7.2 \,\mu\text{g/L}$$

If all PAH was adsorbed the concentration on suspended matter would be

$$C_{s,PAH} = \frac{C_{PAH}}{C_{Suspenderet}} = \frac{7.2 \,\mu\text{g/L}}{1 \text{mg/L} \times 10^{-6} \text{ kg/mg}} = 7.2 \text{ g/kg}$$

By an iteration starting with this concentration is found c, which is used to find a better a and so on, we find that $c = C_s = 5.1\,g/kg$ and $a = C_w = 2.1\,\mu g/L$.

c.
$$BCF = 10^{0.8 \times 5.2 - 0.25} = 10^{3.91}\,L/kg$$

The concentration in fish can be found from

$$BCF = \frac{C_{fisk}}{C_W}$$
$$\Downarrow$$
$$C_{fisk} = BCF \times C_w = 10^{3.91}\,kg^{-1} \times 2.1\,\mu g/L = 17\,mg/kg$$

10.8 Biological Concentration Factor

The distribution between the biosphere and the hydrosphere is also of importance. BCF is the ratio between the concentrations in an organism and in water. It can be found for many compounds and for some organisms in the literature. BCF may also be estimated, see Figure 10.5, where two log–log

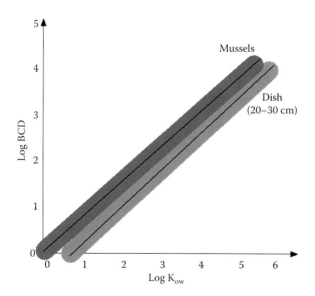

FIGURE 10.5
Relationships between log BCF (fish 20–30 cm long) or mussels and log K_{ow}. The two relationships are almost parallel. The gray zones shown around the straight lines indicate the bands corresponding to 95% of the observations on which the correlation is based. The correlations are based on 142 observations for fishes and 82 for the mussels. The observations are taken from several sources among which the most important is Geyer et al. (1982).

plots between BCF and K_{ow} are shown for mussels and fish (length 20–30 cm) with an average lipid fraction of 5%–12%.

H (or rather K_H), K_{oc}, K_D, and BCF all express a ratio between two equilibrium concentrations in two different spheres. A transfer of a compound from one sphere to another will take place until the equilibrium concentrations have been attained. The rate of transfer will usually be proportional to the distance from equilibrium and dependent on the diffusion coefficient of the compounds and of the resistance at the boundary layer between the two spheres. The reaeration of aquatic ecosystems follows this pattern; see Section 2.6. The rate of reaeration is proportional to the difference between the oxygen concentration at saturation and the actual oxygen concentration.

Example 10.5

Compare the BCF found in Example 10.4 with a BCF value found from Figure 10.5.

Solution

From the figure is found log BCF = 3.55 ± 0.4 and the equation presented in Example 10.4 gave log BCF = 3.91. The difference between the two results is not significant. The equation presented in Example 10.4 had no indication of fish size, while Figure 10.5 is valid for 20–30 cm fish.

References

Geyer, H. et al. 1982. Estimation of the biological concentration factor. *Chemosphere* 11:1121–1130.

Jørgensen, S.E. 2000. *Principles of Pollution Abatement*. Elsevier, Amsterdam, the Netherlands, 520pp.

Jørgensen, S.E. 2012. *Fundamentals of Systems Ecology*. CRC Press, Boca Raton, FL, 320pp.

Jørgensen, S.E. and Bendoricchio, G. 2001. *Fundamentals of Ecological Modelling*, 3rd edn. Elsevier, Amsterdam, the Netherlands, 628pp.

Jørgensen, S.E.; Halling-Sørensen, B.; and Mahler, H. 1997. *Handbook of Estimation Methods in Ecotoxicology and Environmental Chemistry*. Lewis Publishers, Boca Raton, FL, 230pp.

Jørgensen, S.E.; Nors Nielsen, S.; and Jørgensen, L.A. 1991. *Handbook of Ecological Parameters and Ecotoxicology*. Elsevier, Amsterdam, the Netherlands. Published as CD under the name ECOTOX, with Jørgensen, L.A. as first editor in year 2000.

van Leeuwen, C.J. and Hermens, J.L.M. 1995. *Risk Assessment of Chemicals. An Introduction*. Kluwer Academic Publishers, Dordrecht, the Netherlands, 288pp.

Weber, W.J. and DiGiano, F.A. 1996. *Process Dynamics in Environmental Systems*. John Wiley & Sons Inc., New York, 942pp.

11

Application of Aquatic Chemistry in Environmental Management II: Equilibrium Calculations of the Four Types of Reactions

11.1 Double Logarithmic Diagrams Applied on Acid–Base Reactions

The chemistry of the aquatic environment involves always many simultaneous reactions, which can be difficult to overview unless we use computer programs or double logarithmic diagrams that allow us quickly to assess the approximate concentrations. It includes a determination of which components we do not need to consider because they have relatively small concentrations. The double logarithmic diagram is applied in this handbook because it is relatively easy to construct and is very illustrative. The double logarithmic diagram for acid–base reactions plots the logarithmic of the concentration of the various species versus pH.

If we consider an acid–base reaction, $HA \Leftrightarrow A^- + H^+$, the logarithmic form of the equilibrium constant expression is

$$pK = -\log \{A^-\} - \log \{H^+\} + \log \{HA\} = -\log \{A^-\} + pH + \log \{HA\} \quad (11.1)$$

or in the form of the so-called Henderson–Hasselbalch's equation

$$pH = pK + \log \frac{\{A^-\}}{\{HA\}} \quad (11.2)$$

Notice that "p" is a general abbreviation for "−log." K is the equilibrium constant (see Chapter 10).

From (11.2), it is clear that when pH $\ll$ pK is $\{HA\} \gg \{A^-\}$. As the acid–base system only has the two forms HA and A^-, $\{HA\} \approx [HA] \approx C$, where C is the total concentration of the acid–base system. Log $\{A^-\}$ at low pH-values can be derived from (11.1): log $\{A^-\}$ = pH − pK + C ≈ log $[A^-]$. This will correspond to

TABLE 11.1

HA and A⁻ as Function of pH

	Log [HA]	Log[A⁻]
pH << pK	log C	pH − pK + C
pH = pK	log (C/2)	log (C/2)
pH >> pK	pK − pH + C	log C

a straight line with the slope + pH in a logarithmic diagram. For pH = pK, the line will take the value C. A straight line through the point (pK, C) with a slope of +1 represents therefore log [A⁻] in a diagram of log c_i (c_i symbolizing various species) versus pH.

At pH >> pK, {A⁻} ≈ [A⁻] ≈ C. Log {HA} = pK − pH + C ≈ log [HA]. This implies that log [HA] in a double logarithmic diagram for high pH-values, and it is represented by a straight line with the slope − 1 going through the point (pK, C).

For pH = pK, we know from Equation 11.2 that {HA} = {A⁻} = C/2 ≈ [HA] = [A⁻]. Table 11.1 summarizes these results.

Figure 11.1 shows a double logarithmic diagram for a 0.01 M acid (log C = −2) with pK = 6.0. The diagram is drawn by the use of the straight lines for log [HA] and log [A⁻] at low and high pH and the point (pK, C/2), which is valid for both log [HA] and log [A⁻]. The gap between low pH and pH = pK can easily be drawn and correspondingly the gap between pH = pK and high pH.

The double logarithmic diagram represents two equations: the mass equation expression and the information that the total concentration is 0.01 M. We have, however, four that are unknown [HA], [A⁻], [H⁺], and [OH⁻]. We need therefore two more equations. Water's ion product [H⁺] [OH⁻] = 10⁻¹⁴ or, in logarithmic form, log [H⁺] + log [OH⁻] = −14 = pK_w can be used as the third equation. The fourth and last equation is the information that a solution always will be uncharged—that the sum of the concentrations of positive charged ions times their charge = the sum of the concentrations of the negative charged ions times their charge. In most cases it is, however, more beneficial to use the fact that the concentrations resulting from dissociation of hydrogen ions is equal to the concentrations of ions that have taken up hydrogen ions. It is called the proton balance. To use this fourth equation to assess the composition of a solution including pH, it is recommended to write up the components that are in the solution before any reaction.

For instance, if an acid HB in water is considered, the components before any reaction are HB and H_2O. These two components can therefore not result from dissociation or uptake of hydrogen ions. Dissociation of hydrogen ions from these two components can only result in the formation of B⁻ and OH⁻, and the uptake of hydrogen ions is only possible for water forming the oxonium ions, which is often just written as H⁺ only. The proton balance yields therefore the following equation:

$$[B^-] + [OH^-] = [H^+] \tag{11.3}$$

As we have an acidic solution, pH is relatively low, and it is therefore assumed—at least in the first hand—that + [OH⁻] ≈ 0. Therefore, the proton balance gives

$$[B^-] = [H^+] \tag{11.4}$$

This equation corresponds to the point 1 in Figure 11.1. This point represents the composition of the 0.01 M HB solution. It is of course necessary to check the assumption that the hydroxide ions are negligible. As seen $[OH^-] = 10^{-9}$ at point 1. It was therefore fully acceptable to consider $[OH^-]$ as negligible compared with $[B^-]$ and $[H^+]$.

With the same argument, it can be found that the proton balance for a solution of NaB yields $[OH^-] = [HB]$ corresponding to point 2 in Figure 11.1.

Figure 11.2 shows a double logarithmic diagram of the system H_2B, HB^-, and B^{2-}. The total concentration is 0.01 M. The acid in this case is able to dissociate two hydrogen ions, and it has therefore two pK values: pK_1 and pK_2. The H_2B-curve will at $pH \geq pK_2$ have the slope -2, as seen, because the actual process is $H_2B \Leftrightarrow B^{2-} + 2H^+$. B^{2-} will correspondingly have the slope $+2$ at $pH \geq pK_1$.

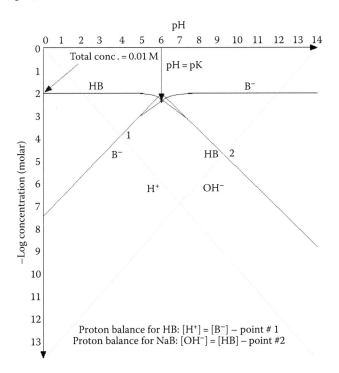

FIGURE 11.1
Double logarithmic diagram for an acid–base system with a total concentration of 0.01 m and pK = 6.0. The proton balance for HB and NaB is shown. The pH for the two cases are found as point 1 (pH = 4.0) and 2 (pH = 9.0), respectively. Notice that the total composition can be read at the diagram. B⁻ at point 1 is 0.0001 and HB at point 2 is 10^{-9}.

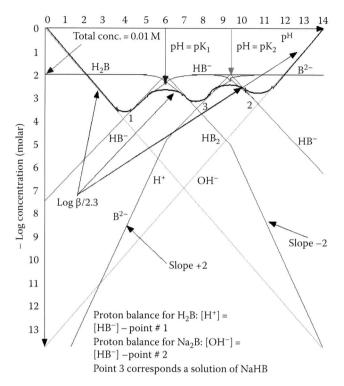

FIGURE 11.2
Double logarithmic diagram of the system H_2B, HB^-, and B^{2-}—total concentration = 0.01 M. Notice that the slopes of the curves for H_2B and B^{2-} are +2 above pK_2 and below pK_1, respectively. The proton balance after approximations were considered are shown for the two cases 0.01 M H_2B and 0.01 M Na_2B. The composition of the two solutions can be read from the figure point 1 and 2, respectively. A 0.01 M solution of NaHB has the composition corresponding to point 3.

The proton balance for the ampholyte NaHB will with good approximations yield the equation $[H_2B] = [B^{2-}]$ as $[OH^-]$ and $[H^+]$ both $\ll [H_2B] = [B^{2-}]$. The point 3 in Figure 11.2 corresponds to the composition of the various species in a 0.01 M NaHB solution with the pK values as shown in Figure 11.2, that is, 6.1 and 9.5. Point 1 corresponds to a 0.01 M H_2B solution, where the concentration of B^{2-} and hydroxide ions are considered negligible in the proton balance. The approximated proton balance is shown in the figure. Point 2 in the Figure 11.2 corresponds to a 0.01 M B^2 solution. The proton balance is shown in the figure. H_2B and hydrogen ions are in this case considered negligible.

Phosphoric acid is a medium strong acid and has therefore a low pK_1 value, namely, 2.12. It implies that point 1 corresponding to the composition of a 0.4 M phosphoric acid solution yields an undissociated phosphoric acid concentration different from the total concentration—compare with Figures 11.1 and 11.2—where a weak acid solution clearly gives an undissociated acid close to the total concentration. If, in the case of phosphoric acid, we have to read the concentration of phosphoric acid in a 0.4 M solution, it is advantageous to read the concentration of dihydrogen phosphate and deduct it from the total concentration. The concentration of dihydrogen phosphate is found from the diagram to be antilog (−1.4) corresponding to a concentration of about 0.04 M. pH is 1.4. The concentration of phosphoric acid becomes therefore $0.4 − 0.04 = 0.36$ M.

11.2 Molar Fraction, Alkalinity, and Buffer Capacity

Introduction of molar fractions makes it possible to set up equations to compute the composition of acid–base solutions. As mentioned earlier, the application of a double logarithmic diagram corresponds to find x unknown concentrations from x equations. It is therefore of course possible to find a composition of a complex acid–base mixture by solving the equations. It is however in most cases faster and sufficiently accurate to use a double logarithmic diagram. Molar fractions may also be used in the double logarithmic diagram instead of concentrations. The concentrations for a number of different cases (different total concentrations) of the same components can easily be found by multiplying the molar fractions found on the diagram with the total concentration.

The molar fractions are shown in Figure 11.3 for the system H_2B, HB^-, and B^{2-}. The three shown equations are found from the two mass equations and from the equation that expresses that the sum of the three molar fractions is one.

The buffer capacity, ß, is defined as

$$\text{ß} = \frac{dC}{dpH} \tag{11.5}$$

where
dC is strong acid or base added to the considered solution
dpH is the corresponding change of pH

Acids and Bases—important equations:

Molar fraction $H_2B = 1/(1 + K_1[H^+] + K_1K_2[H^+]^2)$

Molar fraction $HB^- = 1/(1 + [H^+]/K_1 + K_2/[H^+])$

Molar fraction $B^{2-} = 1/(1 + [H^+]/K_2 + [H^+]^2/K_1K_2)$

ß = buffer capacity = $2.3\,([H^+] + [OH^-] + [HA][A^-]/([HA] + [A^-]))$

$[Alk] = [HCO_3^-] + 2[CO_3^{2-}] - [H^+] + [OH^-] + \sum \text{other base ions}$

NB!! Alkalinity is conserved.

FIGURE 11.3
Important definitions: molar fraction, buffer capacity and alkalinity.

When ß is high, relatively much acid or base is needed to change pH, while a low ß value indicates that pH is changed by addition of a minor amount of acid or base. It can be shown by differentiation according to the definition given in (11.5) that

$$ß = 2.3 \left(\frac{[H^+] + [OH^-] + [HA][A^-]}{([HA] + [A^-])} \right) \tag{11.6}$$

ß can be found directly from this expression. The various concentrations can be found by calculations or from the double logarithmic diagram. It is also possible to find and draw on the double logarithmic diagram the equation log ß/2.3 = log (([H^+] +[OH^-] + [HA][A^-]/([HA] + [A^-])). At very low pH, [H^+] is dominating the expression (11.6), and a log (ß/2.3) line in the double logarithmic diagram will therefore follow the line for log [H^+]. At slightly higher pH, where [H^+] = [A^-] and [HA] ≈ C = [HA] + [A^-] (see Table 11.1), log ß/2.3 = log (2 [H^+]) = log (2 [A^-]) = 0.3 + log ([H^+]) = 0.3 + log [A^-]). Where the two lines for [H^+] and log ([A^-]) intersect, log (ß/2.3) value will therefore be 0.3 above the intersection. At pH = pK, the ß-expression is dominated by [HA][A^-]/([HA] + [A^-] = C/2 C/2/C = C/4. It means that the log (ß/2.3) value will be 0.3 below the intersection of [HA] and [A^-] at pH = pK. At high pH, the expression is dominated by [OH^-], and log [ß/2.3) will therefore follow the line of log [OH^-]. Where log [OH^-] = log [HA], log [ß/2.3) = 0.3 + log [OH^-] = 0.3 + log [HA].

Figure 11.4 shows how a log (ß/2.3) plot is found for the acid–base system in Figure 11.2. The double logarithmic diagram is applied to find which concentrations we have to consider in Equation 11.6 and which concentrations we can eliminate because they are negligible.

The two most important acidic components in the oceans are the carbon dioxide system and boric acid. As pH in the oceans is 8.1, the buffer capacity of the water in the oceans can easily be found. The buffer capacity of the oceans is, however, much higher than the value obtained from the buffer capacities of the ions due their content of suspended clay minerals, which are able to buffer due to the following reaction:

$$3Al_2Si_2O_5(OH)_2 + 4SiO_2 + 2K^+ + 2Ca^{2+} + 12H_2O \longleftrightarrow$$

$$2KCaAl_3Si_5O_{16}(H_2O)_6 + 6H^+ \tag{11.7}$$

The pH-dependence is indicated by the corresponding equilibrium expression in logarithmic form:

$$\log K = 6 \log(H^+) - 2 \log K^+ - 2 \log Ca^{2+} \tag{11.8}$$

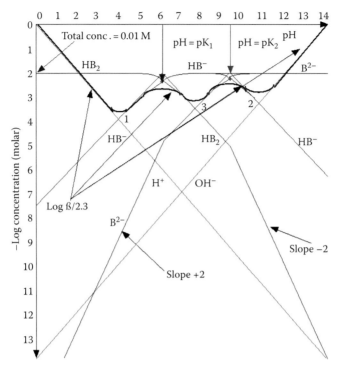

FIGURE 11.4

Log (ß/2.3)–line for Figure 11.2 is shown. At point 1, 2, and 3 the line is 0.3 units above the intersections and at pK_1 and pK_2 the log (ß/2.3) line is 0.3 below the intersection.

Sillen (1961) *estimated the buffering capacity of these silicates to be about 1 mol/L or approximately 2000 times the buffering capacity of carbonates.*

Alkalinity is the sum of all alkaline components minus the sum of all acidic components. In aquatic chemistry, we are usually particularly interested in the amount of hydrogen ions that we have to add to a considered aquatic solution to obtain a pH corresponding to an aquatic solution of carbon dioxide. This is the alkalinity defined in Figure 11.3.

Alkalinity of many natural aquatic systems is often (but not always) with good approximation equal to the hydrogen carbonate concentration. Hydrogen carbonate is at pH between 5 and 8.5, the dominating component yielding alkalinity. Other ions such as chloride and sulfate do not contribute to the alkalinity.

Notice that we cannot find the pH for a mixture of two solutions with known pH, by calculation of the weighted average of the two pH-values, because the resulting number of free hydrogen ions depends on the composition of the two solutions: which ions would be able to react with the free hydrogen and hydroxide ions? We can therefore only find pH for a mixture when we know the concentrations of alkaline and acidic components in the two solutions and can calculate the possible neutralization reactions. The alkalinity can, however, be found easily for a mixture of two solutions, because alkalinity is based on a "book keeping" of *all* alkaline and acidic components, that is, all components that can participate in acid–base reactions in the actual pH-range.

If, for instance, the alkalinity is 4 meq/L for one solution and 8 meq/L for another solution and we mix equal volumes of the two solutions, the resulting alkalinity will be 6 meq/L, which can be used to find the resulting pH for the mixture.

11.3 Dissolved Carbon Dioxide

Open aquatic systems have, in addition to the acid–base reactions of the dissolved components, an equilibrium between carbon dioxide in the atmosphere and dissolved in water. Equation 11.3 can be applied to find the concentration in water of carbon dioxide. This equation implies that the carbon dioxide concentration is constant and independent of pH. A part of the dissolved carbon dioxide (in the order of 1%) reacts with water and forms carbon acid. It is however in most calculations convenient not to distinguish the dissolved carbon dioxide and the carbon acid but to consider the total amount of carbon dioxide and carbon acid. The usually applied pK values are based on this assumption. Henry's constant and the pK values for carbon acid and hydrogen carbonate are dependent on the temperature; see Table 11.2.

Figure 11.5 shows a double logarithmic diagram for an open aquatic system in equilibrium with the carbon dioxide in the atmosphere. The figure illustrates the system during 1999–2002 where the partial pressure of carbon dioxide was about 0.000385 atm corresponding to a concentration of 385 ppm on volume/volume basis. Year 2012/2013 the concentration is expected to be approximately 400 ppm. Room temperature is presumed in the diagram.

TABLE 11.2

Equilibrium Constants of the Various Carbon Dioxide–Carbonate Equilibria at I = 0

Type of Constant	5°C	10°C	15°C	20°C	25°C	40°C	100°C
Solubility product of $CaCO_3$	8.35	8.36	8.37	8.39	8.42	8.53	
pK_1 for H_2CO_3	6.52	6.46	6.42	6.38	6.35	6.35	6.45
pK_2 for H_2CO_3	10.56	10.49	10.43	10.38	10.33	10.22	10.16
pK_w	14.73	14.53	14.34	14.16	14.00	13.53	11.27
K_H	1.20	1.27	1.34	1.41	1.47	1.64	1.99

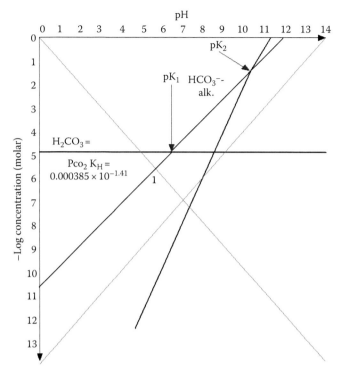

FIGURE 11.5
Double logarithmic diagram for an open aquatic system in equilibrium with the carbon dioxide in the atmosphere. A concentration of 385 ppm is presumed. It corresponds to the carbon dioxide concentration year 1999/2000.

Total carbonic acid = carbonic acid + dissolved carbon dioxide, denoted C_T, will therefore in a double logarithmic diagram correspond to a horizontal line at log $C = 0.000385 \times 10^{-1.41}$. The line representing hydrogen carbonate has a slope of +1 and intersects the carbon acid line at pH = pK_1. The carbonate line will correspondingly have a slope of +2 and intersect the hydrogen carbonate line at pH = pK_2.

The composition of an open aquatic system can easily be found by the application of the diagram Figure 11.5, provided that the alkalinity is known and other alkaline components can be omitted. If for instance the alkalinity is found to be 0.001 M and it is assumed that pH is below 9, the alkalinity will with good approximation be equal to the hydrogen carbonate concentration. It is seen in Figure 11.5 that a hydrogen carbonate concentration of 0.001 corresponds to a pH of 8.0—the assumption that pH was below 9.0 was therefore correct. The concentrations of carbon acid and carbonate can easily be read from the figure in this case.

Example 11.1

Construct a double logarithmic diagram for carbonic acid–hydrogen carbonate in a freshwater system at 10°C in equilibrium with the atmosphere containing 450 ppm carbon dioxide.

    a. Mark on the diagram a stream with alkalinity 0.0025 eqv/L and indicate the corresponding pH.
    b. Mark on the diagram a stream with alkalinity 0.0005 eqv/L and indicate the corresponding pH value.
    c. What is the resulting pH in a lake receiving a mixture of equal volumes of the two streams?

Solution

At moderate pH, the alkalinity is $[HCO_3^-]$. On the diagram, see Figure 11.6, is shown where $[HCO^{3-}] = 0.0025 M$ (point 1) corresponding to pH = 8.5. Point 2 indicates $[HCO^{3-}] = 0.0005 M$ corresponding to pH = 7.8.

The alkalinity is conserved when two streams are mixed. Therefore, the alkalinity in the lake $\approx [HCO^{3-}] = 0.0015 M$ at pH = 8.3, (point 3).

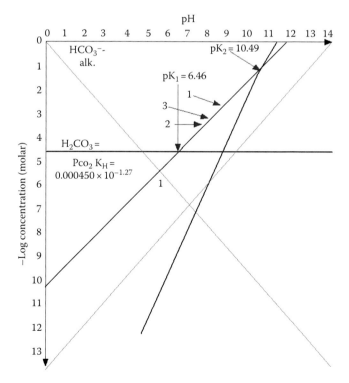

FIGURE 11.6
Example 11.1. See the solution.

In addition to the equilibrium between carbon dioxide in the atmosphere and carbon acid in water, solid carbonate may be present as suspended matter and/or in sediment and be in equilibrium with the calcium and carbonate ions in the water according to the solubility product:

$$[Ca^{2+}][CO_3^{2-}] = K_S = 10^{-8.4} \tag{11.9}$$

It is possible to include also this equation in the double logarithmic diagram as shown in Figure 11.7. The calcium ion line gets a slope of +2 and intersects the carbonate line at $\log C = -4.2$. The composition under these circumstances is found by a charge balance. The sum of the negative and positive ions must balance. In most aquatic systems at pH < 9.0 the dominant cations will be the calcium ions and the dominant anions will be hydrogen carbonate ions. It implies that the following equation is valid:

$$2\,[Ca^{2+}] = [HCO_3^-] \tag{11.10}$$

This corresponds, in Figure 11.7, to pH = 8.4. The carbonate ion concentration is negligible at this pH-value. Notice that in this case where two

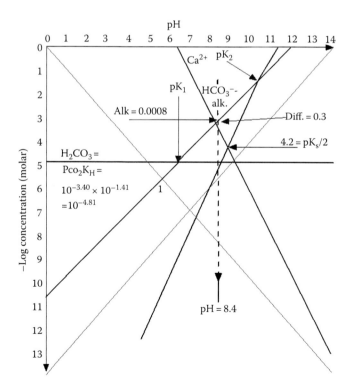

FIGURE 11.7
Double diagram for an aquatic system is simultaneously in equilibrium with carbon dioxide in the atmosphere and solid calcium carbonate.

equilibria are imposed on the system (equilibrium with carbon dioxide in the atmosphere and equilibrium with solid calcium carbonate), the composition of the water is given and no information of alkalinity is needed. The concentration of the carbonate ions must be included in the charge balance at higher pH, which is possible by an iteration.

Example 11.2

What is the composition when an open freshwater system is in equilibrium with solid calcium carbonate and carbon dioxide at $10^{-3.41}$ atm?

Solution

The diagram in Figure 11.7 can be applied.
  The process determining the dissolution of calcium carbonate:

$$CaCO_3(s) + CO_2 + H_2O \rightarrow Ca^{2+} + 2HCO_3^{-}$$

The charge balance:

$$2[Ca^{2+}] + [H^+] = [HCO_3^{-}] + 2[CO_3^{2-}] + [OH^-]$$

With good approximation:

$$2[Ca^{2+}] = [HCO_3^{-}]$$

$$pH \approx 8.4;\ [HCO_3^{-}] \approx 10^{-3}M;\ [H_2CO_3^*] \approx 10^{-5}M;\ [Ca^{2+}] \approx 5 \cdot 10^{-4}M\ og$$

$$[CO_3^{2-}] \approx 10^{-5}M$$

11.4 Precipitation and Dissolution: Solubility of Hydroxides

The solubility products of hydroxides, oxides, and carbonates have particular interest in aquatic chemistry because these anions are present in high concentrations in natural aquatic systems. The hydroxides are furthermore of interest, because they are for most heavy metals very insoluble, which is utilized in the treatment of industrial wastewater containing heavy metals. As it has been shown in Figure 11.7, it is also possible to apply the double logarithmic plot to represent a solubility product. This is also illustrated in Figure 11.8, where the solubility of several heavy metal ions as function of pH due to precipitation of hydroxides. The solubility products for several heavy metal hydroxides are shown in Table 11.3.

  In many cases, the cations are able to form hydroxo-complexes with hydroxide ions and thereby change the solubility that is found directly from the solubility product. The actual reactions for aluminum ions are shown in Figure 11.9. It is

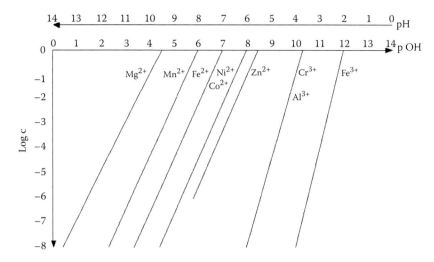

FIGURE 11.8
Solubility of several heavy metal cations as function of pH.

TABLE 11.3

pK$_s$ Values at Room Temperature for Metal Hydroxides[a]

Hydroxide	z = Charge of Metal Ions	pK$_s$
AgOH(1/2 Ag$_2$O)	1	7.7
Cu(OH)$_2$	2	20
Zn(OH)$_2$	2	17
Ni(OH)$_2$	2	15
Co(OH)$_2$	2	15
Fe(OH)$_2$	2	15
Mn(OH)$_2$	2	13
Cd(OH)$_2$	2	14
Mg(OH)$_2$	2	11
Ca(OH)$_2$	2	5.4
Al(OH)$_3$	3	32
Cr(OH)$_3$	3	32

[a] pK$_s$ = $-$ log K$_s$, where K$_s$ = [Mez$^+$][OH$^-$]z.

illustrated in the figure how to cope with the solubility as a function of pH if hydroxo-complexes can be formed. The double logarithmic diagram corresponding to the presented procedure is shown in Figure 11.10.

The solubility of iron(III) hydroxide, zinc oxide, and copper oxide can be found by the same method. The diagram for iron(III) hydroxide is shown in Figure 11.11.

How to make a double logarithmic diagram for the solubility as influenced by hydrolysis/formation of hydroxo complexes?

1. Set up a diagram indicating the initial components and how they can react to form various species. Example:

Species	$Al(OH)_3$	H^+	logK*
Al^{3+}	1	3	8.5
$Al(OH)^{2+}$	1	2	3.53
$Al(OH)_2^+$	1	1	−0.8
$Al(OH)_3$	1	0	−6.5
$Al(OH)_4^-$	1	−1	−14.5

2. Find the equilibrium constants (log $^*K_1$, log $^*K_2$, log $^*K_3$, log $^*K_4$) for the reactions between the aluminum hydroxide and one or more hydrogen ions, here exemplified by $Al(OH)_3 + 3H^+ <-> Al^{3+} + 3H_2O$.

   $^*K_1$ can be found from the solubility product, $K_s = [Al^{3+}] [OH^-]^3$:

   $$K_s = [Al^{3+}] [OH^-]^3 [H^+]^3/[H^+]^3 = K_1 K_w^3$$

   where K_w is the ion product of water = 10^{-14} at room temperature.

3. Plot in a double logarithmic diagram log (species) versus pH by use of the equilibrium constants for the formation the considered species. The solubility of Al(III) is the sum of all the soluble species, Al^{3+}, $Al(OH)^{2+}$ etc.

FIGURE 11.9
It is shown how to utilize a double logarithmic diagram, when hydroxo-complexes may be formed.

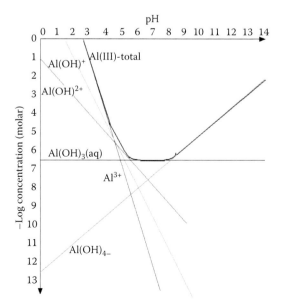

FIGURE 11.10
Solubility of aluminum (III) considering formation of several hydroxo-complexes as function of pH.

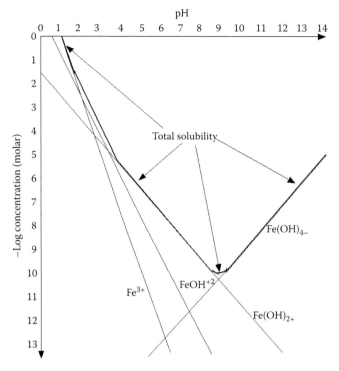

FIGURE 11.11

The solubility of amorphous iron(III) hydroxide as function of pH is shown. The construction of the diagram is as explained for aluminum(III) hydroxide. The solubility of iron(III) hydroxide is determined by formation of iron(III) ions and formation of complexes with one, two, or four hydroxide ions.

11.5 Solubility of Carbonates in Open Systems

The solubility of calcium carbonate in open aquatic systems has already been treated in Section 11.3. Figure 11.12 gives for strontium carbonate, iron(II) carbonate, cadmium carbonate, and zinc carbonate the same diagram as shown in Figure 11.7 for calcium carbonate.

11.6 Solubility of Complexes

Complex formation will be covered later in this chapter, but it will be shown here how a simple complex formation under simultaneous reaction with a compound with very little solubility can enhance the solubility considerably.

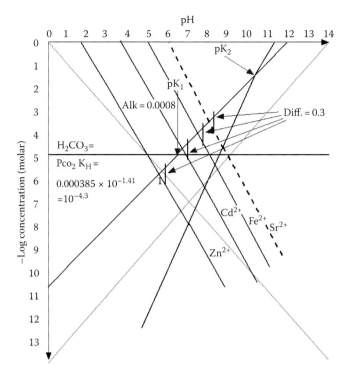

FIGURE 11.12

Double logarithmic representation of an equilibrium of solid strontium carbonate, iron(II) carbonate, cadmium carbonate, and zinc carbonate in an open aquatic system. The composition will correspond to the lines where the difference between the concentration of the metal ion and hydrogen carbonate is 0.3, corresponding to the equation $2[Me^{2+}] = [HCO_3^-]$.

The calculations are made by adding two suitable reactions and determining the equilibrium constant for the total process as the product of the equilibrium constants of the two added processes. If the equilibrium constant for the total process is high, the solubility will with very good approximation be very pronounced. If the equilibrium constant is not very high, the solubility is made unknown in a suitable equation.

The method is shown by use of an example; see Example 11.3.

Example 11.3

Find the solubility of iron(III) in a freshwater lake with and without presence of $10\,\mu mol$ citrate/L. The ion strength of the lake water is 0 and pH = 9.0. The solubility product of iron(III) hydroxide is 10^{-24} mol^4/L^4. The formation constant for the complex between citrate and iron(III) at pH = 9.0, where citric acid has dissociated all hydrogen ions with good approximation, is 10^{28}.

Solution

From the solubility product, it is found that the solubility at pH = 9.0 (the hydroxide concentration is 10^{-5} M) is 10^{-9} M. Solubility of iron(III) when citrate is present is determined by the reaction

$$Fe(OH)_3 + citrat^{3-} \leftrightarrow Fe(III)citrat- + 3OH^-$$

and the mass equation yields

$$[Fecitrat-][OH^-]^3/[citrat^{3-}] = 10^{+4} = X * 10^{-15}/(0.00001 - X)$$

$$X/(0.00001 - X) = 10^{+19}$$

X must be $10\,\mu M/L$ with good approximation, which of course also cor-responds to the concentrations of citrate as it, due to the equilibrium constant (10^{+19}), determines how much iron(III) will dissolve.

11.7 Stability of the Solid Phase

It is not possible, by a comparison of the numerical values of the equilibrium constant, to decide which of the two or more solid phases control the solubil-ity of an ion. It is necessary to determine by calculation which solid phase that gives the smallest concentration of the consider ion and it will be the solid phase with the highest stability. Example 11.4 illustrates a case study: Is it iron(II) carbonate or iron(II) hydroxide that determines the solubility of iron(II) in natural water containing carbonate?

Example 11.4

Is it iron(II) carbonate or iron(II) hydroxide that determines the solubil-ity of iron(II) in natural anoxic water at pH = 6.8 and with an alkalin-ity of 10^{-4}? The solubility products are $10^{-10.7}$ mol^2/L^2 and $10^{-14.7}$ mol^3/L^3, respectively. The pK for hydrogen carbonate is 10.1.

Solution

The iron(II) concentration resulting from solubility of iron(II) carbonate and iron hydroxide is determined:

$$FeCO_3(s) = Fe^{2+} + CO_3^{2-} \quad \log K_s = -10.4$$

$$H^+ + CO_3^{2-} = HCO_3^- \quad -\log K_2 = 10.1$$

For the process,

$$FeCO_3 + H^+ = Fe^{2+} + HCO_3^-$$

the equilibrium constant is found by adding the processes, which implies that log $*K_s$ = −0.3. It means that log $[Fe^{2+}]$ = log $*K_s$ − pH − log $[HCO_3^-]$ = −0.3 − 6.8 − (−4) = −3.1.

$$Fe(OH)_2(s) = Fe^{2+} + 2OH^- \quad log K_s = -14.5$$

$$2H^+ + 2OH^- = 2H_2O \quad -2 log K_w = +28.0$$

For the process,

$$Fe(OH)_2(s) + 2H^+ = Fe^{2+} + 2H_2O$$

the equilibrium constant is found by adding the two processes, which implies that log $*K_s$ = 13.5 and log $[Fe^{2+}]$ = log $*K_s$ − 2pH = 0.1.

The calculations show that iron(II) will be present under the given condition in the concentration $10^{-3.1}$, determined by a solid phase consisting of iron(II) carbonate, with the mineralogical name siderit.

It is possible under the given conditions to find the pH value at which iron(II)-hydroxide will become the most stable solid and thereby determine the iron(II)-solubility. The two expressions for the iron(II)-concentration yields −0.3 − pH −(−4) = 13.5 − 2pH. At pH = 13.5 + 0.3 − 4 = 9.8, will the two solid phases have the same stability, but above this pH value iron(II)-hydroxide will determine the iron(II) -solubility. At pH = 9.8 the solubility of iron(II) is $10^{-6.1}$.

11.8 Complex Formation

Any combination of cations with molecules or anions containing free pairs of electrons is named coordination formation. The metal ion is denoted the central atom and the anions or the molecules are called the ligands. The atom responsible for the coordination is called the ligand atom. If a ligand contains more than one ligand atom and thereby occupies more than one coordination position in the complex, it is referred to as multidentate. Ligands occupying one, two, three, four, and so on are named unidentate, bidentate, tridentate, tetradentate, and so on. Complex formation with multidentate ligands is called chelation and the complexes are called chelates. Typical examples from aquatic chemistry are

Oxalate and ethylenediamine are bidentate

Citrate is tridentate

Ethylenediamine tetraacetate is hexadentate

If there is more than metal ion in the complex, we are talking about polynuclear complexes.

Complex formation takes place according to the following reaction scheme:

$$Me^{n+} + L^{m-} \rightarrow MeL^{(n-m)+} \tag{11.11}$$

where

Me is a metal

LK is a ligand

The mass equation gives the following expression:

$$\frac{[MeL^{(n-m)+}]}{[Me^{n+}][L^{m-}]} = K \tag{11.12}$$

K is named the stability constant, complexity constant, or formation constant.

The coordination number for a metal is the number of coordination position on the metal–positions where the free electron pairs can attach the ligand to the metal. Even coordination numbers are most frequently found. Coordination numbers for the metal ions most frequently found in aquatic environment are listed in Table 11.4.

TABLE 11.4

Coordination Numbers for Metals of Interest in Aquatic Chemistry

Metal	Coordination Number
Cu^+	2
Ag^+	2
Hg_2^{2+}	2, 4
Li^+	4
Be^{2+}	4, 6
Al^{3+}	4, 6
Fe^{3+}	4, 6
Cu^{2+}	4, 6
Ni^{2+}	4, 6
Hg^{2+}	2, 4
Fe^{2+}	6
Mn^{2+}	4, 6

11.9 Environmental Importance of Complex Formation

Complex formation has a great influence on the effects of metal ions on aquatic organisms and ecosystems. Complex formation has the following effects:

1. *The solubility can be increased*

$$MeY(S) + L = MeL + Y \tag{11.13}$$

    Me represents a metal
    Y a sediment
    L a ligand that is able to react with the metal ion

2. *The oxidation stage of the metal ion may be changed. The mass equation constants for the following two processes may differ*

$$Me^{n+} + me^{-} \rightarrow Me^{(n-m)+} \tag{11.14}$$

$$MeL^{n+} + me^{-} \rightarrow MeL^{(n-m)+} \tag{11.15}$$

if the stability constant (mass equation) for the two complexes in Equation 11.15 are different.

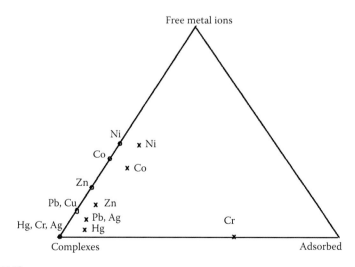

FIGURE 11.13
Triangle diagram for the form of heavy metal ions in freshwater and salt water. Presence of ligands in aquatic systems will generally enhance the transfer of metal ions from sediment, soil, and suspended matter to water; but the toxicity will generally be reduced. x, Fresh water; o, Salt water.

3. *The metal toxicity may be changed*, because the complexes have another bioavailability than the metal ions.
4. *The ion exchange and adsorption processes of the metal ions may be changed.* The adsorption isotherms and ion exchange equilibrium constant will in far most cases be different for the complexes and the metal ions due to the bigger size of the complexes among other properties.

In Figure 11.13, the triangle diagram shows which forms of metal ions are usually found in freshwater and saltwater. The difference is due to the high salinity (chloride concentration) and ion strength in salt water and the presence of organic ligands and a higher concentration generally of suspended matter in freshwater.

11.10 Conditional Constant

The complex formation in aquatic systems is often very complicated because a number of side reactions (acid–base reactions, precipitations, redox reactions) are possible in addition to the main reaction and the complex formation. Reactions between metal ions and ligands are often determining the release of metal ions from sediment, pH is often determining the strength of a complex, and the bioavailability of toxic substances is highly dependent on the form of the compound—complex bound or not, base or acid form, and so on. It is therefore of importance to be able to find the concentrations of the various forms in aquatic systems under given conditions (Alloway and Ayres 1993).

It is possible to consider the side reactions by the application of what is called conditional constants. Let us consider the reaction

$$Me^{n+} + L^m \leftrightarrow MeL^{(n-m)+} \tag{11.16}$$

It is assumed that the following side reactions are possible:
A precipitation

$$pMe^{n+} + nY^{p-} \leftrightarrow Me_pY_n(s) \tag{11.17}$$

An acid–base reaction

$$L^{m-} + H^+ \leftrightarrow HL^{(m-1)-}; \quad HL^{(m-1)-} + H^+ \leftrightarrow H_2L^{(m-2)-} \tag{11.18}$$

f_{Me}, defined as the ratio between the metal ion concentration and the metal concentration including other forms determined by the side reactions, can be found from the solubility product, K_S, if $[Y^{P-}]$ e is known. If concentrations are used, we get

$$[Y^{P-}]^n[Me^{n+}]^P = K_S \qquad (11.19)$$

$$\frac{[Me^{n+}]}{[Me']} = f_{Me} = \frac{K_S^{1/P}}{[Y^{P-}]^{n/P}} * \left[\frac{1}{[Me']} \right] \qquad (11.20)$$

where the symbol Me′ represents all the metal forms except the complex formed by what is considered the main reaction (11.16).

f_L, defined as the ratio between L^{m-} and the ligand concentration including other forms determined by side reaction bs, can be found in a similar manner:

$$\frac{[L']}{[L^{m-}]} = \frac{1}{f_L} = 1 + \frac{[H^+]^m}{K_1 * K_2, \ldots, K_m} + \frac{[H^+]^{m-1}}{K_2 * K_3, \ldots, K_m} + \frac{[H^+]^{m-2}}{K_3 * K_4, \ldots, K_m} + \cdots + \frac{[H^+]}{K_m}$$

$$(11.21)$$

where L′ symbolizes all ligand forms except the complex formed by the main reaction (9.6). K_1 to K_m are the dissociation constants for the step-wise dissociation of hydrogen ions for the acid H_mL.

A reformulation takes place

$$\frac{[MeL^{(n-m)+}]}{[Me^{n+}][L^{m-}]} = K = \frac{[MeL^{(n-m)+}]}{[Me'] * f_{Me} * [L'] * f_L}$$

$$(11.22)$$

$$\frac{[MeL^{(n-m)+}]}{[Me'][L']} = K * f_{Me}f_L = K_{cond}$$

K_{cond} is denoted the conditional equilibrium constant. If K_{cond} is known, it is possible to find the concentrations of free metal ions and metal complexes by the use of the Equations 11.21 and 11.22. K_{cond} is of course only valid under the conditions given by pH and $[Y^{P-}]$. If however these two concentrations are known—the conditions are known—then the usual mass constant calculations can be carried out using K_{cond} instead of K.

Complexation constant, stability constants, formation constant.

$$K_1 = [ML]/[M] [L], \quad K_2 = [ML_2]/[ML] [L], \quad K_3 = [ML_3]/[ML_2] [L], \text{ etc.}$$

$$\text{ß}_1 = K_1, \quad \text{ß}_2 = K_1 K_2, \quad \text{ß}_3 = K_1 K_2 K_3, \text{ etc.}$$

K* and ß* symbolize the constants where HL is the reactant (a ligand with only one H is considered in his example)

If f_L is the fraction of the L-form at a given pH of the total concentration, that is, L + HL = L', the conditional constant

$$K_{1cond} = [ML]/[M] [L'] = f_L K_1$$
$$K_{2cond} = [ML_2]/[ML] [L'] = f_L K_2$$
$$K_{3cond} = [ML_3]/[ML_2] [L'] = f_L K_2$$

Side reactions by M is accounted for in a similar manner.

FIGURE 11.14
Important equations by equilibrium calculations of complex formations.

The most important equations for calculations of complex formation are summarized in Figure 11.14. Notice that K is used for the mass equation constant where one ligand is attached to the metal ion at the time, while ß_n is used for the reaction between the metal ion and n ligand.

Example 11.5

The biological concentration factor, BCF (BCF = C_{Alge}/C_{vand}), for cadmium in brown algae is 890 L/kg, while the BCFs for cadmium chloro-complexes are negligible with good approximation. In the Little Belt in the mid-1980s, a concentration of cadmium of 32 µg/L was found in the most contaminated areas. The salinity is 1.8% and can with good approximation be considered to be entirely sodium chloride. Cadmium forms mono-chloro, di-chloro, tri-chloro, or tetra-chloro complexes. The β-values for formation of these complexes are $25 M^{-1}$, $50 M^{-2}$, $32 M^{-3}$, and $11 M^{-4}$, respectively, by infinite dilution. The temperature is presumed 25°C and $M_{NaCl} = 58.44$ g/mol.

 a. Find the ß-values by the application of Davies equation.
 b. What is the concentration of cadmium in the brown algae?

Solution

$M_w(NaCl) = 58.45$ g/mol
$[Na^+] = [Cl^-] = m(NaCl)/M_w(NaCl) = 18$ g/L$/58.45$ g/mol $= 0.3080$ M
$I = 0.5 \,(0.3080 M * 1^2 + 0.3080 M * 1^2) = 0.3080$ eqv/L
$z = 1 \Rightarrow f_1 = 10^{-0.1477}$
$z = 2 \Rightarrow f_2 = 10^{-0.5606}$
$\text{ß}_1 =$
$\text{ß}_2 =$
$\text{ß}_3 =$
$\text{ß}_4 =$
$[Cd^{2+}] =$
Cd- in brown algae = BCF $[Cd^{2+}] = 890$

Example 11.6

The total concentration of mercury in a lake at pH = 6.5 is determined to be 0.05 mM. The equilibrium constants for formation of
$Hg(OH)^+$ and $Hg(OH)_2^0$ by hydrolysis are $10^{-3.7}$ M and $10^{-2.6}$ M. M_{Hg} =
200.59 g/mol.

a. At which concentrations can the various forms of mercury (II)
   be found under these circumstances?
   The LC_{50} value for Hg^{2+} ions is 1.2 mg/L for daphnia.
b. Is the LC_{50} value exceeded when it is assumed that the hydroxo-
   complexes are 100 times less toxic than Hg^{2+} ions?

Solution

a. The two reactions forming hydroxo-complexes

$$Hg^{2+} + H_2O \rightarrow Hg(OH)^+ + H^+; \quad K_1 = 10^{-3.7} M. \quad pK_{s1} = 3.7.$$

$$Hg(OH)^+ + H_2O \rightarrow Hg(OH)_2^0 + H^+; \quad K_2 = 10^{-2.6} M. \quad pK_{s2} = 2.6.$$

At pH = 6.5 is $Hg(OH)_2^0$ the dominating form of mercury (II)
according to the two equilibrium constants, that is, $[Hg(OH)_2^0] \approx$
0.05 mM.
   The concentrations of $[Hg(OH)^+]$ or $[Hg^{2+}]$ are found from
$[Hg(OH)_2^0] \approx 0.05$ mM and the equilibrium constants

$$[Hg(OH)^+] = [Hg(OH)_2^0] \cdot 10^{-pH}/K_2 = 10^{-8.2} M$$

$$[Hg^{2+}] = [Hg(OH)^+] \cdot 10^{-pH}/K_1 = 10^{-11} M$$

b. The hydroxo-complexes are 100 times less toxic; that is, they
   have an LC_{50}-værdi or 0.12 g/L = 0.55 mM
   The ratio between concentrations and LC_{50} values are found:

$$\frac{(10^{-8.2} M + 0.05 \text{ mM})}{0.55 \text{ mM}} + \frac{10^{-11} M}{0.0055 \text{ mM}} = 0.09 < 1$$

The LC_{50} for daphnia is not exceeded.

11.11 Application of Double Logarithmic Diagrams to Determine the Conditional Constants for Complex Formation

Let us consider a complex formation according to the following reaction scheme:

$$Me^{n+} + L^{m-} = MeL^{(n-m)+} \quad \log K = F$$

exemplified by (Glycinate = Gly)

$$Fe^{3+} + Gly -^- = FeGly^{2+} \quad \log K = 10.8$$

The conditional constant, $K_{cond} = [FeGly^{2+}]/[Fe'][Gly']$

The following equations can be applied to find the conditional constant as function of pH:

$$[Fe'] = [Fe(III)_{total}] - [FeGly^{2+}] \tag{11.23}$$

$$[Fe'] = [Fe^{3+}] + [FeOH^{2+}] + [Fe(OH)_2{}^+] + [Fe(OH)_3] + [Fe(OH)_4{}^+] \tag{11.24}$$

$$[Fe'] = [Fe^{3+}] (1 + {}^*K_1/[H^+] + {}^*\beta_2/[H^+]^2 + {}^*\beta_3/[H^+]^3 + {}^*\beta_4/[H^+]^4) \tag{11.25}$$

$$[Fe'] = [Fe^{3+}]/f_{Fe} \tag{11.26}$$

$$[Gly'] = [Gly_{total}] - [FeGly^{2+}] \tag{11.27}$$

$$[Gly'] = [H_2Gly^+] + [Hgly] + [Gly^{--}] \tag{11.28}$$

$$[Gly'] = [Gly^{--}] (1 + [H^+]/K_2 + [H^+]^2)/K_1K_2 \tag{11.29}$$

$$[Gly'] = [Gly^{--}]/f_{gly} \tag{11.30}$$

Based upon these equations, it is possible to calculate f_{Fe} and f_{Gly} as functions of pH, provided of course that the various applied mass equations constants are known: $\log {}^*K_1 = -3.05$, $\log {}^*\beta_2 = -6.31$, $\log {}^*\beta_3 = -13.8$, and $\log {}^*\beta_4 = -22.7$. The pK values for H_2Gly^+ are 3.1 and 9.9.

Figure 11.15 shows the calculation of f_{Fe}, f_{Gly}, and the product of the two, $f_{Fe} f_{Gly}$, calculated as a function of pH. Equations 11.25, 11.26 and 11.29, 11.30 are applied. f_{Fe}, f_{Gly}, and the product $f_{Fe} f_{Gly}$ are found at a given pH and the conditional constant is calculated as $K_{cond} = K f_{Fe} f_{Gly}$, K_{cond} has a maximum around pH = 3.2. At this pH $\log K_{cond} = \log K + \log f_{Fe} f_{Gly} = 10.8 - 7.0 = 3.8$.

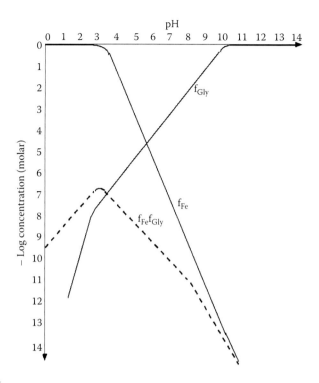

FIGURE 11.15

f_{Fe} and f_{Gly} and the product $f_{fe}f_{Gly}$ are found as function of pH by the Equations 11.23 through 11.30.

Example 11.7

a. By the application of the diagrams shown next, answer the following questions:
   What is the dominant form of cadmium(II) in water with different salinities? $M_{NaCl} = 58.44\,g/mol$.

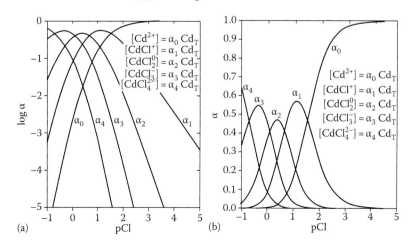

b. What is the LC_{20} value for blue mussels for total cadmium in mg/L, when the following information is available: LC_{20} for Cd^{2+} ions is 2.5 mg Cd/L, 25 mg Cd/L for the monochloro-complex, 60 mg Cd/L for the dichloro-complex, 75 mg Cd/L for the trichloro-complex, and 120 mg Cd/L for the tetrachloro-complex. It is presumed that no synergistic or antagonistic effect will occur. The salinity can with good approximations be referred to sodium chloride, and the specific density of sea water is independent of the salinity 1.0 kg/L.

Use the following table to answer the questions:

Salinity	5‰ (Baltic Sea 200 km North of Stockholm)	12‰ (Baltic Sea at the Island Møn)	20‰ (Kattegat by Anholt)	40‰ (The Mediterranean Sea)
Dominating form of Cd(II)				
Approximate LC_{20} value in mg/L				

Solution

a. Use the diagrams to answer the questions. The answer can be found in the table.
b. For the calculations of $LC_{20,Cd(II),pCl}$ for the four pCl-values read on the diagram the α-values. Calculate $LC_{20,Cd(II),pCl}$ as

$$1/LC_{20,Cd(II),pCl} = \sum_{n=0}^{n=4} \alpha_n / LC_{20,Cd[Cl]^{n/2-n}}$$

Salinity	5‰ (Baltic Sea 200 km North of Stockholm)	12‰ (Baltic Sea at the Three Islands Møn)	20‰ (Kattegat by Anholt)	40‰ (The Mediterranean Sea)
pCl	1.07	0.69	0.47	0.16
Dominating form of Cd(II)	$CdCl^+$	$CdCl^+$ og $CdCl_2^0$	$CdCl_2^0$	$CdCl_2^0$ og $CdCl_3^-$
Approximate LC_{20} value in mg/L	9.4 mg/L	17.6 mg/L	30.4 mg/L	45.9 mg/L

11.12 Redox Equilibria: Electron Activity and Nernst's Law

There is an analogy between acid–base and reduction–oxidation reactions. Acid–base reactions exchange protons. Acids are proton donors and bases proton acceptors. Redox process reactions exchange electrons. Reductants are electron donors and oxidants are electron acceptors. As there are no free hydrogen ions, or protons, there are no free electrons. It implies that every oxidation is accompanied by a reduction and vice versa. Like pH has been introduced as the proton activity, we may introduce an electron activity defined as

$$pe = -\log\{e^-\} \tag{11.31}$$

where e^- is the electron and p as usually is an abbreviation of $-\log$. As free electrons do not exist pe should be considered a concept that is introduced to set up a parallel to pH and facilitate the calculations of redox equilibria. The relationship between pe and the redox potential, E, is

$$pe = \frac{F*E}{2.3RT} \tag{11.32}$$

where
  F is Faraday's number $= 96485\,C/mol =$ the charge of 1 mol of electrons
  R is the gas constant $= 8.314\,J/mol\,K = 0.082057\,L\,atm/mol\,K$

Nernst law

$$E = E° + \frac{RT}{nF}\log\frac{\{ox\}}{\{red\}} \tag{11.33}$$

may be rewritten by the use of pe^0

$$pe^0 = \frac{F*E°}{2.3RT} \tag{11.34}$$

to

$$pe - pe^0 + \left(\frac{1}{n}\right)\log\frac{\{ox\}}{\{red\}} \tag{11.35}$$

This yields furthermore the following relationship between the free energy and pe, as $\Delta G = -En$:

$$pe = \frac{-\Delta G}{n*2.3RT} \quad pe^0 = \frac{-\Delta G°}{n*2.3RT} \tag{11.36}$$

Notice that these equations are applied on the half reaction that an oxidant takes up an electron, which is transferred to a reductant. A redox process, however, requires, as pointed out previously, that another reductant is available to deliver the electron and thereby this reductant is changed to the corresponding oxidant.

Equation 11.36 implies that the following relationship between pe and the equilibrium constant, K, for the half reaction, in which the electron takes part, is valid:

$$pe^0 = \frac{1}{n} \log K \qquad (11.37)$$

As illustration of the application of these equations, the following reaction is considered:

$$Fe^{3+} + e^- = Fe^{2+} \qquad (11.38)$$

The standard redox potential corresponding to the {ox} = {red} = 1 for this process can be found in any table of standard redox potentials to be 0.77 V by room temperature. It means

$$E^0 = 0.77; \quad pe^0 = \frac{F * 0.77}{2.3 * R * 298} \quad (t = 25°C) \qquad (11.39)$$

$$\log K = n * pe^0 = 1 * \frac{0.77}{0.059} = 13.0 \qquad (11.40)$$

where $\log K = \log (\{Fe^{2+}\}/\{Fe^{3+}\}\{e^-\})$.

If, for instance, in an acidic solution $[Fe^{3+}] = 10^{-3}$ and $[Fe^{2+}] = 10^{-2}$, we get, applying concentrations instead of activities as a reasonable good approximation,

$$pe = pe^0 + \left(\frac{1}{n}\right) \log \frac{\{ox\}}{\{red\}} = 13 + 1 * \log 0.1 = 12.0 \qquad (11.41)$$

The equations that are applied to make the redox calculations are summarized in Figure 11.16.

Example 11.8

Find the pe value for a natural aquatic system at pH = 7.0 and in equilibrium with the atmosphere (partial pressure of oxygen = 0.21 atm). What is the ratio $\{Fe^{2+}\}$ to $\{Fe^{3+}\}$ in the water?

Solution

The following process determines the redox potential:

$$0.5O_2 + 2H^+ + 2e^- = H_2O \qquad (11.42)$$

In tables, it is possible to find that log K for this process is = 41.56—or the standard redox potential can be found and log K calculated based upon E^0.

It means that $pe^0 = 41.55/2 = 20.78$

Redox equations:

$$E = E_o + 2.3\ RT/nF\ (\log\ \{ox\}/\{red\})$$

or

$$pe = pe_o + 1/n\ (\log\ \{ox\}/\{red\})$$

$$\Delta G_o = -E_o\ nF = -RT\ \ln K = 2.3\ RT\ \log K = 2.3\ RT\ n\ pe_o$$

Notice that when the composition, that is, $\{ox\}/\{red\}$ is given, the redoxpotential E or pe can be found.

Freshwater in equilibrium with the atmosphere (oxygen partial pressure 0.21 atm) will have the following pe:

$$pe = 20.78 + 0.5 \log (0.21)^{0.5}\ \{H+\}^2)$$

From the pe value the ratio between any redox pair in freshwater is determined.

FIGURE 11.16
Summary of the most important equations to calculate redox potential, pe and the equilibrium constant.

$$pe = pe^0 + 0.5 \log (\sqrt{p_{O2}}\{H^+\}^2) = 20.78 + 0.5 \log (\sqrt{0.21 \times 10^{-14.0}})$$

$$= 20.78 + 0.5\ (-14.34) = 13.61$$

This pe value is determining all ratios between oxidants and reductants, because the oxygen concentration in water in equilibrium with the atmosphere will inevitably have a constant concentration determined by the partial pressure of 0.21 atm. Therefore,

$$13.61 = pe = pe^0 + \frac{1}{n} \log \frac{[Fe^{3+}]}{[Fe^{2+}]} = 13.0 + \log 10^{0.61} \tag{11.43}$$

The ratio $\{Fe^{2+}\}$ to $\{Fe^{3+}\} \approx 4.0$ (Figure 11.17).

11.13 pe as Master Variable

Equation 11.35 can be used to set up a graphical presentation in a double logarithmic diagram of a redox process. How is for instance the ratio $\{Fe^{2+}\}$ to $\{Fe^{3+}\}$ changed with changing redox potential or pe? Obviously, at low pe the reductant will be dominating and at high pe the oxidant will be dominating. When pe = pe^0 the two forms will be equal. The double logarithmic diagram is therefore very similar to the double logarithmic diagram applied for acid–base reactions. A double logarithmic diagram for $\{Fe^{2+}\}$ and $\{Fe^{3+}\}$ is shown in Figure 11.17.

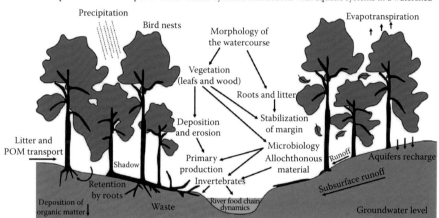

Conceptual scheme of a riparian forest and its dynamic interactions with aquatic systems in a watershed

FIGURE 2.3

Role of riparian forest as a control and regulating factor in a watershed. (Modified from Likens, G.E., The ecosystem approach: Its use and abuse, in: Kinne, O. (Series Ed.), *Excellence in Ecology*, Ecology Institute, Oldendorf/Luke, Germany, 166pp., 1992; Paula Lima, W. and Zakia, M.J.B., Hidrobiologia de Matas Ciliares, in: Rodrigues, R.R. and Leitão Filho, H. (Eds.), *Matas Ciliares: Conservação e recuperação*, EDUSP, FAPESP, São Paulo, Brazil, pp. 33–44, 320pp., 2001.)

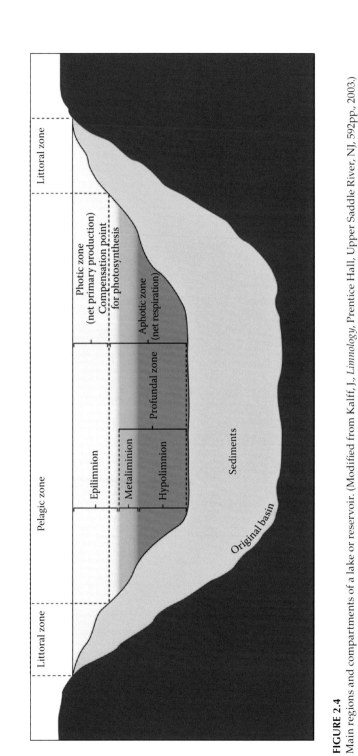

FIGURE 2.4
Main regions and compartments of a lake or reservoir. (Modified from Kalff, J., *Limnology*, Prentice Hall, Upper Saddle River, NJ, 592pp., 2003.)

FIGURE 2.13
Lake Taihu suffers from eutrophication and frequent bloom of blue-green algae.

FIGURE 2.14
Eutrophication in Dianchi Lake, China.

FIGURE 2.15
Western lake in Hangzhou. By a management that has considered all the sources to the eutrophication problems, it has been possible to obtain a reasonable water quality in spite of a previous hypereutrophication. Many of the tool boxes that are mentioned in Chapter 14 have been applied.

FIGURE 2.16
Lake Konstanz, where the management of the eutrophication has consequently considered all sources and has reduced the eutrophication significantly. The phosphorus concentration in 1980 was more than 80 mg/m³ while it is about 13 mg/m³ today.

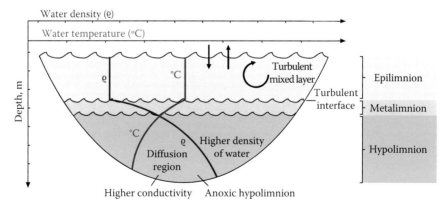

FIGURE 3.5
Vertical structure of a stratified lake and its physical and chemical features (Original Degani and Tundisi 2011.).

FIGURE 3.8
Bara Bonita reservoir, S. Paulo State, Brazil.

FIGURE 3.9
Broa reservoir, S. Paulo State, Brazil.

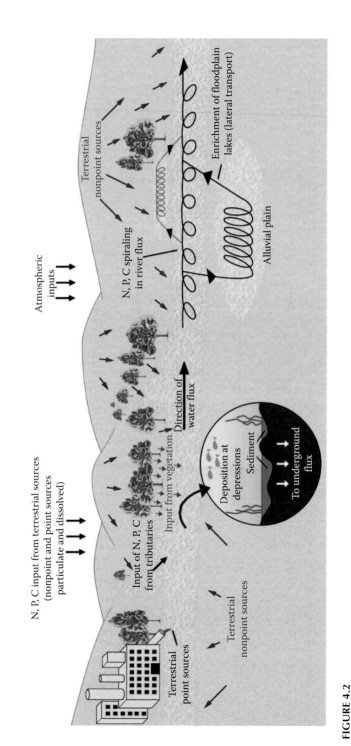

FIGURE 4.2

Sources of point and nonpoint inputs of nutrients and organic matter to the river; N, P, C spirals in river and floodplain lakes. (Credit to Degani and Tundisi, 2011.)

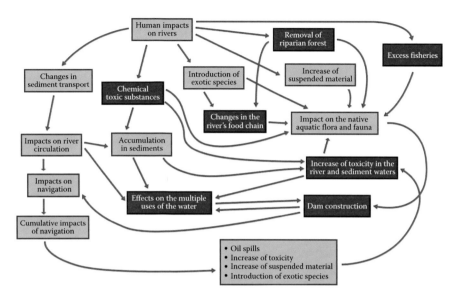

FIGURE 4.4
Human impacts on rivers systems: a synthesis.

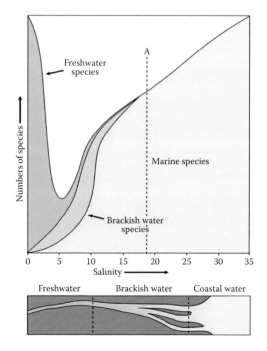

FIGURE 5.3
Reciprocal biological gradient in estuaries. (Modified from Emery, K.O. and Stevenson, R.E., Estuaries and lagoons, in Hedgpeth, J.W., Ed., *Treatise on Marine Ecology and Paleoecology*, Geological Society of America, Boulder, CO, Memoir 67, Chapter 23, pp. 673–750, 1227pp, 1957.)

FIGURE 6.1
Wetland at a Brazilian reservoir.

FIGURE 6.3
Constructed wetland in Tanzania. A high density of plants are selected in the design phase to ensure a high treatment efficiency. The selected plants species are local wetland plants.

FIGURE 6.5
Discharge of wastewater for the treatment on natural tundra in northern Canada.

FIGURE 6.6
Typical tundra applied for wastewater treatment in northern Canada.

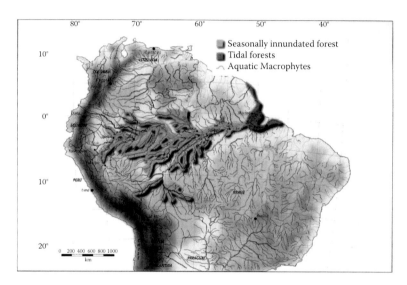

FIGURE 7.9
Extension of the macrophyte growth in Amazon River and its tributaries. (From Barthem, R.C. and Goulding, M., *Um ecossistema inesperado. Amazônia revelada pela pesca*, ACA SCM, Lima, Peru, p. 241, 2007.)

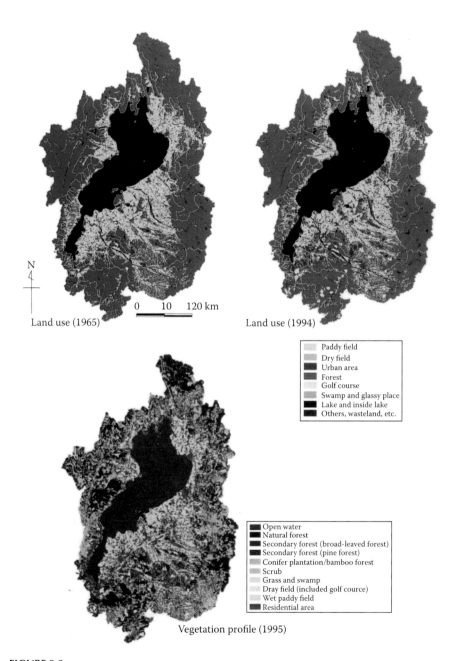

Land use (1965)

Land use (1994)

N

0   10   120 km

Paddy field
Dry field
Urban area
Forest
Golf course
Swamp and glassy place
Lake and inside lake
Others, wasteland, etc.

Open water
Natural forest
Secondary forest (broad-leaved forest)
Secondary forest (pine forest)
Conifer plantation/bamboo forest
Scrub
Grass and swamp
Dray field (included golf cource)
Wet paddy field
Residential area

Vegetation profile (1995)

FIGURE 8.3
Changes in vegetation cover and land use in the Lake Biwa Watershed. (From *LBRI, Lake Biwa Study Monographs*, Lake Biwa Research Institute, Otsu, Japan, 118pp, 1984.)

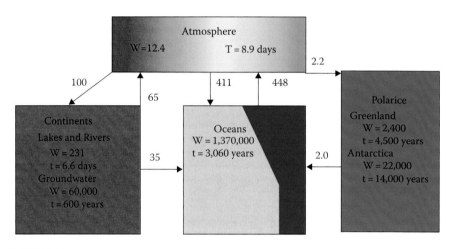

FIGURE 9.11
Water cycle is shown. Water in the compartment is indicated in 1000 km³ and the retention time is indicated in years. Number of fluxes represents 1000 km³/year. The estimation of the ground water volume is to a depth of 5 km of the earth's crust; much of this water is not actively exchanged.

The main impacts on watersheds and aquatic ecosystems and their consequences and links

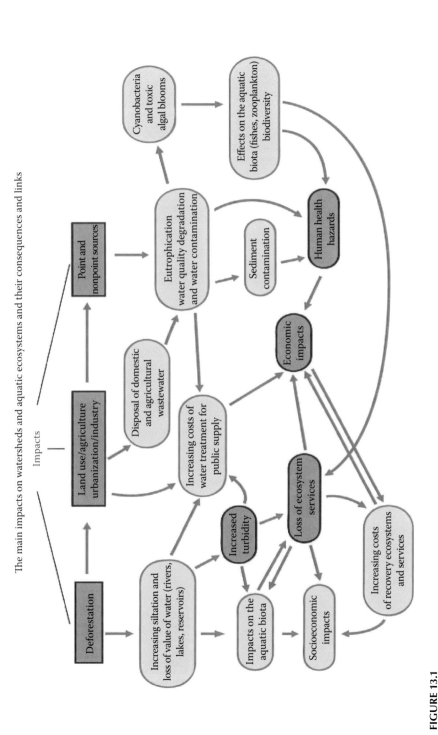

FIGURE 13.1

Synthesis of the main sources of impact and their consequences and links (Original Degani and Tundisi 2011).

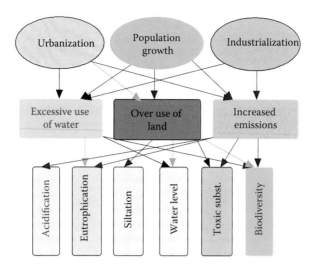

FIGURE 13.2
ILEC has formulated six problems for lakes, rivers, and reservoirs, which are due to an excessive water use, overuse of land, and increased emissions. The problems are as shown in the figure rooted in the increased urbanization, population growth, and industrialization.

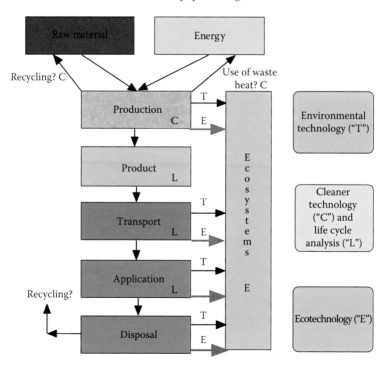

FIGURE 16.1
Arrows show mass and energy flows. The thin black arrows are point sources and the thick grey arrows are nonpoint sources. The letters by the arrows indicate the possibilities to use environmental technology, T; cleaner technology, C; the life cycle approach, L; and the ecotechnology, E. Recycling possibilities are indicated—they belong to clear technology. Life cycle analyses are able to reveal where in the history of the product the waste actually takes place and it will make it possible to change the production and the product the recorded quantities of waste.

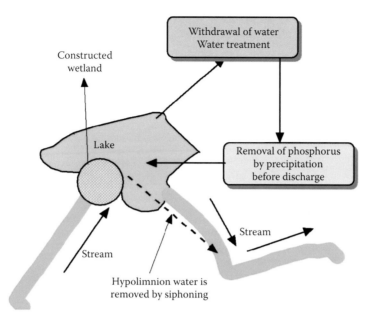

FIGURE 17.2
Control of lake eutrophication, illustrating a combination of chemical precipitation for phosphorus removal from wastewater (environmental technology), a wetland to remove nutrients from the inflow (ecotechnology, Class 1 or 2), and siphoning of nutrient-rich hypolimnetic water downstream (ecotechnology Class 3).

FIGURE 17.3
Riparian wetland in Malaysia. The dense vegetation along the river is adsorbing much of the pollution coming from land.

FIGURE 17.4
Siphoning of hypolimnic water in the Lake Bled, Slovenia.

FIGURE 17.7
Shallow lake with dominance of submerged vegetation is named a clear water stage.

FIGURE 18.2
Lake Fure.

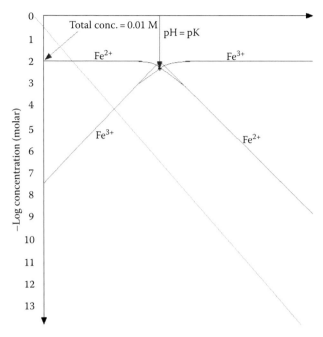

FIGURE 11.17
Double logarithmic diagram for the redox process $Fe^{3+} + e^- = Fe^{2+}$.

11.14 Examples of Relevant Processes in the Aquatic Environment

An aqueous solution at a given pH and a given pe determines the partial pressure of hydrogen and oxygen according to the following redox processes:

$$2H^+ + 2e^{--} = H_2(g) \quad \log K = 0 \text{ (the process is reference to the redox potential scale)} \quad (11.44)$$

$$2H_2O + 2e^- = H_2(g) + 2OH^- \quad \log K = -28 \text{ (combine water's ion product } 10^{-14} \text{ with process (11.44))} \quad (11.45)$$

$$O_2(g) + 4H^+ + 4e^- = 2H_2O \quad \log K = 83.1 \quad (4 \cdot 20.78;) \quad (11.46)$$

$$O_2(g) + 4e^- + 2H_2O = 4OH^-$$

It implies that we obtain the following relationships between the partial pressure and pH and pe:

$$\log pH_2 = 0 - 2pH - 2pe \tag{11.47}$$

$$\log pO_2 = -83.1 + 4pH + 4pe \tag{11.48}$$

If on the other side the partial pressure of oxygen can be considered constant and in equilibrium with aqueous solution and pH is given, the pe value can be found from Equation 11.48.

Manganese is present in aquatic systems as manganese dioxide (s) and manganese ions according to the following redox potential:

$$MnO_2(s) + 4H^+ + 2e^- \rightarrow Mn^{2+} + 2H_2O \tag{11.49}$$

The concentration of manganese ions is determined by the redox potential for the aqueous solution. If the redox potential is determined by the partial pressure of oxygen in equilibrium with the aqueous solution, we obtain the following equation to determine the manganese ion concentration:

$$pe = 13.6 = 20.42 + 0.5\log\left(\frac{[H+]^4}{[Mn^{2+}]}\right) \tag{11.50}$$

$$\Downarrow$$

$$\log [Mn^{2+}] = 2 \cdot (20.42 - 13.6) - 7.4 = -14.4 \tag{11.51}$$

Iron(II) is often under anaerobic conditions present in sediment and soil as iron(II) sulfide. If the sediment or soil is exposed to air, the following oxidation process takes place:

$$2FeS_2(s) + 2H_2O + 7O_2 = 2FeSO_4 + 2H_2SO_4 \tag{11.52}$$

$$4FeSO_4 + O_2 + 2H_2SO_4 = 2Fe_2(SO_4)_3 + 2H_2O \tag{11.53}$$

$$Fe_2(SO_4)_3 + 6H_2O = 2Fe(OH)_3(s) + 3H_2SO_4 \tag{11.54}$$

These processes involve formation of sulfuric acid, and extreme low pH values may occur as a result of these processes. Simultaneously, the solid iron(II) sulfide is replaced by the solid iron(III) hydroxide.

Formation of iron(II) sulfide under anaerobic conditions may also mobilize phosphate stored in the sediment of aquatic systems. If sulfide is formed as a result of a low redox potential, the following process will take place:

$$FePO_4(s) + HS^- + e = FeS(s) + HPO_4^{2-} \tag{11.55}$$

Chlorine is widely used as a disinfectant in aqueous solutions. Chlorine disproportionates (the same compound goes up and down in oxidation state)

$$Cl_2 + H_2O = HOCl + H^+ + Cl^- \tag{11.56}$$

The effect of chlorine is associated with the formation of HOCl which is a strong oxidant that is able to damage the enzyme system of bacteria. The hypochlorite ion OCl⁻ does not have this effect, which implies that pH is a determining factor for the disinfection effect of an aquatic chlorine solution. pK for HOCl at room temperature is 7.2. A pH above 8.0 is therefore not recommendable when chlorine is applied as a disinfectant.

11.15 Redox Conditions in Natural Waters

It is often convenient to consider the redox potential in natural aquatic systems, where it is presumed that pH = 7.0. A symbol $pe^0(w)$ is applied for these calculations. a is analogous to pe^0, except that the activities of protons and hydroxide ions correspond to natural water at pH = 7.0. This implies that the following expression between $pe^0(w)$ and pe^0 is valid:

$$pe^0(w) = pe^0 + 0.5 \, n \log K_w \tag{11.57}$$

where n is the number of moles of protons exchanged per mole of electrons.

$0.25 \, O_2(g) + H^+ + e^- = 0.5 \, H_2O$ has a pe^0 value of 13.75. The corresponding $pe^0(w)$ is therefore 20.75.

A table of $pe^0(w)$ values may be used to determine whether a system will tend to oxidize equimolar concentrations of any other system, which would require that it would have higher $pe^0(w)$. Sulfate/sulfide has, for instance, a $pe^0(w) = -3.5$, while CO_2/CH_2O has $pe^0(w) = -8.20$. Sulfate is accordingly able to oxidize CH_2O in natural water.

Example 11.9

The air in immediate contact with a wetland has a partial pressure of carbon dioxide and methane on 100 and 250 Pa, respectively. There is equilibrium between the water in the wetland and the air above the wetland. pH in the wetland is 7.0. The temperature is 25°C.

1. Find pe and the redox potential for the water in the wetland, when $pe^0(w)$ for the carbon dioxide/methane redox equilibrium is −4.13
2. What is the ratio iron(III) to iron(II) in the water? pe^0 for Fe^{3+}/Fe^{2+} is 13.0
3. How much will it change the redox potential (pe) if pH is changed from 7.0 to 7.6?
4. At a later stage it is found that pH is 7.2 and the redox potential is −0.23 V. Is methane produced under these conditions in detectable amounts?

Solution

1. $pe = pe^0(W) + 1/8(\log pCO_2/pCH_4) = −4.13 + 0.125 \log (100/250)$
   $= −4.18$ EH $= 0.05895 * (−4.18) = −0.246$
2. $−4.18 = 13.0 + \log Fe^{3+}/Fe^{2+}$; $[Fe^{3+}]/[Fe^{2+}] = 10^{-17.2}$
3. $0.125 \log (10^{-7.6}/10^{-7.0})^8 = −0.6$
4. The redox potential implies that $1/8 (\log pCO_2/pCH_4) \gg 0$

Example 11.10

In the lagoon of Venice, sulfate may be transformed to free sulfur in hot weather with no or little wind. pe for this redox reaction is 6.03.

1. What is pe under these conditions? pH is measured to be 7.5. A 10 mM concentration of sulfate is assumed.
2. What is the ratio $[Fe^{3+}]/[Fe^{2+}]$ under these conditions? pe^0 for Fe^{3+}/Fe^{2+} is 13.0.

Solution

1. $pe = 6.03 + \log [SO_4^{2-}]^{1/6}[H+]^{4/3} = − 4.3$
2. $−4.3 = 13.0 + \log [Fe(III)]/[Fe(II)]$ $[Fe(III)]/[Fe(II)] = 10^{-17.3}$

Figure 11.18, named as redox staircase, can be used in the same manner to decide which oxidants are able to oxidize which reductants. pe^0 values for various redox pairs at four different pH are shown. The oxidants placed higher in the figure can oxidize the reductants placed lower in the figure. At the same time, the figure indicates the pe value corresponding to the present at an equilibrium of a given redox pair. If we know for instance that oxygen is present at a normal concentration around 8–10 mg/L in an aquatic ecosystem, it will inevitably imply that pe must be around 13.75 and that the other redox pair must adjust their ratio according to this pe value. Notice that pe^0 for Fe^{3+}/Fe^{2+} is independent of pH, because no protons participate in the redox reaction.

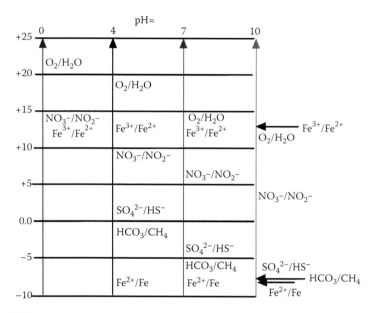

FIGURE 11.18

pe^0 values for the most common redox pair in aquatic systems are shown for four different pH-values.

Finally, the redox staircase indicates which oxidant will be used for oxidation under given circumstances. If pe is around 13–15 and pH = 7.0, oxygen will be applied as oxidant before iron(III), nitrate, etc. When oxygen has been used Fe^{3+} will be used before nitrate and sulfate, etc. This is a parallel to the titration of a mixture of acids with a base. The strongest acid will be neutralized first, then the second strongest acid, and so on.

It is interesting that the sequence of oxidants (oxygen is used as oxidant before iron(III), which is used before nitrate, which is used before sulfate, which is used before carbon dioxide) also is the sequence that gives the highest energy efficiency (most kJ for formation of ATP and thereby most exergy; see also Jørgensen 2012). This is shown in Table 11.5.

11.16 Construction of pe–pH Diagrams

The equations for pe obtained from Equations 11.47 and 11.48

$$pe = -pH - 0.5 \log pH_2 \qquad (11.58)$$

$$pe = 20.78 - pH + 0.25 \log pO_2 \qquad (11.59)$$

can be plotted in a pH–pe diagram; see Figure 11.19. The diagram gives the following important information: above the upper line, water is an effective

TABLE 11.5

Yields of kJ and ATP's per mole of Electrons, Corresponding to 0.25 mol of CH_2O Oxidized

Reaction	kJ/mol e$^-$	ATP's/mol e$^-$
$CH_2O + O_2 \rightarrow CO_2 + H_2O$	125	2.98
$CH_2O + 0.8NO_3^- + 0.8H^+ \rightarrow CO_2 + 0.4N_2 + 1.4H_2O$	119	2.83
$CH_2O + 2MnO_2 + H^+ \rightarrow CO_2 + 2Mn^{2+} + 3H_2O$	85	2.02
$CH_2O + 4FeOOH + 8H^+ \rightarrow CO_2 + 7H_2O + Fe^{2+}$	27	0.64
$CH_2O + 0.5SO_4^{2-} + 0.5H^+ \rightarrow CO_2 + 0.5HS^- + H_2O$	26	0.62
$CH_2O + 0.5CO_2 \rightarrow CO_2 + 0.5CH_4$	23	0.55

The released energy is available to build ATP for various oxidation processes of organic matter at pH = 7.0°C and 25°C.

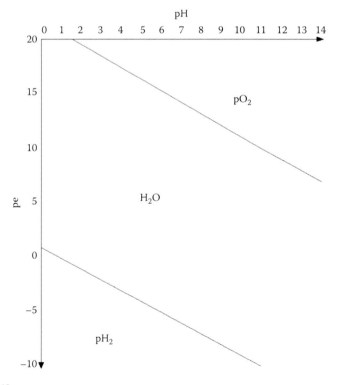

FIGURE 11.19
Diagram represents the equilibria between water and oxygen (upper line) and between water and hydrogen (lower line).

reductant (forming oxygen), and below the lower line, water is an effective oxidant, producing hydrogen. Aquatic systems in equilibrium with the oxygen in the atmosphere will have a pe = f(pH) as the upper line. Under these normal aerobic conditions, oxygen is an oxidant and hydrogen is a reductant.

pe–pH diagrams illustrate which are the components that will prevail under given conditions, that is, at given pe and pH. It is therefore advantageous to construct pe–pH diagrams for the most commonly found elements in aquatic systems to be able to determine quickly which form of the considered element is stable under the prevailing conditions. At the same time, the pe–pH diagram gives a good overview of the possible processes where both redox processes and acid–base reactions can take place. Figure 11.19 gives the information for the two elements oxygen and hydrogen. Example 11.11 gives the pe–pH diagram for the sulfur system.

Example 11.11

Construct a diagram considering that sulfur can be found in natural aquatic ecosystem as sulfate, as sulfur (s), and as hydrogen sulfide (g) in equilibrium with aqueous solution of hydrogen sulfide. The following redox reactions can be found in the literature:

$$SO_4^{2-} + 8H^+ + 6e^- = S(s) + 4H_2O \qquad pe^0 = 6.03$$

$$SO_4^{2-} + 10H^+ + 8e^- = H_2S(aq) + 4H_2O \quad pe^0 = 5.12$$

$$S(s) + 2H^+ + 2e^- = H_2S(aq) \qquad pe^0 = 2.40$$

$$HSO_4^- + 7H^+ + 6e^- = S(s) + 4H_2O \qquad pe^0 = 5.70$$

$$SO_4^{2-} + 9H^+ + 8e^- = HS^- + 4H_2O \qquad pe^0 = 4.25$$

Hydrogen sulfate has pK = 2.0 and hydrogen sulfide has pK = 7.0

The total concentration of the soluble sulfur species is 0.01 M.

Solution

Based upon the shown reactions, it is possible to obtain the following equations for the pe–pH diagram:

$$pe = 6.03 + 1/6 \log [SO_4^{2-}] - 8/6 \cdot pH$$

$$pe = 5.12 + 1/8 \cdot \log ([SO_4^{2-}]/[H_2S(aq)]) - 10/6\ pH$$

$$pe = 2.40 - 1/2 \cdot \log [H_2S] - pH$$

$$pe = 5.70 + 1/6 \cdot \log [HSO_4^{2-}] - 876\ pH$$

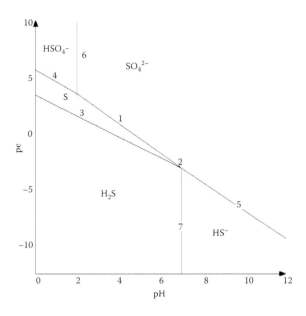

FIGURE 11.20
pe–pH diagram for the sulfate-sulfur-hydrogen sulfide system. 1; pe = 6.03 + 1/6 log $[SO_4^{2-}]$ – 8/6 pH; 2; pe = 8.01 + 1/8 log $[SO_4^{2-}]/[H_2S]$ – 10/8 pH; 3; pe = 2.4 – pH-0.5 log $[H_2S]$; 4; pe = 5.7 + 1/6 log $[HSO_4^-]$ – 7/8 pH; 5; pe = 4.25 + 1/8 log $[SO_4^{2-}][HS^-]$ – 9/8 pH; 6; log $[SO_4^{2-}]/[HSO_4^{2-}]$ –pH = 2.0; and 7; log $[HS^-]/[H_2S]/$–pH = 7.0.

$$pe = 4.25 + 1/8 \cdot \log\, ([SO_4^{2-}]/[H_2S(aq)]) - 9/8\, pH$$

$$\log\, ([SO_4^{2-}]/[HSO_4^{2-}]/[H_2S(aq)]) - pH\ = -2.0$$

$$\cdot\log\, ([HS^-]/[H_2S(aq)]) - pH\ = -7.0$$

These equations can easily be plotted in a pe–pH diagram; see Figure 11.20.
   Where only one soluble species is included in the equation the con-centration 0.01 M is applied. Where two soluble species are included, for instance, equation number 2, they are both assumed to be 0.005 M. The resulting diagram is shown in Figure 11.20.
   Figure 11.21 shows another example, namely, the pe–pH diagram for iron. It is constructed by the same method as Figures 11.19 and 11.20.

11.17 Redox Potential and Complex Formation

Figure 11.22 shows how it is possible to calculate the pe^0 value for two dif-ferent oxidation states of metal complexes. It is possible to go from Me^{2+} to MeL^{2+} by the processes (1) + (3) or by the processes (2) + (4) and these two

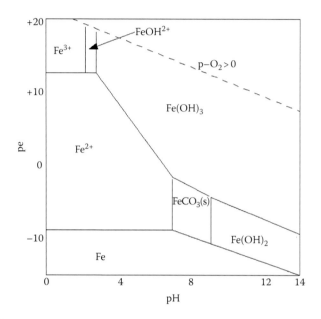

FIGURE 11.21
pe–pH diagram for iron.

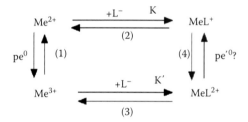

$$npe^0 + \log K = \log K' + npe'^0$$
when pe^0, $\log K$ and $\log K'$
are known, pe'^0 can be determined.

FIGURE 11.22
It is possible to go from Me^{3+} to MeL^{2+} by the processes (1) + (3) or by the processes (2) + (4), and the two pathways must necessarily give the same result with respect to the equilibrium between Me^{3+} and MeL^{2+}. It implies that $npe^0 + \log K = \log K' + npe'^0$.

pathways must necessarily give the same result with respect to the overall equilibrium between Me^{2+} and MeL^{2+}. The equilibrium constants for the two pathways are therefore equal. The equilibrium constant for two successive processes are the product of the equilibrium constants for the two processes:

$$npe^0 + \log K = \log K' + npe'^0 \tag{11.60}$$

As seen from this equation, if the highest oxidation state has the strongest complex, pe^0 will be reduced by addition of the corresponding ligand, and vice versa, if the lowest oxidation state has the strongest complex, the pe^0 will be increased by addition of the ligand.

Example 11.12

The complexity constants for formation of Fe^{3+}-hydroxo-complexes by hydrolysis at room temperature are, for mono-, di-, tri-, or tetra- hydroxo-complexes, the so-called *ß-values, $10^{-2.2}$, $10^{-5.7}$, $10^{-15.6}$, and $10^{-21.6}$, respectively.

The complexity constants for formation of fluoride complexes for Fe^{3+} at room temperature are

$$Fe^{3+} + F^- = FeF^{2+} \qquad \log \text{ß}_{1F} = 5.2$$

$$Fe^{3+} + 2F^- = FeF_2^+ \qquad \log \text{ß}_{2F} = 9.2$$

$$Fe^{3+} + 3F^- = FeF_3^0 \qquad \log \text{ß}_{3F} = 11.9$$

Iron(II) does not form complexes with fluoride and the possible hydroxo-complexes are much weaker than the corresponding complexes for Fe^{3+}, and they can therefore be neglected. pe^0 for the following process

$$Fe^{3+} + e^- = Fe^{2+}$$

is 13.00 at the room temperature.

    a. Find the pe value at equal concentrations of iron(II) and iron(III), pH = 5.0 and a fluoride concentration of 0.01 M.
    b. Which iron(III) complexes are strongest by these conditions?

Solution

    a.

$$a_0 = [Fe^{3+}]/Fe(III)$$

$$= 1/(1 + {}^*\text{ß}_1/[H^+] + {}^*\text{ß}_2/[H^+]^2 + {}^*\text{ß}_3/[H^+]^3$$

$$+ {}^*\text{ß}_4/[H^+]^4 + \text{ß}_{1F}[F^-] + \text{ß}_{2F}[F^-]^2 + \text{ß}_{3F}[F^-]^3)$$

$$= 1/(1 + 10^{2.8} + 10^{4.3} + 10^{-0.6} + 10^{-1.6} + 10^{3.2} + 10^{5.2} + 10^{5.9})$$

$$= 1/9.7 \times 10^5 = 1.03 \times 10^{-6}$$

$$pe = 13.00 + \log [Fe^{3+}]/[Fe^{2+}] = 13.00 + \log 10^{-6}/1 = 7.00$$

    b. From the expression for $a_0 = [Fe^{3+}]/Fe(III)$, it can be seen that the dihydroxo, difluoride, and trifluoride complexes are contributing most to the formation of complexes.

References

Alloway, B.J. and Ayres, D.C. 1993. *Chemical Principles of Environmental Pollution.* Blackie Academic and Professional, London, U.K., 394pp.

Jørgensen, S.E. 2012. *Fundamentals of Systems Ecology.* CRC Press, Boca Raton, FL, 320pp.

Sillen, L.G. 1961. *Oceanography.* Publication Number 67. AAAS, Washington, DC, pp. 549–581.

12

Future of Limnology and Aquatic Ecology as a Tool for Management of Inland Waters

12.1 Our Demand to Limnology and Aquatic Ecology

Limnology as all ecological sciences is an inclusive science, incorporating into the studies of ecological dynamics of lakes, rivers, reservoirs, wetlands, and ponds the physical, chemical, and biological data. The sustainability of the inland water resources; the recovery of the water quality of degraded lakes, rivers, and reservoirs; and the protection of critical freshwater ecosystems are dependent on the profound knowledge of limnology and aquatic ecology. The future of limnology meets two basic challenges: (1) to incorporate vigorously the watershed approach into the study of inland water ecosystems in order to evaluate the dynamics of the terrestrial/aquatic interactions in space and time Likens (1992) and (2) to integrate social, economic, and environmental data into the limnological database in order to broaden the scope of the analysis and help to create an integrated approach to water management.

One of the most important developments to enhance the role of limnology in this integrated water management is the building up of capacity to promote predictions and to provide a basis to build predictive models (Håkanson and Peters 1995). All the main problems of today, related to the degradation of inland freshwater ecosystems, such as eutrophication, acidification, metal contamination, and wastewater (industrial and domestic) disposal need a scientific understanding, a technical approach, and a predictive capacity in order to anticipate changes in species composition, evaluate the loss of ecosystem services, and estimate the economic consequences of the impacts. By introducing the limnological research in a broader and dynamic ecosystem problem, sensitivity analysis can be consolidated and the fluxes of nutrients and contaminants can be predicted. For example, limnological studies were a key component in the environmental impact assessment of tropical and subtropical reservoirs (Tundisi and Straskraba 1999, Tundisi et al. 2008, Tundisi 2010). The changes in engineering projects for reservoirs taking into account limnological data (Kennedy 1999, Kennedy et al. 2003) are examples of this use of limnology in the design and operation of reservoirs. The several water quality problems and technological efforts to control and operate reservoirs in South America and North America were, in part, solved with the application of limnological data and by long-term studies (Tundisi and Matsumura-Tundisi 2003).

The evaluation of recovery costs of lakes, rivers, and reservoirs can also require considerable inputs of limnological data when the desirable water quality of the freshwater ecosystem is based on limnological information.

To conclude, limnological studies can not only have a scientific value and usefulness, but, if the scope of the science is broadened, its use should be much more effective for societal needs and economic measures. A better understanding of the ecosystem networks allows access to a better understanding of the behavior of the ecosystems and their reactions to perturbations that are very important for development of a better environmental management. Thereby, it is feasible to develop a predictive capability on the response of the ecosystems to changes in forcing functions (Jørgensen and Svirezhev 2004). Limnology can also have a strong role on science education and education of general public, since public participation is one of the bottlenecks to the improvement of inland water correction and recovery (Tundisi 2010).

Chapter 19 presents the use of models in environmental management of inland waters and here it is emphasized that the use of models acts as a tool to synthesize our knowledge—both the theoretical knowledge and our observations. With a good synthesis in hand we can make better management decisions. Consequently, the more ecological knowledge we have about aquatic ecosystems, the better models will we be able to develop and the more predictive power will the model be able to offer the environmental management.

References

Hakanson, L. and Peters, R.H. 1995. *Predictive Limnology: Methods for Predictive Modeling*. SBP Academic Publishing, Amsterdam, the Netherlands, 464pp.

Jørgensen, S.E. and Svirezhev, Y. 2004. *Towards a Thermodynamic Theory for Ecological Systems*. Elsevier, Amsterdam, the Netherlands, 366pp.

Kennedy, R. 1999. Reservoir design and operation: Limnological implications and management opportunities. In: Tundisi, J.G. and Straskraba, M. (Eds.), *Theoretical Reservoir Ecology and Its Applications*. BAS, Backhuys Publishers, Leiden, the Netherlands, pp. 1–28, 585pp.

Kennedy, R.H.; Tundisi, J.G.; Straskrabova, V.; Lind, O.T.; and Hejlar, J. 2003. Reservoirs and the limnologists growing role in sustainable water resource management. *Hydrobiologia* 504:XI–XII. Special issue: Reservoir Limnology and Water Quality. Straskrabova, V.; Kennedy, R.H.; Lind, O.T.; Tundisi, J.G.; and Hejlar J., 325pp.

Likens, G.E. 1992. The ecosystem approach: Its use and abuse. In: O. Kinne (series Ed.), *Excellence in Ecology*. Ecology Institute, Oldendorf/Luhe, Germany, 166pp.

Tundisi, J.G. 2010. The advocacy responsibility of the scientist. In: Moore, K.D. and Nelson, M.P. (Eds.), *Moral Ground: Ethical Action for a Planet in Peril*. Trinity University Press, San Antonio, TX, pp. 448–451, 478pp.

Tundisi, J.G. and Straskraba, M. 1999. Theoretical reservoir ecology and its applications. Backhuys Publishers/Brazilian Academy of Sciences/International Institute of Ecology. 585pp.

Tundisi, J.G. and Matsumura–Tundisi, T. 2003. Integration of research and management in optimizing uses of reservoirs: The experience of South America and Brazilian case studies. *Hydrobiologia*, Baarn, the Netherlands, 500: 231–242.

Tundisi, J.G.; Matsumura–Tundisi, T.; and Abe, D.S. 2008. The ecological dynamics of Barra Bonita reservoir: Implications for its biodiversity. *Braz. J. Biol.*, 68 (4): 1079–1098.

Part II

Holistic, Environmental and Ecological Management

13

Impacts on Watersheds
and Inland Aquatic Ecosystems

13.1 Environmental Problems, Their Sources, and Evaluation of Impacts

Continental aquatic ecosystems, coastal lagoons, and estuaries are subject to several environmental problems and impacts as a consequence of human activities. These impacts can be permanent and cumulative, periodic, or as a pulse due to accidental occurrences such as oil spills or discharges of contaminants from industrial and agricultural activities. There are two main sources of impacts: nonpoint sources, originating from several processes in the watersheds, or point sources such as effluents of industries or urban discharges of insufficiently treated domestic wastewater. Another nonpoint source of impacts is the atmospheric contribution as particulate matter or dissolved substances in rainwater. For example, acid rain originates from the processes of accumulation of H_2S or SO_x in the atmosphere (Jørgensen et al. 2005). Besides the intensification of the sources and the magnitude of impacts on continental aquatic ecosystems, coastal lagoons, and estuaries, it is fundamental to understand the fate of the pollutants and contaminants in the aquatic ecosystems including their possible accumulation in the food chain (see the use of mass balances in Chapter 9 to quantify this important process). It is also relevant in the identification and study of impacts to clarify effects on the ecosystem services. For example, the eutrophication or toxic metal contamination can affect water supply and other (multiple) uses of water. The impact on ecosystem services has indirect or direct effects on the regional economy; for example, by accumulation of toxic metals in the sediment or in the water, fisheries may be affected. Recreation and tourism, which are, in many regions of the world, a major source of economic and social development, may also be impaired by eutrophication. For loss of ecosystem services and general degradation of rivers, reservoirs, estuaries, and coastal waters see Straskraba and Tundisi (1999).

This evaluation of the impacts (and the resilience or buffer capacity of the ecosystems to cope with them) should include not only the biogeochemical aspects but also the ecosystem services. The evaluation of impacts from the economic point of view should also include the cost of recovery of the ecosystem function or ecosystem service.

Thus, the following criteria can be applied to impact assessment (quantification and qualification of impacts):

- Nature of the impact
- Location and spatial scale of impact
- Temporal scale of the impact (short, medium, and long term)
- Reversibility/irreversibility
- Relevance for the applied ecosystem services—low/medium/high
- Magnitude—low/medium/high

Tools for impact assessment need to be implemented at regional scale and it is also necessary to develop predictive models that develop scenarios about the effects of impacts. Indirect and direct effects have in this context to be considered. It is also important to consider the links of the impacts with biogeophysical, economic, and social processes. One of the recent trends in environmental impact assessment is to analyze the human health hazards associated with pollution; see Jørgensen (2000), where the ERA (environmental risk assessment) and the human health risk assessment are presented in parallel. Chemical pollution is a threat to public health and this has to be included in the impact assessment of water quality degradation. Water quality and sanitation are intrinsically linked to patterns of land use such as massive urbanization and food production by more or less intensive agricultural developments with application of pesticides and herbicides and deforestation (Confalonieri et al. 2010).

13.2 Impacts

Figure 13.1 gives an overview of the most important impacts on watersheds and aquatic ecosystems, including their consequences and links.

The impacts on the watersheds and continental aquatic ecosystems are therefore diverse with several origins and many consequences (short, medium, and long term). Table 13.1 shows a list of impacts and their consequences on watershed and aquatic ecosystem.

All the impacts in Table 13.1 have economic consequences for the watershed, imply qualitative or quantitative changes of the water resources, and increase the health hazards for the human population. The aquatic ecosystems and the water quality are the ultimate recipients of these impacts.

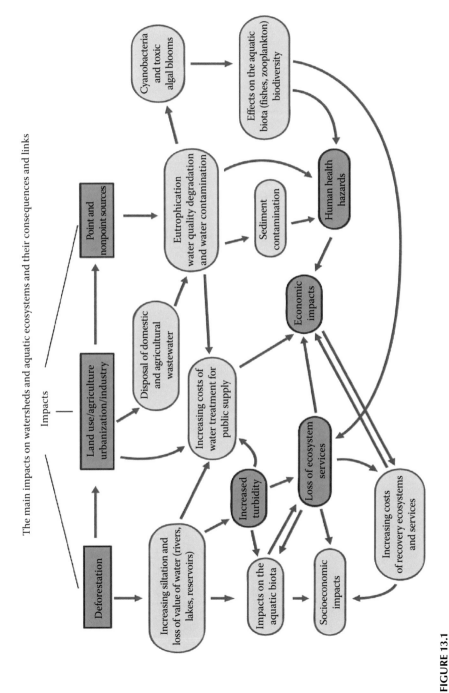

FIGURE 13.1
(See color insert.) Synthesis of the main sources of impact and their consequences and links (Original Degani and Tundisi 2011).

TABLE 13.1

Impacts and Consequences on Watershed and Aquatic Ecosystems

Deforestation

Nutrient input (nitrogen, phosphorus) and eutrophication from agriculture

Sediment load from runoff

Extensive agriculture and soil use

Air pollution and atmospheric contamination

Toxic metal contamination

Acidification

Biodiversity changes due to toxicity

Biodiversity changes due to introduction of exotics species

Reservoir construction

Organic pollution (persistent organic pollutants)

Removal of wetlands

River degradation (channel construction, reservoir construction)

Flood plain degradation

Thermal pollution

Fish stock depletion

Oil pollution

Discharge of untreated urban wastewater and eutrophication

Industrial waste disposal

Toxic waste disposal

Salinization (in semiarid reservoirs)

Increased turbidity

Loss of volume of water by increased siltation and mechanical filling of lake basin

Increased bacterial contamination and formation of clay organic bacteria aggregates (Lind and Dayalos Lind 1999, Jørgensen et al. 2005)

Waterborne diseases

Increased health risk for human population

Sources: Tundisi, J.G. and Straskraba, M., *Theoretical Reservoir Ecology and Its Applications*, IIE, São Carlos, Brazil, 585 pp., 1999; Jimenez, B.E.C., *La contaminacion ambiental en México*, Limusa-Noruega Editores, Inst. Eng. UNAM, México, 913 pp., 2001; Jørgensen, S.E. et al., *Lake and Reservoir Management*, Development in Water Sciences, Vol. 54, Elsevier, Amsterdam, the Netherlands, 502 pp., 2005; Tundisi, J.G. and Matsumura-Tundisi, T., *Limnologia*, Oficina de textos, São Paulo, Brazil, 632 pp., 2008.

Therefore, a monitoring procedure that detects the intensity and relevance of the impacts and their persistence is recommended. Real-time monitoring is an important innovation because it allows for rapid assessment anticipating impacts on the physics and chemistry of the water. Biological monitoring with the use of biological indicators is a very important additional procedure with special importance for bottom macroinvertebrates (Tundisi and Matsumura-Tundisi 2008) and diatoms (Bere and Tundisi 2009, 2011).

ILEC, International Lake Environment Committee, has defined six water quality and quantity problems for lakes, rivers, and reservoirs (Jørgensen et al. 2005):

1. *Acidification*: pH gets too low, often due to acidic rain—it means that air pollution is not controlled sufficiently.
2. *Eutrophication*: Very high concentration of phytoplankton due to very high discharge of nitrogen and phosphorus from both point and nonpoint sources.
3. *Siltation*: Water becomes too turbid due to very high erosion in the watershed.
4. *Water level changes*: Very high consumption of water or it can be said no respect for the mass conservation principle.
5. *Very high concentrations of toxic substances*: Concentrations of toxic organic substances and heavy metals increase due to an uncontrolled discharge of toxic substances from both point and nonpoint sources.
6. *Loss of biodiversity*: Reduction of the biodiversity due to pollution or introduction of exotic species.

These six problems are caused by an excessive use of water, overuse of land, and/or increased and uncontrolled emissions, which again are rooted in the increasing urbanization, population growth, and industrialization; see Figure 13.2.

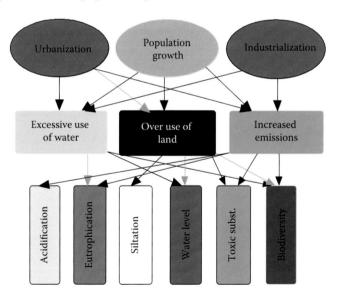

FIGURE 13.2
(See color insert.) ILEC has formulated six problems for lakes, rivers, and reservoirs, which are due to an excessive water use, overuse of land, and increased emissions. The problems are as shown in the figure rooted in the increased urbanization, population growth, and industrialization.

This means that the problems are direct consequences of the development. It implies that a complete solution of the problems requires that we shift to a more sustainable development, including the use of more holistic and integrated environmental management; see Chapter 14.

The identification and applications of ecological models and indicators (see Chapters 18 and 19) are a first step in the process of evaluating and assessing impacts (including their intensity and persistence). However, ecotoxicological studies are necessary in order to further evaluate the consequence of the impacts and their direct and indirect effects on organisms and communities (see Jørgensen et al. (2005) and Jørgensen and Fath (2011)).

13.3 Impacts of Climatic Change on Aquatic Ecosystems

The Intergovernmental Panel on Climatic Change (IPCC) concludes in their third evaluation report (IPCC 2001a) that the average temperature of the atmosphere has increased by 0.6°C ± 0.2°C during the twentieth century.

The global models of IPCC have shown that between 1900 and 2100 the global temperature can increase between 1.4°C and 5.8°C, dependent on the political decisions about the out phasing of fossil fuel. A higher temperature rise is therefore expected for the present century than the one detected for the twentieth century (Marengo 2006). There are evidences that extreme events such as droughts, flooding, large-scale and continuous rainfall, and cyclones will happen more frequently and that it will affect many terrestrial and aquatic ecosystems around the world with consequences on the human health, death losses, and the economy (IPCC 2001b).

In the aquatic ecosystems, several impacts as consequences of the climatic changes will inevitably occur. The water temperature of lakes and reservoirs can increase and maintain for longer periods the thermal stratification and the chemical stratification. This can stimulate the cyanobacteria blooms that will be more intense and have longer duration (Paerl and Hussmann 2008). Deterioration of surface water sources can increase the costs of treatment for production of potable water (Tundisi et al. 2010). The aquatic biota and the biodiversity will also be affected by these changes. Particularly, the increase in water temperature, and the increase of the discharge will have a significant effect. Increase in the discharge due to changes in the hydrological cycle can increase the turbidity of rivers, lakes, and reservoirs, affecting the primary productivity of phytoplankton, fish production, and dissolved oxygen concentration. For example, in the La Plata basin the impact of the cold fronts on lakes and reservoirs can be changed with consequences on the phytoplankton and zooplankton succession, the organization of the food chain, and the sediment–water interactions (Tundisi et al. 2004, 2010). The synergy between the impacts of a higher temperature of the atmosphere and water due to global changes and the transformation of the watersheds under the influences of the impact (deforestation, extensive soil use for food production, industrialization, and urbanization) have to be considered. Therefore, future studies have to demonstrate and quantify the links and the ecosystem dynamics under the joint influences of all the forcing functions at atmospheric and watershed levels (Tundisi and Matsumura-Tundisi 2008, Bicudo et al. 2010).

References

Bere, T. and Tundisi, J.G. 2009. Weighted average regression and calibration of conductivity and pH of benthic diatoms is streams, influenced by urban pollution, São Carlos, SP. Brazil. *Acta Limnol. Bras.* 21:317–325.

Bere, T. and Tundisi, J.G. 2011. Influence of ionic strength and conductivity on benthic diatom communities in a tropical river (Monjolinho), S. Carlos-Brazil. *Hydrobiologia* 661:261–276.

Bicudo, C.E.; Tundisi, J.G.; and Schementrue, M.C. (Eds.). 2010. *Water in Brazil Strategic Analysis.* Brazilian Academy of Sciences, Rio de Janeiro, Brazil, 222pp.

Confalonieri, U.; Heller, L.; and Azevedo, S. 2010. Water and health: Global and national problems. In: Bicudo, C.E.; Tundisi, J.G.; and Schementsul, M.C.B. (Eds.), *Water in Brazil. Strategic Analyses.* Brazilian Academy of Sciences, São Paulo, Brazil, pp. 27–42, 222pp.

IPCC. 2001a. Climate change 2001. Synthesis report: A contribution to Working Group I, II and III, to the Third Assessment Report of the Intergovernmental panel on climate change. Watson, R.T. and the core writing team (Eds.), Cambridge University Press, U.K., New York, 398pp.

IPCC. 2001b. Climate change 2001: Impacts, adaptation, and vulnerability. A contribution of the Working Group II to the Third Assessment Report, Cambridge University Press.

Jimenez, B.E.C. 2001. *La contaminacion ambiental en México.* Limusa-Noruega Editores. Inst. Eng. UNAM, México, 913pp.

Jørgensen, S.E. 2000. *Principles of Pollution Abatement.* Elsevier, Amsterdam, the Netherlands, 520pp.

Jørgensen, S.E. and Fath, B. 2011. *Fundamentals of Ecological Modelling,* 4th edn. Elsevier, Amsterdam, the Netherlands, 400pp.

Jørgensen, S.E.; Loffler, H.; Rast, W.; and Straskraba, M. 2005. *Lake and Reservoir Management.* Development in Water Sciences, Vol. 54. Elsevier, Amsterdam, the Netherlands, 502pp.

Lind, O.T. and Davalos Lind, L. 1999. Suspended Clay: Its role on reservoir productivity. In: Tundisi, J.G. and Straskraba, M. (Eds.), *Theoretical Reservoir Ecology and Its Applications.* Backhuys Publishers/Brazilian Academy of Science/International Institute of Ecology, pp. 45–56, 585pp.

Marengo, J.A. 2006. *Mudanças climáticas globais e seus efeitos sobre a biodiversidade.* IBAMA, Brasília, Brazil, 212pp.

Paerl, H.W. and Hussmann, J. 2008. Blooms like it hot. *Science* 320:57–58.

Strakraba, M. and Tundisi, J.G. 1999. *Reservoir Water Quality Management.* Guidelines of Lake Management, Vol. 9. ILEC, Kusatsu, Japan, 229pp.

Tundisi, J.G. and Matsumura-Tundisi, T. 2008. *Limnologia.* Oficina de textos, São Paulo, Brazil, 632pp.

Tundisi, J.G.; Matsumura-Tundisi, T.; Arantes, J.D. Jr.; Tundisi, J.E.M.; Manzini. N.F.; and Ducrot, R. 2004. The response of Carlos Botelho (Lobo/Broa) reservoir to the passage of cold fronts as reflected by physical, chemical and biological variables. *Braz. J. Biol.* 64(1):177–189.

Tundisi, J.G. and Straskraba, M. 1999. *Theoretical Reservoir Ecology and Its Applications.* IIE. São Carlos, Brazil, 585pp.

Tundisi, J.G. et al. 2010. Cold fronts and reservoir limnology: An integrated approach forwards the ecological dynamics of freshwater ecosystems. *Braz. J. Biol.* 70(Suppl. 3):815–824.

14

Integrated Ecological and Environmental Management

14.1 Introduction

Integrated ecological and environmental management means that the environmental problems are viewed from a holistic angle considering the ecosystem as an entity and considering the entire spectrum of solutions including all possible combinations of proposed solutions. The integration in this context means that all the available tools and possibilities are taken into account and that all the problems are considered simultaneously. The experience gained from environmental management in the last 40 years has clearly shown that it is important not to consider solutions of single problems but to consider *all or at least all major* problems associated with a considered ecosystem simultaneously and evaluate *all* the solution possibilities proposed by the relevant disciplines at the same time or, expressed differently, to observe the forest and not the single trees. The experience has clearly underlined that there is no alternative to an *integrated* management at least not on a long-term basis. Fortunately, as it will be presented in this chapter, new ecological subdisciplines have emerged and they offer tool boxes to perform an integrated ecological and environmental management.

The presentation in this chapter is based on Jørgensen and Nielsen (2012), where a general procedure to perform integrated environmental and ecological management has been proposed. This chapter presents this procedure with indications of how to use it particularly for management of inland water ecosystem. The application of the procedure is therefore focusing mainly on lakes, ponds, and reservoirs (abbreviation L); rivers and streams (abbreviation R); wetlands (abbreviation W); and lagoons and estuaries (abbreviation E).

14.2 Ecological and Environmental Management Procedure

Present-day integrated ecological and environmental management consists of seven steps:

1. Define the problem.
2. Determine the ecosystems involved.
3. Find and quantify all the sources to the problem. It is often beneficial to do so by the use of mass balances.
4. Set up a diagnosis to understand the relation between the problem and the sources.
5. Determine all the tools we need to implement to solve the problem.
6. Implement the selected solutions.
7. Follow the recovery process for instance by the use of a monitoring program.

When an environmental problem has been detected, it is necessary to determine and quantify the problem and all the sources to the problem. It requires the use of chemical, biological, physical, and analytical methods coupled eventually with a monitoring program. To solve the problem, a clear diagnosis has to be developed: what are actually the problems that the ecosystems are facing? and what are the relationships between the sources and their quantities and the determined problems? Or expressed differently, to what extent do we solve the problems by reducing or eliminating the different sources to the problems? The erection of mass or energy balances for the relevant components and elements is beneficial in this context. Chapter 9 gives the physical considerations that are needed to set up the balances.

A holistic integrated approach is needed in most cases because the problems and the corresponding ecological changes in the ecosystems are most often very complex, particularly when several environmental problems are interacting. It is therefore essential to know the focal ecosystem, its processes, and its reactions to changes. It is furthermore necessary to draw on freshwater ecology/limnology when an environmental management strategy has to be developed. The first nine chapters of this handbook give the background knowledge that is needed to understand freshwater ecosystems and their reactions to pollutants. The environmental problems of the main inland water ecosystems are listed. Only the main problems and the main sources are mentioned to get a proper overview of the environmental problems. The main environmental problems are

For lakes and reservoirs (L) (supplement with the illustration in Figure 13.2),

1. Eutrophication is caused by very high nutrient concentrations—often phosphorus but could also be nitrogen. The main source is

mostly wastewater but drainage water from agricultural areas is often contributing significantly to the problem.

2. Acidification is caused by industrial pollution or more often by acidic precipitation. It was a major problem is northern Europe and in the northeastern part of United States 25–40 years ago, but the problem is reduced significantly today due to air pollution abatement. The problem is, however, increasing in the industrial parts of China. Environmental chemical calculations are needed to find proper solutions to this problem. Chapters 10 and 11 give the chemical basis for these calculations, which are relatively easy to carry out by the use of aquatic environmental chemistry.

3. Siltation is caused by a very high concentration of suspended matter in the inflowing tributaries due to a very high erosion upstream.

4. Changing water levels is caused by a very high withdrawal of water from the lake or reservoir or their rivers. The classical example is the Aral Lake. It is a question about respecting the mass balance principle, which makes the abatement strategy very simple, although it often has very significant social implications.

5. Discharge of toxic substances is either due to insufficient treatment of industrial wastewater or due to discharge of pesticides from agriculture. The latter is often caused by an extensive use of pesticides or by drainage of wetlands, which otherwise could have been used as buffer zones. Heavy metals can occur in many different forms that have different toxicity. It is therefore important to be able to find the concentrations of the various heavy metal forms to be able to manage the problem properly. Chapters 11 and 12 give the basic environmental chemistry that present how the calculations of the relevant concentrations can be made.

6. Overfishing, which is caused by a lack of fishing regulations. In principle, a model should be used to give information about a sustainable fishery.

7. Introduction of alien species. It is often caused by an accidental introduction of alien species but can also be a result of a wrongly managed plant, for instance, when the Nile Perch was introduced to Lake Victoria.

For rivers and streams (R),

1. Oxygen depletion is most often caused by discharge of insufficiently treated wastewater, by discharge of polluted drainage water from industries or agriculture, or by an uncontrolled erosion. High fish mortality may be a consequence. The biodiversity is also reduced due to a low oxygen concentration.

2. Too much and/or too dense vegetation, which is a result of discharge of too much nutrients either from wastewater or from agriculture drainage water.

3. Toxic substances—the sources are the same as for lakes and reservoirs, see the aforementioned Point 5.

4. Drainage of riparian wetlands, which implies that the buffer zones along the shore of the rivers and streams are removed, significantly reduced or damaged. The result is that all the human activities on land get a significant (negative) influence on the water quality.

5. Drainage of riparian wetlands means usually that the river or stream eliminate their meanders and get a straight flow. It implies often significantly reduced diversity.

For wetlands (W),

1. Drainage of wetlands means often more or less complete elimination of the wetlands and thereby loss of the many ecosystem services that wetlands can offer. It is often due to wrong political decisions, where the idea is to provide more agriculture land. These decisions are often shortsighted and do not consider the long-term negative effects of eliminating the wetlands, including their ability to reduce the probability for flooding.

2. Toxic substances—the source is the same as for lakes and reservoirs.

3. Impact from urban activities will often partially destroy wetlands and reduce the possibilities to apply the wetland services to the benefit of the society.

For lagoons and estuaries (E),

1. Eutrophication is caused by very high nutrient concentrations—could be both nitrogen and phosphorus or both simultaneously or shifting between the two nutrients. The main sources are wastewater and drainage water from agricultural or urban areas.

2. Oxygen depletion is most often caused by discharge of insufficiently treated wastewater, by discharge of polluted drainage water from industries or agriculture, or by an uncontrolled erosion. High fish mortality may be a consequence. The biodiversity is also reduced due to a low oxygen concentration in lagoons and estuaries.

3. Toxic substances—the source is the same as for lakes and reservoirs.

When the first green wave started in the mid-1960s, the tools to answer the mentioned questions associated with the seven steps of the management procedure were not yet developed. We could carry out the first three points on the previously shown list, but had to stop at Point 4 and could at that time only recommend to eliminate the source completely or closed to completely by the methods that were available at that time—it means by environmental technology. This tool box has of course more to offer today—a wider spectrum of methods and more effective methods. The tool box of environmental technology will be presented in the next chapter—Chapter 15.

Due to the development of several new ecological subdisciplines, it is today, however, possible to accomplish all the seven points. Next, we will present the tool boxes that we can apply today to carry out particularly the Points 4 and 5. They are the result of the emergence of six new ecological subdisciplines: For a better diagnosis we have developed: ecological modeling, ecological indicators, ecological services. Ecological modeling is presented in Chapter 19, and Chapter 18 covers ecological indicators. Ecological service can be considered an indicator as shown in Costanza et al. (1992), Jørgensen (2010) and in Jørgensen et al. (2004 and 2010). For a better solution we have today ecological engineering (also denoted ecotechnology), cleaner production, and environmental legislation. These three tool boxes will be presented. Chapter 16 covers cleaner technology and Chapter 17 gives an overview of ecological engineering. Environmental legislation will not be covered in this volume, because it is different from country to country and it would therefore be difficult to give a good overview in relatively few pages.

14.3 Tool Boxes Available Today to Develop an Ecological–Environmental Diagnosis

A massive use of ecological models as an environmental management tool was initiated in the early seventies. The idea was to answer the question what is the relationship between a reduction of the impacts on ecosystems and the observable, ecological improvements. The answer could be used to select the pollution reduction that the society would require and could effort economically. Ecological models were developed already in the 1920s by Steeter-Phelps and Lotka Volterra (see for instance Jørgensen and Fath [2011]), but in the 1970s, a much more consequent use of ecological models started and many more models of different ecosystems and different pollution problems were developed. Today we have practically for all combinations of ecosystems and environmental problems at least a few models available. The journal *Ecological Modeling* was launched in 1975 with an annual publication of 320 pages and about 20 papers. Today, the journal publishes 20 times as many papers. It means that ecological modeling has been adopted as a very powerful tool in ecological–environmental management to cover particularly Point 4 in the integrated ecological and environmental management procedure proposed in the introduction of this chapter. An overview of the models available for management of inland water ecosystems will be given in Chapter 19, but for those interested in a more detailed information about the appropriate models to be applied can refer to Jørgensen (2011) and Jørgensen and Fath (2011).

Ecological models are powerful management tools but they are not always easily developed. They require in most cases good data, which are resource and time consuming to provide. About 20 years ago, it was therefore proposed to use another tool box that required less resources to provide a diagnosis, namely, ecological indicators (see for instance Costanza et al. [1992]). Ecological indicators can be classified as shown in Table 14.1 according to the spectrum from a more detailed or reductionistic view to a system or holistic view (see Jørgensen [2002]). The reductionistic indicators can for instance be a chemical compound that causes pollution or specific key species. A holistic indicator could for instance be a thermodynamic variable or the biodiversity. The indicators can either be measured or they can be determined by the use of a model. In the latter case, the time consumption is of course not reduced by the use of indicators instead of models, but the models get a more clear focus on one or more specific state variable, namely, the selected indicator, which best describes the problem(s). In addition, indicators are usually associated with very clear and specific health problems of the ecosystems, which of course is beneficial in environmental management. Jørgensen et al. (2004 and 2010) give a comprehensive overview of ecological indicators including ecological services that can be applied to assess the ecosystem health. Chapter 18 gives a brief overview of the ecological indicators most frequently applied to assess the ecosystem health of inland-water ecosystems. Furthermore, the use of ecosystem services as diagnostic tool is also discussed in the following text.

The last 10–15 years, the services offered by the ecosystems to the society have been discussed and it has been attempted to calculate the economic values of these services (Costanza et al. 1997). A diagnosis could be developed that would focus on the services actually reduced or eliminated due to environmental problems. Another possibility of using ecological services to assess the environmental problems and their consequences could

TABLE 14.1

Classification of Ecological Indicators

Level	Example
Reductionistic (single) indicators	PCB, Species present/absent
Semiholistic indicators	Odum's attributes
Holistic indicators	Biodiversity/ecological network
"Superholistic"	Thermodynamic indicators as ecoexergy and emergy

TABLE 14.2

Work Capacity Used to Express the Ecosystem Services for Various Types of Ecosystems[a]

Ecosystem	Biomass (MJ/m² year)	Information Factor (β-Value)	Work Capacity (GJ/ha year)
Desert	0.9	230	2,070
Open sea	3.5	68	2,380
Coastal zones	7.0	69	4,830
Coral reefs, estuaries	80	120	960,000
Lakes, rivers	11	85	93,500
Coniferous forests	15.4	350	539,000
Deciduous forests	26.4	380	1,000,000
Temperate rainforests	39.6	380	1,500,000
Tropical rainforests	80	370	3,000,000
Tundra	2.6	280	7,280
Croplands	20.0	210	420,000
Grassland	7.2	250	18,000
Wetlands	18	250	45,000

[a] It is calculated as biomass *the information factor.

be to determine the economic values of the overall ecological services offered by the ecosystems and then compare them with what is normal for the type of ecosystems considered. Jørgensen (2010) has determined the values of all the services offered by various ecosystems by the use of the ecological holistic indicator eco-exergy expressing the total work capacity. It is a good measure of the total amount of ecological services as all services require a certain amount of free energy, that is, energy that can do work, to be carried out. The values published in Jørgensen (2010) are shown in Table 14.2 and can be used for the previously indicated comparison. The eco-exergy is found as the sum of the β-values times the biomass. The β-values express the information the different ecosystem components are carrying (see Jørgensen et al. [2005] and Jørgensen [2012]). The average β-values for the various ecosystems are based upon the living components that are present in typical ecosystems representing the various types of ecosystems.

Assessment of the values of the ecosystem services may also be coupled to sustainability, because it is crucial to maintain the many ecosystem services on which the society is dependent. Environmental management scenarios, preferably developed by ecological models, are tested by means of the ecological sustainability trigon (EST); see Marques et al. (2009). The use of eco-exergy as indicator to find the value of the ecosystem services in this context is beneficial, because the development of sustainability can be determined as maintenance of the total work capacity that is at our disposal (Jørgensen 2006).

Assessment of the ecosystem services is frequently using ecological indicators. The indicators are followed by the use of models and models can determine the reduced or lost ecological services of ecosystems. The three diagnostic tool boxes, ecological models, ecological indicators, and ecological services are, in other words, closely related and obviously the use of all three tool boxes will give the most complete diagnosis and image of the health of an ecosystem. On the other hand, the resources available for environmental management are always limited, which means that it is hardly possible to apply all three tool boxes in all cases, but it is necessary in many cases to make a choice. If an ecological model is developed, anyhow, to be able to give more reliable prognoses, it is of course natural to apply the developed model and it may be beneficial in addition to select one or a few indicators to focus more specifically on a well-defined problem. If a model is not available but a monitoring program is applied, it would naturally direct the observations to encompass the state variables that can be applied to assess the indicators that are closely related to the defined health problems. If the society is dependent on specific ecological services of the ecosystem, it would be natural to assess to what extent these services are reduced or lost, maybe supplemented with health indicators that are particularly important for the maintenance of these services. The choice of tool boxes is therefore a question about the available resources and the specific case and problem.

14.4 Tool Boxes Available Today to Solve the Environmental Problems

The tool box environmental technology was the only methodological discipline available to solve the environmental problems 45 years ago when the first green waves started in the mid-1960s. This tool box was only able to solve the problems of point sources but sometimes but not always at a very high cost. Today, this tool box has more tools and often more cost moderate tools than it had 30–40 years ago Fortunately, we have today additional tool boxes that can solve the problems of the diffuse or nonpoint pollution or find alternative solutions at lower costs when the environmental technology would be too expensive to apply. As for the diagnostic tool boxes, these new tool boxes are developed on basis of new ecological subdisciplines.

To solve environmental problems, we have today four tool boxes:

1. Environmental technology
2. Ecological engineering, also denoted as ecotechnology
3. Cleaner production (technology) and under this heading we would also in this context include industrial ecology
4. Environmental legislation

Environmental technology was available by the emergence of the first green waves about 45 years ago. Since, several new environmental–technological methods have been developed and all the methods have been streamlined and are generally less expensive to apply today. Chapter 15 will give an overview of the environmental technological methods that are available to solve the aquatic ecosystem problems. We have today a wide spectrum of applicable methods, as the chapter will demonstrate. It is possible in principle to solve all water problems associated with point sources.

There is and has been, however, an urgent need for other alternative methods to be able to solve the entire spectrum of environmental problems at an acceptable cost. The environmental management today is more complicated than it was 45 years ago because of the many more tool boxes that should be applied to find the optimal solution and because global and regional environmental problems have emerged. The use of tool boxes and the more complex situation today is illustrated in Figure 14.1.

The tool box containing ecological engineering methods has been developed since the late seventies. Ecological engineering is defined as design of sustainable ecosystems that integrate human society with its natural environment to the benefit of both (Mitsch and Jørgensen 2004). It is an engineering discipline that operates in ecosystems, which implies

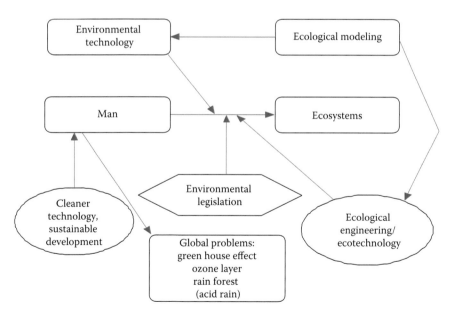

FIGURE 14.1
Solutions of environmental problems can today apply four tool boxes: environmental technology, ecotechnology, cleaner technology, and environmental legislation. Ecological models can be applied to select the needed methods of environmental technology and ecotechnology. The environmental management today is further complex due to global pollution problems such as the decrease of the ozone layer and the greenhouse effect.

that it is based on both design principles and ecology. The tool box contains four classes of tools:

1. Tools that are based on the use of natural ecosystems to solve environmental problems (for instance, the use of wetland to treat agricultural drainage water)
2. Tools that are based on imitation of natural ecosystems (for instance, construction of wetlands to treat wastewater)
3. Tools that are applied to restore ecosystems (for instance, restoration of lakes by the use of biomanipulation)
4. Ecological planning of the landscape (for instance, the use of agroforestry)

The introduction of ecological engineering has made it possible to solve many problems that environmental technology could not solve, first of all nonpoint pollution problems and a fast restoration of deteriorated ecosystems. Chapter 17 will give an overview of the ecological engineering methods that we can apply to solve the aquatic environmental problems, particularly the freshwater problems.

Some environmental problems, however, cannot be solved without a more strict environmental legislation, and for some problems, a global agreement may be needed to achieve a proper solution, for instance, by out phasing the use of freon to stop or reduce the destruction of the ozone layer. Notice that also environmental legislation requires an ecological insight to assess the required reduction of the emission that is needed through introduction of the environmental legislation.

As the environmental legislation has been tightened, it has been more and more expensive to treat industrial emissions, and the industry has of course considered whether it was possible to reduce the emission by other methods at a lower cost. That has led to development of what is called cleaner production, which means the idea to produce the same product on a new method that would give a reduced emission and therefore less costs for the pollution treatment. New production methods have been developed by the use of innovative technology that has created a completely new method to produce the same product with less environmental problems. Other emission reductions have been developed by the use of ecological principles on the industrial processes, for instance, recycling and reuse. In many cases it has also been possible to achieve a reduction of the environmental problems by identification of unnecessary waste. Industrial ecology could after the author's opinion be defined as the use of ecological principle in the production such as recycling, reuse, and holistic solutions to achieve a high efficiency in the general use of the resources. Industrial ecology is today, however, used to cover the use of waste from one production in another production. Some use even the word industrial ecology, when trees are planted in industrial areas, which of course is not very innovative.

Today, we have by the four tool boxes a possibility to solve any environmental problem and often at a moderate cost and sometimes even at a cost that makes it beneficial to solve the problem properly. As is the case for the diagnostic tool boxes, also the tool boxes with problem solution tools are rooted in recently developed ecological subdisciplines that are named after the following tools: ecological engineering, environmental legislation, and cleaner technology.

14.5 Follow the Recovery Process

Environmental management is only complete if the environmental problem and the ecosystem are followed carefully after the tool boxes have been applied. It is usually not a problem because it is a question of providing the observations needed to follow the prognoses of the

1. Eventually developed ecological model
2. The selected ecological indicators
3. The recovery of the ecological services of the ecosystem (which can be done by focusing on a specific service or on the values of all the ecological services offered by the ecosystem)

14.6 Implementation of the Presented Ecological and Environmental Management Procedure in Practice on Inland Water Ecosystems

It is strongly recommended to apply the presented procedure in practical management of aquatic ecosystems. A few remarks are, however, needed to consider in this context. These considerations are presented in this section.

Ecosystems are living systems, and do not react in the same predictable manner as rigid physical systems. In fact, ecosystems have often exhibited surprising reactions to changed impacts, which obviously complicates environmental management considerations. A deeper understanding of systems ecology, therefore, is a prerequisite for a more ecologically sound environmental management approach, which is one of threads unifying this chapter. The state of an ecosystem is obviously dependent on the forcing functions that result in impacts on it. The history of the ecosystem, however, is also important in assessing the possible reactions to a changed water quality or other type of impact. It can be shown that the same impact on, for instance, a lake (e.g., the same total phosphorus concentration in lake water) may result in two different responses, depending on the structure of the system, which again is determined by the history of the system. This important reaction of ecosystems is touched upon in the following text and will also be further treated in Chapter 19, where the use of structurally dynamic models often is needed to be able to capture the possible reactions of ecosystems.

Many failures in freshwater management can be explained by the omission of the ecological consideration of the water body. As an example, an ecosystem resists changes, with a tendency to try to maintain its existing structure. A reduction in external forcing functions (e.g., a pronounced reduction in the phosphorus load entering a lake from wastewater discharges) is often not sufficient to restore a lake or reservoir. If the structure of the food web in the water body is adapted to a high phosphorus input, it will require very pronounced reductions in the phosphorus load to shift to a mesotrophic or oligotrophic condition. Thus, biomanipulation is an alternative to support the system in its efforts to shift to another food web structure. In this context it is absolutely necessary to consider the storage of nutrients and other pollutants during the implementation of a management strategy. Significant quantities of nutrients and other pollutants may be stored in the lake bottom sediments. If less-polluted water flows into a lake as a result of proper wastewater treatment, nutrients stored in the sediments may be released if appropriate chemical conditions exist in the water layers at the bottom of the lake and can significantly delay its restoration. This internal pollutant source often may be very significant, dominating for an extended period over external pollutant.

The importance of internal pollution sources is closely related to the proper timing of effective management strategies. Because an ecosystem has the ability to reduce the impacts of pollutant inputs by various accumulation processes, accumulation in the sediment being one of the most significant, it becomes very important to reduce the pollution sources at an early stage. Environmental management efforts in Denmark clearly illustrate how a good management strategy, applied at the wrong time, can lead to significant increases in pollution abatement costs. An ambitious environmental management strategy was launched in Denmark in 1976, with the goal of letting the ecosystem determine the best restoration strategy. Consequently, it was necessary to make plans about the use of all aquatic ecosystems. Could discharge of wastewater be tolerated? What were the recreational values of the various aquatic ecosystems? Did some ecosystems have a particular scientific value? Which ecosystems were valuable as water resources? Many more questions became evident. To answer these crucial questions required more than 7 years of interactions between the different levels in the political hierarchy, from the communities to the government, which resulted in excellent environmental plans, but no pollution abatement. The Gordian Knot was cut in 1986, when it was decided that all Danish wastewater treatments should meet certain standards, including a biochemical oxygen demand ($BOD_5 \leq 10 \, mg/L$, total phosphorus concentration

$\leq$1.5 mg P/L, and a total nitrogen concentration $\leq$8 mg N/L). In addition, it was possible to assess even more strict requirements for particularly important or vulnerable ecosystems (e.g., lakes of important recreational value). This goal was subsequently widely applied in Danish lake management efforts. The result was that the entire environmental strategy in Denmark was quite successful—except for the timing. These measures should have been taken 7–9 years earlier! The consequences were that 7–9 additional years of phosphorus (and nitrogen) discharges accumulated in the Danish lakes, making it considerably more difficult to find a management strategy able to return the lakes to the conditions prevailing 30–50 years earlier. A specific example of this failure is Arresø in Zealand, Denmark's largest lake. The lake is very shallow, with a maximum depth of 4 m, which obviously accentuates the importance of nutrients accumulated in the lake sediments. If a proper environmental management had been implemented in the mid-1970s, the condition of the lake today would have been close to its condition in the 1950s. However, because of the aforementioned delay of 7–9 years, so much additional phosphorus had accumulated in the sediments that it will take more than 100 years to bring the lake back to its previous conditions based solely on reduction of its phosphorus load. In fact, it will be necessary to remove the upper 0.5 m of the lake bottom sediment to significantly reduce the internal phosphorus load, which will cost as much as U.S.$ 60–80 million based on a sediment surface area of 41 km^2.

As this example illustrates, ecosystem restoration can be very expensive. It is usually best, therefore, to develop and implement good environmental planning at an early stage, in order to be able to also take preventive measures that would eliminate the problems of phosphorus accumulation, as well as heavy metals and persistent organic chemicals, in freshwater sediments. A proper environmental planning also would help identify the pollution priorities that a good strategy should address.

Many developing countries lack the economic resources to undertake appropriate environmental management programs. Nevertheless, the reality is that proper environmental planning can be carried out at a fraction of the costs of major pollution abatement programs. It is strongly recommended, therefore, that developing countries undertake environmental planning efforts at an early stage. This will facilitate their assigning a proper priority to the various steps in the environmental management planning process, and also to make better use of preventative measures, which will work to enhance the possibilities of moderate-cost solutions over the long term.

As with other ecosystems, inland water ecosystems are "open systems." Thus, it is not sufficient to consider only the water basin in a management regime. Indeed, it is also essential to consider the *entire* watershed in order to include all the pollution sources in the management considerations, which is very consistent with the idea behind integrated environmental management. It is often much more cost-effective, for example, to reduce the concentration of a pollutant at its source—a solution that can only be identified by considering the watershed as one combined, holistic system. Further, there are usually many more possible solutions to the environmental problems when we attempt to optimize a large, combined lake–watershed system, than focusing only on the lake itself. Including the entire watershed in environmental management considerations is of particular importance for developing countries, which generally have fewer resources to devote to pollutant abatement efforts than do the developed countries.

The use of ecological modeling has become much more important over the last 35 years, due to a need for quantification in ecology and environmental management. This is probably the only significant tool available for obtaining a quantitative overview of complex ecosystems such as lakes, rivers, wetlands, lagoons, or reservoirs, which is a prerequisite for selecting an optimum solution to the complex problems facing them. Because solutions to the entire spectrum of environmental problems require quantitative estimation methods to assist in identifying a realistic trade-off between ecological and economic concerns, it is not surprising that ecological models have been used increasingly in environmental management efforts. Chapter 19 provides an overview of models in environmental management of freshwater systems.

From this introductory discussion, several important recommendations may be deduced, as summarized in the following six points:

- Expect that a proper environmental strategy will require a wide spectrum of approaches and techniques.
- Expect that proper environmental management will require the application of a combination of end-of-the-pipe technologies (environmental technology), ecotechnology, cleaner technology, and environmental legislation.
- Correct timing in applying the various steps in environmental management efforts is extremely important; thus, it is recommended that a comprehensive environmental management plan be developed at a very early stage, in order to be able to use the available resources in the most optimal manner.
- It is usually very beneficial, particularly from an economic perspective, to consider prevention, rather than correction, primarily because it is often very costly to restore heavily-degraded lakes.
- Because of the complexity of ecosystems and their problems, proper ecological knowledge about ecosystems is a prerequisite for ecologically sound environmental management programs; this is the only reasonable method for avoiding unexpected ecosystem responses.
- Optimum solutions to environmental management problems are best obtained if the entire lake–watershed ecosystem is taken into consideration in developing and implementing management actions.

14.7 Conclusions about Integrated Environmental and Ecological Management of Inland Water Ecosystems

From the review of the up-to-date integrated environmental management, it is possible to conclude the following:

1. It is recommended to use the three diagnostic tool boxes to understand the environmental problems properly.

2. The three diagnostic tool boxes can also be applied to follow the development of an environmental and ecological problem—including an eventual recovery process.

3. Eco-exergy is a useful indicator as it expresses sustainability and the total amount of ecological services offered by an ecosystem. See also Chapter 18, where an overview of applicable ecological indicators in freshwater management will be presented.

4. Integrated environmental management based on the use of the three diagnostic tool boxes and considering all sources of the problem requires the use of all four "problem solving" tool boxes:
   a. Environmental technology
   b. Ecotechnology
   c. Cleaner technology, including industrial ecology
   d. Environmental legislation

   A combination of the tools from all the four tool boxes should always be considered.

5. The integrated up-to-date environmental management requires the use of the seven presented tool boxes and would not be possible if these tool boxes were not developed as a result of recently emergent ecological subdisciplines: ecological modeling, ecological engineering, application of ecological indicators, cleaner technology, and industrial ecology. These ecological subdisciplines are therefore crucial for the environmental management of today and they form an indispensable bridge between ecology and environmental management—between the basic science of ecology and its application in practical environmental management.

References

Costanza, R.; Norton, B.G.; and Haskell, B.D. 1992. *Ecosystem Health, New Goals for Environmental Management.* Island Press, Washington, DC, 270pp.

Costanza, R. et al. 1997. The value of the world's ecosystem services and natural capital. *Nature* 387:252–260.

Jørgensen, S.E. 2002. *Integration of Ecosystem Theories: A Pattern*. Kluwer, Dordrecht, the Netherlands, 386pp.

Jørgensen, S.E. 2006. *Eco-Exergy as Sustainability*. WIT, Southampton, U.K., 220pp.

Jørgensen, S.E. 2010. Ecosystem services, sustainability and thermodynamic indicators. *Ecol. Complex.* 7:311–313.

Jørgensen, S.E. (Ed.). 2011. *Handbook of Ecological Models Used in Ecosystem and Environmental Management*. CRC Press, Boca Raton, FL, 620pp.

Jørgensen, S.E. 2012. *Introduction to Systems Ecology*. CRC Press, Boca Raton, Florida, 320pp.

Jørgensen, S.E. and Nielsen, S.N. 2012. Tool boxes for an integrated ecological and environmental management. *Ecological Indicators* 20:104–109.

Jørgensen, S.E.R.; Costanza, R.; and Xu, F.L. 2004 and 2010. *Handbook of Ecological Indicators for Assessment of Ecosystem Health*, 2nd edn. CRC Press, Boca Raton, FL, 436pp. (2004) and 482pp. (2010).

Jørgensen, S.E. and Fath, B. 2011. *Fundamentals of Ecological Modelling*, 4th edn. Elsevier, Amsterdam, the Netherlands, 400pp.

Jørgensen, S.E.; Ladegaard, N.; Debeljak, M.; and Marques, J.C. 2005. Calculations of exergy for organisms. *Ecol. Model.* 185:165–176.

Marques, J.C. et al. 2009. Ecological sustainability trigon. *Mar. Pollut. Bull.* 58:1773–1779.

Mitsch, W.J. and Jørgensen, S.E. 2004. *Ecological Engineering and Ecosystem Restoration*. John Wiley, New York, 410pp.

15

Application of Environmental Technology in the Environmental and Ecological Management

15.1 Introduction

It is possible to distinguish between pollution from point sources and from nonpoint sources. Environmental technology, which was already widely applied in environmental management 40–50 years ago, is covering the technological methods that are able to solve almost exclusively pollution problems from point sources. Impact on aquatic ecosystems from point sources is originated from discharge of wastewater. Wastewater discharges into inland water ecosystems are man-controlled forcing functions of crucial importance for the water quality. It is, however, possible in many situations to control it completely, either by water diversion or by wastewater treatment methods. Water diversion, however, results in another downstream water body that has to cope with the pollutant load. Thus, treating the wastewater properly should be considered a generally more acceptable solution to the problem. This gives rise to two questions, namely,

1. Is it possible to solve all pertinent wastewater problems?
2. What is understood by a proper wastewater treatment?

The water pollution problems associated with municipal and industrial wastewaters include their content of

- Nutrients causing eutrophication
- Biodegradable organic matter causing oxygen depletion
- Bacteria and virus effecting the sanitary quality of water, which is of particular importance when the water is used for bathing, swimming, and drinking purposes

- Heavy metals, mainly lead, zinc, and cadmium, from gutters; heavy metals from fungicides and other agricultural chemicals; and a wide range of other heavy metals in minor concentrations
- Refractory organic matter, originating from industries and hospitals, the use of pesticides, and even the use of a wide spectrum of household articles

15.2 Wastewater Treatment: An Overview

Tables 15.1 and 15.2 provide an overview of a wide range of wastewater treatment methods, their efficiencies, and their costs. Clearly, there is with good approximation a method available to virtually solve any of the mentioned problems.

Industrial wastewaters can cause the same water pollution problems as municipal wastewater plus a few more. In addition, they can also contain toxic organic and/or inorganic compounds, (particularly heavy metals and persistent organic pollutants). However, it is necessary to solve at the source the problems associated with industrial wastewater that can hardly be solved with municipal wastewater treatment methods. It is also the general legislation all over the world today that industries are obliged to treat the wastewater before discharge to the public sewage system. In many countries, the practice of the polluter having to pay the system has forced the industries to solve their pollution problems to keep the production costs low. The major portion of toxic substances is therefore today removed by the industries, at least in the industrialized countries. They would only have been partially removed, if at all, at municipal wastewater treatment plants and/or could contaminate the sludge produced at municipal wastewater treatment plants, thereby eliminating the possibility of the use of the sludge as a soil conditioner.

The removal of high concentrations of biodegradable organic matter at the source is strongly recommended, since it is usually much more cost-effective to remove these components, at least partially, when they are present in high concentrations. High concentrations of biodegradable organic matter are found in wastewater from slaughterhouses, starch factories, fish industries, dairies, and canned food industries. The removal of x% of BOD_5 costs as a rule of thumb the same independent on the level of BOD_5. It means that 1 kg of BOD_5 can be removed at a much lower price from wastewater with a high BOD_5, for instance, wastewater from slaughterhouses, starch factories, fish industries, dairies, and canned food industries.

The listed methods often are used in combinations of two or more steps to obtain the overall removal efficiency required by the most

TABLE 15.1

Survey of Generally Applied Wastewater Treatment Methods

Method	Pollution Problem	Efficiency	Costs (US\$/100 m³)
Mechanical treatment	Suspended matter removal	0.75–0.90	3–5
	BOD_5 reduction	0.20–0.35	
Biological treatment	BOD_5 reduction	0.70–0.95	25–40
Flocculation	Phosphorus removal	0.3–0.6	6–9
	BOD_5 reduction	0.4–0.6	
Chemical precipitation $Al_2(SO_4)_3$ or $FeCl_3$	Phosphorus removal	0.65–0.95	10–15
	Reduction of heavy metal concentrations	0.40–0.80	
	BOD_5 reduction	0.50–0.65	
Chemical precipitation $Ca(OH)_2$	Phosphorus removal	0.85–0.95	12–18
	Reduction of heavy metal concentrations	0.80–0.95	
	BOD_5 reduction	0.50–0.70	
Chemical precipitation and flocculation	Phosphorus removal	0.9–0.98	12–18
	BOD_5 reduction	0.6–0.75	
Ammonia stripping	Ammonia removal	0.70–0.95	25–40
Nitrification	Ammonium is oxidized to nitrate	0.80–0.95	20–30
Active carbon adsorption	COD removal (toxic substances)	0.40–0.95	60–90
	BOD_5 reduction	0.40–0.70	
Denitrification after nitrification	Nitrogen removal	0.70–0.90	15–25
Ion exchange	BOD_5 reduction (e.g., proteins)	0.20–0.40	40–60
	Phosphorus removal	0.80–0.95	70–100
	Nitrogen removal	0.80–0.95	45–60
	Reduction of concentrations		10–25
Chemical oxidation (e.g., with Cl_2)	Oxidation of toxic compounds	0.90–0.98	60–100
Extraction	Heavy metals and other toxic compounds	0.50–0.95	80–120
Reverse osmosis	Removes pollutants with high efficiency, but is expensive		100–200
Disinfection methods	Reduction of microorganisms	High, can hardly be indicated	6–10
Ozonation + active carbon adsorption	Removal of refractory compounds	0.5–0.95	100–120

TABLE 15.2

Efficiency Matrix Relating Pollution Parameters and Wastewater Treatment

	Suspended Matter	BOD$_5$	COD	Total Phosphorus	Ammonium Nitrogen	Total Nitrogen	Heavy Metals	Escherichia coli	Color	Turbidity
Mechanical treatment	0.75–0.90	0.20–0.35	0.20–0.35	0.05–0.10	~0	0.10–0.25	0.20–0.40	—E	0.80–0.98	—
Biological treatment[a]	0.75–0.95	0.65–0.90	0.10–0.20	0.05–0.10	~0	0.10–0.25	0.30–0.65	Fair	~0	—
Chemical precipitation	0.80–0.95	0.50–0.75	0.50–0.75	0.80–0.95	~0	0.10–0.60	0.80–0.98	Good	0.30–0.70	0.80–0.98
Ammonia stripping	~0	~0	~0	~0	0.70–0.96	0.60–0.90	~0	~0	~0	~0
Nitrification	~0	~0	~0	~0	0.80–0.95	0.80–0.95	~0	Fair	~0	~0
Active carbon adsorption[a]	—	0.40–0.70	0.40–0.95	~0.1	High[b]	High[b]	0.10–0.70	Good	0.70–0.90	0.60–0.90
Denitrification after nitrification	~0	—	—	~0	—	0.70–0.90	~0	Good	~0	—
Ion exchange	–0.40	0.20–0.50	0.20–0.95	0.80–0.95	0.80–0.95	0.80–0.95	0.80–0.95	Very good	0.60–0.90	0.70–0.90
Chemical oxidation	—	Corresponding to oxidation	~0	~0	~0	~0	~0	~0	0.60–0.90	0.50–0.80
Extraction	—	Corresponding to extraction of toxic compounds	~0	~0	~0	~0	0.50–0.95	~0	~0	~0
Reverse osmosis[a]	See Table 15.1									
Disinfection methods	—	Corresponding to the amount of chlorine or ozone applied	Very high	0.50–0.90	0.30–0.60					

[a] Depends on the composition.
[b] As chloramines.

cost-moderate solution. The methods can also be applied in combination with cleaner technology (Chapter 16) and/or ecotechnology (Chapter 17). Because wastewater treatment often is costly, it is advisable in the planning phase to examine *all* possible combinations of treatment options in order to identify the most feasible and appropriate one.

Many existing municipal wastewater treatment plants were constructed years or decades ago, and may not meet today's higher standards. Nevertheless, upgrading existing wastewater treatment plants is possible, and may be more cost-moderate than building new ones (Novotny and Somlyódy 1995, van Loosdrecht 1998). Because the funding allocated to pollution abatement often is limited, the overall effect of upgrading wastewater treatment plants that can be upgraded with sufficient efficiency will be to the benefit of the environment. An attractive solution is often to introduce *tertiary treatment* by chemical precipitation and flocculation in an existing mechanical–biological treatment plant, with the addition of chemicals and flocculants before the primary sedimentation phase. The installation costs for this solution are minor, and the additional running costs are limited to the costs of chemicals. The result is an 85%–95% removal of phosphorus at low cost. Similarly, nitrification and denitrification, ensuring an 80%–85% removal of nitrogen, can be realized with the installation of additional capacity for biological treatment (the overall water retention time in the plant is increased by 4–12 h, depending on the standards and composition of the wastewater), which is considerably less costly than installation of a completely new treatment plant. For details, see Hahn and Muller (1995) and Henze and Ødegaard (1995).

The second question refers to the selection of the right standards for the treated wastewater. Any removal efficiency of any pertinent parameters (BOD_5, nutrients, bacteria, viruses, toxic organic compounds, color, taste, and heavy metals) is possible to obtain with a suitable combination of the available treatment methods. However, what removal efficiencies are needed in the focal case? Because wastewater treatment is costly, the maximum allowable concentrations should not be set significantly lower than the lake or reservoir receiving the effluents can tolerate. The ban of phosphate detergents to decrease phosphorus concentrations in municipal wastewater treatment plant effluents is a point to consider in this context, as the treatment costs can be reduced considerably by introduction of phosphorus-free detergents. On the other hand, it might be even more expensive to install an insufficient treatment plant. Thus, the potential effects of a wide range of possible pollutant inputs on water quality and on the entire lake or reservoir should be assessed, as the basis for selecting an acceptable option. This will require a quantification of the impacts of various possible pollutant inputs, considering a wide range of solutions. All processes and components affected significantly by the impacts should be included in the quantification. It is usually very helpful to develop a water quality/ecosystem model and use it properly to assist in the selection of specific environmental treatment methods. It is important to emphasize that a model has an uncertainty in all its predictions that must be considered in making a final decision. Thus, it is essential to use safety factors to the benefit of the environment, in order to ensure that the selected treatment methods will have the anticipated effects. If the uncertainty is not taken into account for the sake of economy, as it is unfortunately often done, the investment may be wasted because the foreseen recovery of the freshwater ecosystem will not be realized.

Application of the methods identified in Tables 15.1 and 15.2 gives only approximate results, and the indications should therefore be used with caution. However, first estimates, such as those shown in the tables, are useful for evaluations of various alternative solutions to wastewater pollution problems. The biological treatment may either be an activated sludge plant or a trickling filter.

The cost of treating $100\,m^3$ of wastewater is also based on approximate indications, because they vary from place to place, as the costs of labor are very different in and are highly dependent on the size of the wastewater treatment plant. The costs are calculated as the running costs (electricity, labor, chemicals, and maintenance), plus 10% of the investment to cover interest and annual appreciation. The annual water consumption of one person in an industrialized country corresponds to approximately $100\,m^3$.

A problem in many developing countries is the relatively high cost of wastewater treatment. Although this cost might justify diversion of the wastewater, the application of "soft technology"—"ecotechnology"—also should be considered. Some corresponding methods will be touched on in the following sections, but proper planning at an early phase, and considering all predictable problems, offers the widest range of cost-effective possibilities and may allow prevention of the pollution problems before they will occur.

Corrections at a later stage, when pollution has already degraded the water quality and associated ecosystems, are possible, but will always be more expensive than the costs of proper wastewater treatment at an early stage. This is due in part to the fact that the accumulation of pollutants in a lake over time will always cause additional problems and, therefore, result in additional costs. Thus, pollution prevention at an early stage is better than curing pollution at a later stage. Removal of phosphorus from wastewater at an early stage, for example, is always beneficial since the surplus phosphorus will accumulate in the lake sediments to a large extent and allow its remobilization back into the water column under certain chemical conditions in the water body.

Model studies are able to reveal how long it may require to restore a lake, or how much higher phosphorus removal efficiency will be required to compensate for each year that implementation of an appropriate phosphorus removal technology is postponed. However, it is not unusual that implementation of a phosphorus removal technology a few years later than it was first feasible may delay the restoration of a lake by one or more decades, due to the fact that the additional phosphorus accumulated in the sediments may significantly increase the quantity of phosphorus in the water column.

Important pollution sources may be reduced if the liquid waste and sludge land disposal is avoided or minimized. For municipal areas, two options are used to decrease the hydraulic load of wastewater treatment plants:

1. Decreasing water use, thereby saving water and producing smaller volumes of polluted water.
2. Separating storm water from municipal domestic waste, with a similar result. One result of these options is that the capacity of wastewater treatment plants can be kept smaller, achieving significant cost savings.

New approaches have emerged in regard to sustainable development. For instance, serious consideration is being given to separate toilets in some locations, which collect urine separately from feces, thereby allowing utilization of the septic urine as fertilizer.

The selection of proper wastewater treatment methods for point sources of pollution is summarized in the following points:

- Develop models for the impacts of the wastewater on freshwater ecosystems, considering the impacts on the water quality and the entire lake ecosystem.
- Apply the model to identify the maximum allowable pollutant concentration in the treated wastewater. Any uncertainty associated with the model predictions should be reflected in identifying the lower maximum allowable concentrations.
- Select the combination of available treatment methods able to meet the standards at the lowest costs without impacting the proper operation of the plant.
- If the investment needed for a proper solution to a problem cannot be provided, the application of cost-moderate technology that will reduce the accumulation of pollutants in the lake should be considered. Any measures taken at an early stage will reduce the costs at a later stage.

15.3 Municipal Wastewater

The composition of domestic sewage varies surprisingly little from place to place, although it to a certain extent reflects the economic status of the society.

TABLE 15.3

Typical Composition of Municipal Wastewater (mg/L)

Constituent	Soluble	Particulate	Total
BOD_5	100–200	50–100	150–300
COD	200–500	100–200	200–700
Ammonium-N	20–40	0	20–40
Nitrate-N	5–20	0	5–20
Organic nitrogen	0	5–20	25–60
Suspended matter	—	40–80	40–80
Carbohydrates	20–40	10–15	30–55
Amino acids	10–15	15–25	25–40
Fatty acids	0	50–80	50–80
Surfactants	10–20	5–10	15–30
Creatinine	3–5	0	3–5
Phosphorus	2–4	4–10	6–14

Source: From *Principles of Pollution Abatement*, Jørgensen, S.E., 520 pp., 2000, from Elsevier.

A typical composition of municipal wastewater including BOD_5, COD, suspended matter, and nutrients is shown in Table 15.3.

Most industrialized countries require a treatment of municipal wastewater. The standards vary for the European Union, the United States, Canada, Australia, New Zealand, and Japan, but generally a reduction of BOD_5 to about 10 mg/L, a reduction of nitrogen to about 10 mg/L, and a reduction of the phosphorus concentration to 1–2 mg/L are required. To obtain the required reductions, a combination of the methods presented in Section 15.2. The combinations of methods applied for treatment of municipal wastewater are presented in the next section and it will be demonstrated that it is possible by a suitable combination of methods to meet practically all realistic effluent standards.

15.4 Combinations of Methods for the Treatment of Municipal Wastewater (Reduction of BOD_5)

Models are used increasingly to design and to optimize wastewater treatment methods. For details on the applied models, refer to Jørgensen (2011), which has a comprehensive chapter devoted to models of wastewater treatment systems.

There is a number of different possible designs of biological treatment plants; see Jørgensen (2000).

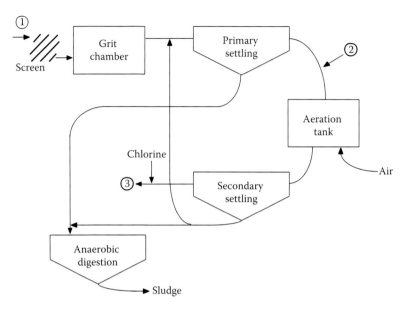

FIGURE 15.1

Flow diagram of a conventional activated sludge plant. It is widely used to treat municipal wastewater. Number 1 is untreated municipal wastewater, which would have a BOD_5 of about 150–300 mg/L, total nitrogen of about 25–45 mg/L, and total phosphorus of about 6–12 mg/L. The mechanical treatment (sample at point 2) will reduce BOD_5 and total N by about 25%–40%, while the reduction of total phosphorus will be minor. The totally treated wastewater (point 3) will have BOD_5 of about 10–15 mg/L, while the nitrogen is reduced 35%–50% only and the phosphorus concentration about 10%–20% only.

The use of mechanical–biological wastewater treatment is a classical method or rather the combination of the two methods. Figure 15.1 gives the flow chart of a classical mechanical-activated sludge plan as it is generally used all over the world. The processes involve screening, a separation of grease and sand in the grit chamber, a primary settling, an aeration step, followed by sedimentation. Chlorine is often added before discharge to the receiving water, particularly when it is used for swimming and bathing. The sludge is most often digested anaerobically, which gives a production of biogas. In the activated sludge plant, a rapid adsorption and flocculation of suspended matter take place. Organic matter is oxidized and decomposed. Sludge particles are dispersed and settled by the secondary settling. The processes are conceptualized in Figure 15.2.

Oxidation ditches, see Figure 15.3, can replace the activated sludge plant and the secondary settling. The rotor provides the aeration to oxidize the organic matter. The influent and the rotor are stopped usually during the night when the inflow is low anyhow to allow settling. The supernatant is withdrawn through an effluent launder. The retention time is usually 2–5 days.

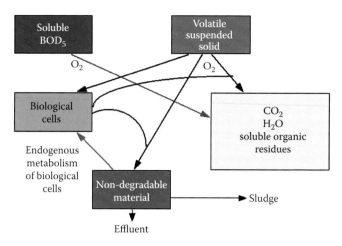

FIGURE 15.2
Processes characteristic for the biological treatment.

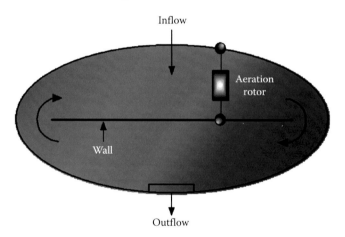

FIGURE 15.3
Oxidation ditch.

A trickling filter is a bed packed with rocks, although recently plastic media (celite pellets; Sorial et al. 1998) or bioblocks with a high surface area due to high porosity are also applied. They require substantially less space than the stone packed trickling filter. They have usually a specific surface on $100\,m^2/m^3$ or more. It is at least two times the specific surface of the trickling filter packed with rocks. The media is covered by a slimy microbiological film. The wastewater is passed through the bed and oxygen and organic matter diffuse into the film, where oxidation occurs. Recirculation improves the removal efficiency. A flow chart of a treatment plant combining a trickling filter and an activated sludge plant is shown in Figure 15.4. The alternative ecotechnological methods are lagoons and

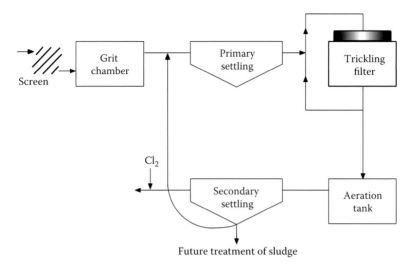

FIGURE 15.4
Treatment of municipal wastewater, combining trickling filter and activated sludge treatment.

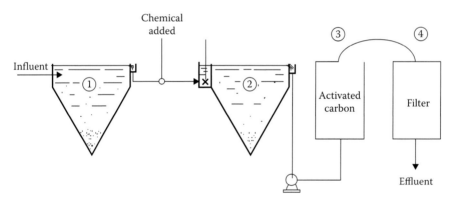

FIGURE 15.5
AWT system consisting of mechanical treatment, chemical precipitation, settling, adsorption on activated carbon, and filtration. The method is not used very much. It has the advantage that it is easier to control but it is also more expensive both with respect to investment and operation.

waste stabilization ponds. Various solutions based on natural and con-structed wetlands are covered in Chapter 6 and 17.

Physical–chemical methods have been proposed to replace mechanical–biological treatment, for instance, the so-called AWT system, which is based on the application of a combination of mechanical treatment, precipitation, adsorption on activated carbon, and sand filtration; see Figure 15.5.

It has been considered to recover proteins and grease to cover at least partially the costs of the treatment of wastewater from slaughterhouses, fish industries, starch factories, and other foodstuff industries. Figure 15.6 shows a method that has been applied in several, but still relatively few cases.

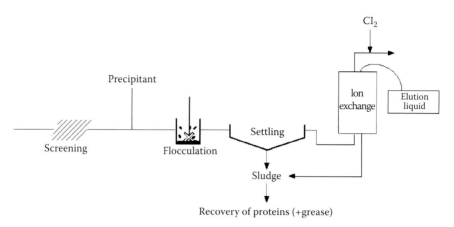

FIGURE 15.6
Recovery of proteins and grease from the wastewater discharge from the foodstuff industries
is possible by a combination of precipitation and ion exchange. The recovery pays partially for
the treatment of the wastewater by this process.

If the wastewater has a high concentration of grease, oil, and fat, it is possible to apply a flotation unit to separate water and suspended matter. It offers therefore an alternative to sedimentation. A portion of water is pressurized by 3–10 atm, and when this water is returned to the normal atmospheric pressure in the flotation unit, air bubbles are created. The air bubbles attach themselves to particles and the air–solute mixture rises to the surface, where it can be skimmed off, while the clarified water is removed from the bottom of the flotation unit. Figure 15.7 shows a flotation unit. Usually the retention time in a flotation unit is three to six times less than for a settling unit, which means that a significant volume reduction is obtained. Flotation has therefore most frequently replaced the sedimentation in slaughterhouses, fish industries, and oil industries.

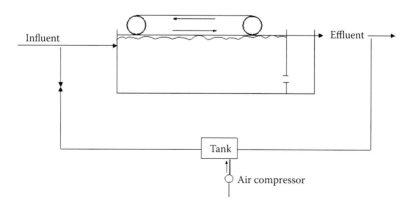

FIGURE 15.7
Flotation unit.

15.5 Methods for the Treatment of Municipal Wastewater (Reduction of Phosphorus Concentration)

Nutrient removal is most frequently carried out by chemical precipitation, often combined with mechanical–biological treatment. The chemical precipitation can be applied at three different points in the mechanical–biological plant as shown in Figure 15.8. Sometimes both direct precipitation and posttreatment is applied, which together with sand filtration makes it possible to obtain a concentration of 0.1 mg P/L or less in the effluent. In many cases where such a low phosphorus concentration is needed for particular sensitive receiving waters (mainly lakes), this double precipitation is a very attractive method, because it is relatively cost moderate compared with other alternatives. Without the sand filtration and only a single precipitation, it is usually relatively easy to obtain a concentration of phosphorus in the effluent between 1.0 and 1.5 mg/L.

Aluminum sulfate, various polyaluminates, calcium hydroxide, and iron (III) chloride can be applied as chemicals for the precipitation. The amount of hydrated lime or calcium hydroxide needed for the precipitation is usually 2.5–6 times higher than for the aluminum and iron compounds, because a high efficiency of the precipitation requires a pH of 10.0

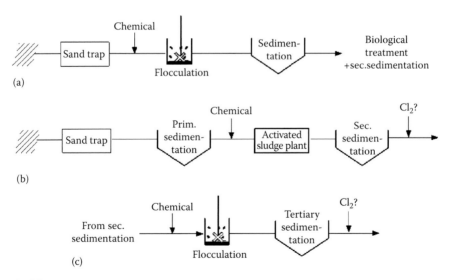

FIGURE 15.8
Precipitation by aluminum sulfate or other aluminum compounds, iron (III) chloride, and calcium hydroxide is able to reduce the phosphorus concentration in wastewater significantly. The precipitation is applied often in combination with mechanical–biological treatment and can be carried out after the sand trap: (a) direct precipitation, before the activated sludge plan; (b) simultaneous precipitation or after the mechanical–biological treatment; and (c) posttreatment.

or higher, which—but it dependents of the hardness of the water—is not possible without the indicated amount of calcium hydroxide. For a hardness of 15–30 hardness degrees, the amount is 100–480 mg calcium hydroxide/L. It would usually give an efficiency of 90%–95% precipitation of the phosphorous compounds, which is slightly better than for most precipitations with aluminum and iron compounds. The disadvantage of precipitation with lime is the high pH, which makes adjustment of the pH necessary. Carbon dioxide produced by incineration of the sludge or solid waste can be used for this purpose. If the sludge is incinerated, it is possible partially to recycle the calcium hydroxide and thereby reduce the costs of precipitation chemicals. Recycling three to five times is possible and afterward it can be applied as fertilizer, as it has a relatively high phosphorus concentration. A flow chart with direct precipitation and recycling of calcium hydroxide is shown in Figure 15.9.

If the sludge after calcium hydroxide precipitation cannot be incinerated, an adjustment of pH is needed before anaerobic digestion or aerobic sludge treatment. Heavy metals are removed more effectively by the use of calcium hydroxide than by aluminum and iron compounds. Lead, copper, and chromium are removed with a very high efficiency by all the mentioned precipitation chemicals, while only calcium hydroxide would give a high removal efficiency for cadmium and zinc.

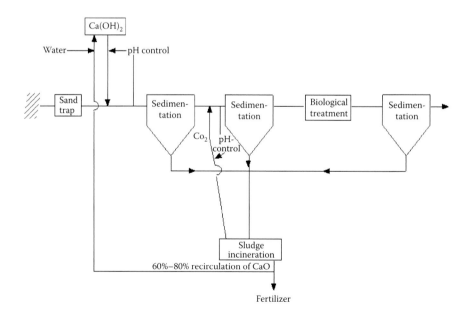

FIGURE 15.9
Chemical precipitation with partial recirculation of calcium hydroxide and use of carbon dioxide from incineration of the sludge for pH adjustment. (From *Principles of Pollution Abatement*, Jørgensen, S.E., 520 pp., 2000, from Elsevier.)

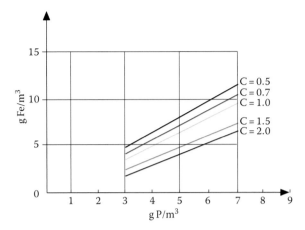

FIGURE 15.10
Addition of iron (III) as f(P-concentrations) at different P-concentrations in the effluent.

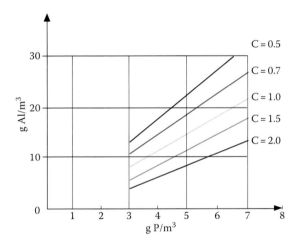

FIGURE 15.11
Addition of aluminum salts as f(P-concentrations) at different P-concentrations in the effluent.

The amount of aluminum and iron compounds can be found from Figures 15.10 and 15.11 or by the use of the following equation based on Freundlich adsorption isotherms:

$$\frac{(C_o - C)}{n} = a * C^b$$

where
   C_o is the initial concentration of phosphorus (mg P/L)
   C is the final concentration
   n is the dose of chemical expressed as mg Al or mg Fe/L
   a and b are characteristic constants that can be found in Table 15.4

TABLE 15.4

Constants in Freundlich Adsorption
Isotherms for Aluminum Sulfate
and Iron (III) Chloride

Precipitation with	a	b
Aluminum sulfate	0.63	0.2
Iron (III) chloride	0.26	0.4

With good approximation the constants
for aluminum sulfate can also be applied
for other aluminum compounds.

A more rapid flocculation, precipitation, and settling can be obtained by addition of synthetic organic polymeric flocculants. They may either be cationic polyelectrolytes, anionic polyelectrolytes, or nonionic polymers. It is hardly possible to indicate which polymeric flocculant would give the best result, as the ionic characteristics of municipal wastewater vary significantly. It is recommended to test at least the various types of polyflocculants from case to case. The optimum design of a flocculator before the settling of the precipitated material should be based on a mathematical model; see Dharmappa et al. (1993) and Thomas et al. (1999).

It is possible to remove phosphorus by biological treatment. Activated sludge systems with anaerobic and aerobic zones in sequence have been developed to achieve a higher phosphorus removal. The system is called EDPR (enhanced biological phosphorus removal). The shift between aerobic and anaerobic conditions activates the microorganisms to take up considerably more phosphorus than under aerobic conditions, particularly if the wastewater contains relatively high concentrations of easily biodegradable organic matter. With a P/BOD_5 ratio of more than 20, a phosphorus removal of 80%–90% can be obtained.

15.6 Methods for the Treatment of Municipal Wastewater (Reduction of Nitrogen Concentration)

A combination of *nitrification and denitrification* can reduce significantly the nitrogen concentration in the effluent. The applied chemical processes are

$$NH_4^+ + 2O_2 \rightarrow NO_3^- + H_2O + 2H^+$$

$$4NO_3^- + 5C + H_2O \rightarrow 5HCO_3^- + 2N_2 + H^+$$

The ammonium is oxidized to nitrate and the nitrate is used to oxidize organic matter under anaerobic conditions—here indicated just as C. Thereby the nitrate is reduced to dinitrogen, which is released to the atmosphere, where there is about 78% dinitrogen and a minor addition of dinitrogen is therefore harmless.

Effective nitrification occurs when the sludge age is greater than the reciprocal rate of constant for the nitrifying microorganisms (Bernhardt 1975). The sludge age is defined as $X/\Delta X$, where X is the mass of sludge in the system and ΔX is the sludge yield per unit of time. Usually the time unit applied is 24 h. The relationship between nitrification efficiency in percentage and the sludge age is shown in Figure 15.12.

The nitrification process is a two-step process. First, ammonia is oxidized to nitrite by *Nitrosomonas*. Second, nitrite is oxidized to nitrate by *Nitrobacter*. The optimum pH range for nitrification is 6.7–7.0 (Groeneweg et al. 1994). The oxygen concentration for nitrification has to be at least 2 mg/L. Heavy metal ions are toxic to nitrification at rather low concentrations. Toxic levels of about 0.2 mg/L are reported for chromium, nickel, and zinc. The nitrification process is sensitive to temperature, and an Arrhenius expression can be applied:

$$Kn \text{ (rate constant } 1/24 \text{ h)} = 0.18 * 1.128 \char94 (\text{temperature} - 15)$$

The denitrification takes place by a number of heterotrophic bacteria present in activated sludge. Anaerobic conditions are absolutely required.

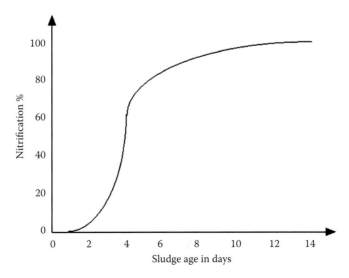

FIGURE 15.12
Nitrification efficiency as a function of sludge age.

The optimum pH for this process is about 7.0. About three times as much BOD_5 as nitrate-N both expressed in mg/L should be applied to ensure adequate denitrification. The process can be realized even at high concentrations of nitrate or at high salinities (Jørgensen 2000). Addition of organic carbon—for instance methanol—has been proposed to ensure a proper denitrification, but it is of course not an attractive method to apply. Alternatives are

1. To switch between aerobic and anaerobic conditions; it is called alternative operation (for more details see Diab et al. 1993, Halling-Sørensen and Jørgensen 1993)
2. To recycle a patty of the wastewater containing nitrate after the treatment is complete

Both methods are working properly. Other possibilities are the use of zeolite ion exchange material to obtain simultaneous nitrification and denitrification (Halling-Sørensen and Jørgensen 1993), the use of electrolysis for denitrification (Islam and Suidan 1998), and the use of activated carbon to enhance the denitrification (Sison et al. 1996).

Ammonia can be removed by blowing air through the wastewater. The process is called *stripping* and the equipment shown in Figure 15.13 can be used. It is of course required that the pH is sufficiently high to ensure that

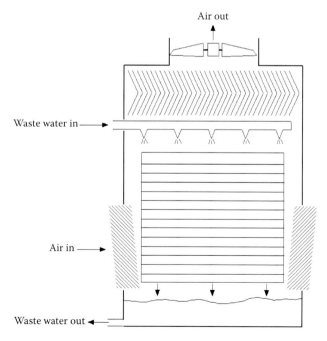

FIGURE 15.13
Sketch of a stripping tower is shown.

it is the gas ammonia that is present. The equilibrium between ammonium and ammonia can be expressed by the following equation:

$$pH = pK + \log \frac{[\text{ammonia}]}{[\text{ammonium}]}$$

where pK is 9.3.

The efficiency of the process is dependent on (1) pH, (2) temperature, (3) amount of air relative to the amount of water, and (4) stripping tower height. The Figures 15.14 and 15.15 show how these factors influence the efficiency.

The cost of stripping is relatively low, but the process has three crucial limitations:

1. It is practically impossible to work at temperatures lower than 7°.
2. Deposition of calcium carbonate can reduce the efficiency or even block the tower.
3. After the ammonia removal, pH is high and must be adjusted to 8.0 or below.

Ammonia can be removed by *chlorination and adsorption on activated carbon*. Chlorine can oxidize ammonia to NH_2Cl, $NHCl_2$, and NCl_3 and the

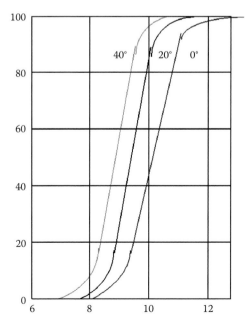

FIGURE 15.14
Stripping efficiency as a function of temperature and pH.

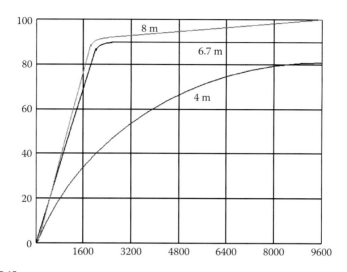

FIGURE 15.15
Efficiency of stripping process versus m³ air/m³ water ratio for three different tower depths and tower height. The three tower depths are 4 m (insufficient), 6.7 m, and 8 m. Tower height of 7–8 m is recommended.

activated carbon is able to adsorb the chloramines formed by these oxidation processes. By the adsorption process, it is most likely that dinitrogen and chloride ions are formed. In the order of ten parts by weight of chlorine are required for each part of ammonium-N. As the wastewater may typically contain 30 mg ammonium-N/L, 300 mg/L of chlorine is required. It would give a cost of at least $0.1/m³ to cover only the chlorine. When the capital costs and the costs of recovery of activated carbon are added, the total costs per cubic meter may easily reach $0.3. The method has therefore not found a wide application due to the high costs. The method has however a high removal efficiency even if the ammonium concentration is low. Chlorination followed by active carbon adsorption can therefore be used as the last treatment for removal of ammonium, where very low concentrations are required in the effluent.

Ion exchange by the use of the natural clay material clinoptilolite can be used for removal of ammonium ions (Halling-Sørensen and Jørgensen 1993). Clinoptilolite has a high selectivity for ammonium ions. About one third to one half of the ion exchange capacity, which is about 1.5 eqv./L, can be obtained for uptake of ammonium ions. It means that 1 L of clinoptilolite will be able to remove $1.5 * 14/3$ g of ammonium-N = 7 g ammonium-N.

Ion exchange has furthermore been applied to remove nitrate from drinking water. A general anion exchanger is applied. It has an ion exchange capacity of about 2.5–3.0 eqv./L, but the selectivity is not very high. It is however not necessary to remove the nitrate ions completely but to reduce the concentrations to the regional standards for nitrate in drinking.

15.6.1 Recycling of Municipal Wastewater

By a combination of the available treatment methods, it is possible to recycle municipal wastewater—it means produce drinking water from municipal wastewater. The recycling takes about 8–10 days from toilet to water tap so to say. All drinking water is in principle recycled. It takes nature normally a couple of thousand years to recycle water in average, which of course much better ensures a good water quality than a recycling with a duration of only 8–10 days. Recycling of wastewater has been applied in Pretoria and South Africa and in Windhoek, Namibia, due to insufficient supply of natural water for production of potable water. In Pretoria, a treatment consisting of the following steps is applied: mechanical–biological treatment, aeration, lime precipitation, ammonia stripping, sand filtration, chlorination, adsorption on activated carbon, and a second chlorination. The plant in Windhoek has the same steps except no ammonia stripping.

References

Bernhardt, E.L. 1975. Nitrification in industrial treatment works. In: *Second International Congress on Industrial Waste Water and WASTES*, Stockholm, Sweden, pp. 272–299, February 1975.

Dharmappa, H.B. et al. 1993. Optimal design of a flocculator. *Water Res.* 27:513–519.

Diab, S. et al. 1993. Nitrification pattern in a fluctuating anaerobic-aerobic pond environment. *Water Res.* 27:1469–1475.

Groeneweg, J. et al. 1994. Ammonia oxidation by *Nitrosomonas* at ammonia concentrations near Km: Effects of pH and temperature. *Water Res.* 28:2561–2566.

Hahn, H.H. and Muller, N. 1995. Factors affecting water quality of (large) rivers past—Experiences and future outlook. In: Novotny, V. and Somlyódy, L. (Eds.), *Remediation and Management of Degraded River Basins with Emphasis on Central and Eastern Europe*. Springer Verlag, Berlin, Germany, pp. 385–426.

Halling-Sørensen, B. and Jørgensen, S.E. 1993. *The Removal of Nitrogen Compounds from Waste Water*. Elsevier, Amsterdam, the Netherlands, 444pp.

Henze, M. and Ødegaard, H. 1995. Wastewater treatment process development in Central and Eastern Europe-strategies for a stepwise development involving chemical and biological treatment. In: Novotny, V. and Somlyódy, L. (Eds.), *Remediation and Management of Degraded River Basins with Emphasis on Central and Eastern Europe*. Springer Verlag, Berlin, Germany, pp. 357–384.

Islam, S. and Suidan, M.T. 1998. Electrolytic denitrification. *Water Res.* 32:528–536.

Jørgensen, S.E. 2000. *Principles of Pollution Abatement*. Elsevier, Amsterdam, the Netherlands, 520pp.

Jørgensen, S.E. (Ed.). 2011. *Handbook of Ecological Models Used in Ecosystem and Environmental Management*. CRC Press, Boca Raton, FL, 620pp.

van Loosdrecht, M. 1998. Upgrading of waste water treatment process for integrated nutrient removal—The BCFS© process. *Water Sci. Technol.* 37:234–256.

Novotny, V. and Somlyódy, L. (Eds.). 1995. *Remediation and Management of Degraded River Basins with Emphasis on Central and Eastern Europe*. Springer Verlag, Berlin, Germany.

Sison, N.F. et al. 1996. Denitrification with external carbon source utilizing adsorption and desorption capability of activated carbon. *Water Res*. 30:217–227.

Sorial, G.A. et al. 1998. Evaluation of trickle-bed air bio filter performance for styrene removal. *Water Res*. 32:1593–1603.

Thomas, N.D. et al. 1999. Flocculation modelling: A review. *Water Res*. 33:1579–1592.

16

Application of Cleaner Technology in Environmental and Ecological Management

16.1 Introduction

Cleaner technology or production implies that changes in the processes are made within a production plant with the result that the emitted pollution is reduced. When the industries are forced to reduce their discharge of pollutants, they are of course considering either to apply a cost-moderate treatment method for their waste or the alternative—to change the production to application of methods that would imply a reduction of the discharge of pollutants. Major advantages are inherent in this approach that can benefit the producer. Aside from reducing the expenses of fees for creating pollution, considerable savings of energy, water, and various materials used in the production process may also be obtained. There are numerous examples on a change of the production methods—both major changes and minor changes—may imply not only reduced discharge of pollutants but also saving of production costs for instance by introduction of recycling. So, the interest in this method for the industries is primarily the possibilities to save considerable amounts of money, although its value in helping solve environmental problems also plays a role.

An even wider approach analyzes the whole production and application process, namely, the life cycle analysis of a product—as it is often expressed from cradle to grave. The production of a product is analyzed, with the goal of minimizing wastes over the entire production cycle, starting with the mining of raw materials through its use and finally to the destruction of the final product after use. This effort usually exceeds the capabilities of a given producer, demanding instead the synchronization of efforts by several enterprises. The environment has profited greatly in recent efforts of this kind, by wise and progressive industry leaders able to increase their competitive capabilities. It is often of great values for industries and enterprises to have a green image.

Industrial enterprises face increasing costs of raw materials, energy, water, and pollution fees and face increased pressure by environmental groups to reconsider their production methods and seek to retain competitiveness while also demonstrating good will toward environmental protection. In addition to automation, this goal can be obtained by the recirculation of materials and water inside the production plants, by energy saving, by better space organization of the process, and by the minimization of transport. All these elements are important for protecting water quality, since the processes of material extraction, transportation, and energy production are causes of pollution, and the quantity and degree of pollution of the water leaving the plant is decisive with regard to the water returned as effluent back to the plant. One scientifically rooted procedure with broad positive consequences for plant efficiency and competitiveness, as well as for water quality, that is favored by large international trusts is *the life cycle evaluation of products* or *environmental life cycle assessment* (Hauschild and Wenzel 2000).

The evaluation process in life cycle analysis consists of following a product throughout its entire life cycle, from its creation to its disposal. The production stages covered with this analysis include mining raw materials, processing for basic chemicals or metals, transportation, production of final materials needed for the product, production of the final product, packaging and distribution, the fate and environmental consequences of the product when it is used, and, finally, the disposal of the remains or unused parts (e.g., the fate of packing materials, nonfunctioning parts, or nonrepairable damaged products). Each step is evaluated with regard to the economy of the product, including the needs for materials, energy, and water, and the environmental consequences of decisions at each step are estimated, as a means of identifying the cheapest, most efficient, and least polluting option. Many factories that have implemented this evaluation have found considerable savings and a resulting increase in competitiveness. This is typically accompanied by a considerable savings in water and energy resources and a corresponding reduction in pollution. Thus, it is hoped that this procedure will become widely used and will benefit water quality. Because of increased water circulation within a factory, the quantity of the effluents is reduced, too. Effective, specialized pretreatment can be used, sometimes with the regeneration of some substances, which further reduces the pollutant load to the effluent or city wastewater treatment plant.

Considerable water quality improvement can also be obtained by citizens in their daily life on the basis of this approach, including the following:

- Water quality managers must stress the usefulness of clean production and product life cycle analysis, working to facilitate such evaluations. Local water management councils can be very helpful in this direction.
- Saving energy has positive consequences for water quality, as energy generation causes environmental degradation, including water pollution.
- Saving of water in households improves water quality, since existing wastewater treatment plants function better with less entering wastes, and there is less need for upgrading, renovation, and new construction.

16.2 Application of Life Cycle Analyses and Cleaner Technology

Figure 16.1 shows the flows of material and energy in the history of a product (Jørgensen 2000), from raw material to final disposal as waste. P indicates the emission of point pollution and NP covers the nonpoint pollution.

The number of products in the modern technological society is not known exactly, but it is probably in the order of 10^8. All these products emit pollutants to the environment during their production, their transportation from producers to users, during their applications, and in the final disposal as waste. The core problem in environmental management is: How can we

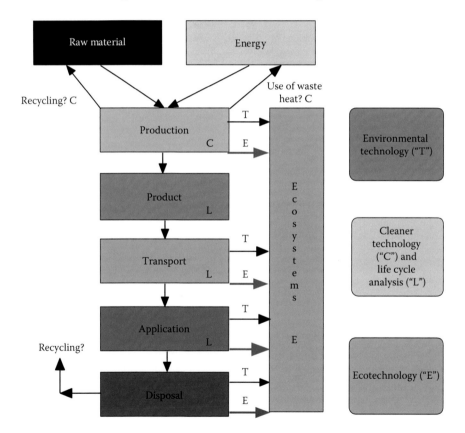

FIGURE 16.1
(See color insert.) Arrows show mass and energy flows. The thin black arrows are point sources and the thick grey arrows are nonpoint sources. The letters by the arrows indicate the possibilities to use environmental technology, T; cleaner technology, C; the life cycle approach, L; and the ecotechnology, E. Recycling possibilities are indicated—they belong to clear technology. Life cycle analyses are able to reveal where in the history of the product the waste actually takes place and it will make it possible to change the production and the product the recorded quantities of waste.

control these pollutants properly? The answer is that we have to use a wide spectrum of methods. Figure 16.1 illustrates where the different technologies can be applied. Environmental technology (Chapter 15) offers a wide spectrum of methods that are able to remove the pollutants from point sources. Cleaner technologies, however, explore the possibilities to recycle by-products and final waste or change completely the production methods to achieve a reduced overall emission. The key question here is: Could we produce our products by more environmentally friendly methods? The ISO 14000 series are among the most important tolls in the application of cleaner technology (Haucshild and Wenzel 2000).

Ecotechnology and its potential for implementation of environmental solution of nonpoint sources pollution will be presented in much more detail in the next chapter. Environmental legislation is hardly possible to cover in this volume, because the legislation is very different from country to country, although some very few general features of this tool were given in Chapter 14. We will, however, in the next section touch on the application of green taxes, because it will encourage the industries to apply cleaner technology to a higher extent. Green taxes could be called a powerful political instrument to reduce pollution.

16.3 Recycling and Reuse: Green Tax

Industries and farms apply recycling and reuse to the extent to which it is profitable, directly by reduction of costs or indirectly by the market value of a green image (Jørgensen 2006). Figure 16.2 illustrates the possibilities.

The profit, P, of the production is

$$P = S - RM - C - W \qquad (16.1)$$

where
  S is the sales price
  RM is the cost of raw material + energy
  C is the production costs
  W is the costs of treating the waste

Reuse is profitable if

$$U - O + W \geq 0 \qquad (16.2)$$

where
  U is the income by reuse sale
  O is the cost of making the reuse possible

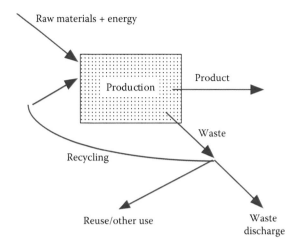

FIGURE 16.2
Production has almost always waste in addition to the product. The waste may be discharged to the environment, reused somewhere else, or recycled, which would save raw material and perhaps energy.

Recycling is profitable if

$$RC - F + W \geq 0 \qquad (16.3)$$

where
RC is the value of the recycled raw material
F is the recycling costs
W is as previously mentioned the costs of discharging the waste

In some cases, it may be possible to add the value of the green image to the left sides of (16.2) and (16.3).

Generally, the costs of discharging waste have increased considerably during the last decades, sometimes due to the introduction of "the polluter pays principle." Therefore, we have seen an increased interest by the companies and the agriculture to recycle or reuse. If we on top of the real costs would add a green tax, the motivation for recycling and reuse could even increase.

Income tax is the dominant tax form in many countries in the industrial world. It was introduced more than 100 years ago, when the state needed more money for the increasing demand for a series of tasks that we found it naturally that the modern state would undertake. The principle was in most cases that the biggest incomes should bear the biggest burden and the tax scale was therefore in most countries progressive. Labor is, however, not the limiting factor for our production and its continuation in the future. On the contrary—in most states, unemployment is a major problem. It is also absurd that the more you are working for the society, the more you are

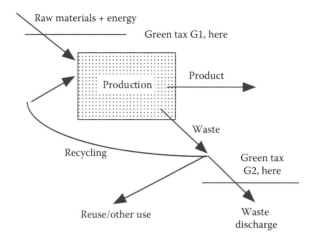

FIGURE 16.3
Production has almost always waste in addition to the product. The total global amount of raw materials, particularly if the raw materials are nonrenewable, are limited. Therefore, introduction of green tax on raw material (G1) and/or on waste discharge (G2) will enhance the motivation to recycle and reuse.

paying in tax, while the tax on capital income in most countries has a lower tax rate. Nonrenewable and partly renewable raw materials are or are going to be the limiting factor in our production. In addition, we have seen that as the gross national product (GNP) is increasing, the environment is gradually deteriorated due to increased discharge of waste. Recycling and reuse would solve these problems and both reuse and recycling require labor in one form or another, which implies that also the unemployment problem would be reduced by introduction of recycling and reuse (Figure 16.3).

We could enhance the application of recycling and reuse by increasing the costs of raw materials and of waste discharge by introduction of a green tax on raw materials and waste discharge as shown in Figure 16.2. If we add a green tax G1 on raw materials including nonrenewable energy sources and on waste discharge G2, the Equations 16.2 and 16.3 would change to the following:

Reuse is now profitable if

$$U - O + W + G2 \geq 0 \qquad (16.4)$$

Recycling becomes profitable if

$$RC - F + W + G1 + G2 \geq 0 \qquad (16.5)$$

Clearly, reuse and recycling can easily be made attractive by a suitable green tax G1 and G2.

Green tax should of course not be considered a new additional tax, but should replace the income tax. That may even reduce O and F in (16.4) and (16.5) and make it even more attractive to recycle and reuse, which is the motivation behind the introduction of the green tax. It is possible to find numerous examples of the positive effect of green tax or even of "the polluter pays principle." All industries in Denmark have reduced their relative water consumption considerably during the last decades due to a high cost for discharge of wastewater. Particularly the breweries have been able successfully to reduce their water consumption. Carlsberg uses, for instance, about three times less water per volume of beer brewed today than 30 years ago.

Europe and most countries all over the world have introduced a significant green tax on gasoline. The result is that the gasoline consumption per kilometer is considerably lower in these countries than in the United States where there is no corresponding green tax on gasoline. Denmark has, for instance, for a long time had a green tax on oil and electricity and the result is that the energy consumption per capita in Denmark is half the energy consumption in the United States although the United States and Denmark have approximately the same GNP/capita. So, clearly introduction of green tax implies a reduction of the consumption.

It is, however, necessary to introduce a green tax on *all* raw materials and on *all* discharges of waste to provoke a major increase of the sustainability. Many European countries have lately discussed the possibilities to decrease the income tax, because the income tax will inevitably decrease the employment. The effect on employment and sustainability would be even more pronounced by a transfer of tax from income tax to green tax. Most governments are, however, conservative and hesitate to make changes. The reduction of income tax and the increase of green taxes can, however, be made gradually over a period of 10–15 years, for instance, to avoid the changes that would impose significant net income changes for a part of the population, which may be the result of very major changes in the tax structure.

When it becomes more attractive to recycle or reuse, industrial cooperative network will inevitably be formed to reduce the costs. Self-organization and self-regulation by the industries and the enterprises will lead to recycling, reuse, and formation of industrial networks with increased possibilities of reuse and recycling.

When the society has reached a certain level of development, it is important that the further development is focusing on development of better communication and cooperation networks, more information and knowledge, and better exchange of the information and the know-how. This is characteristic for what we denote the information society.

So, the conclusion is that cleaner technology has an enormous potential as pollution reducing factor and the potential is reinforced considerably by introduction of green taxes—it means by a political involvement using economic factors is slightly better. The industries will inevitably use the possibilities of all available technologies, environmental technology, ecotechnology, and cleaner technology, provided they can see the advantages of these applications by the reduction of the production cost, by reduction of waste fees, or by achievement of a more green image.

The possibilities of using cleaner technology in any scale in the industry are almost unlimited. The experience has shown that a systematic assessment of the mass and energy balance for an industry inevitably will lead to pollution reductions at no cost or even sometimes with economic benefit for the industry. Due to the wide spectrum of possibilities, it will not be possible to cover here all the possibilities of using cleaner technology in the industry.

The most significant possibilities to use cleaner technology in agriculture should, however, be mentioned:

1. Analysis of the nitrogen and phosphorous compounds in the soil to determine more accurately how much fertilizer is necessary to use and when it is most beneficial to use it

2. Use of biological methods to replace the use of pesticides

3. General use of organic farming

4. Controlled use of particularly selected medicine for domestic animals, for instance, antibiotics that will decompose fast in the soil and therefore be less harmful

5. Adjust the fodder composition to the optimum for the domestic animals and for the optimum composition of the manure to be used as natural fertilizer

The possibilities to recycle and reuse are enhanced if industrial or agricultural productions cooperate in a network. The important property known from ecosystem, operation in networks, can be implemented in industries by building a network of several industries. This would facilitate of course—as always for larger systems—the possibilities to find a better matter and energy (or rather work energy or exergy; see Chapter 18) efficiency, understood as the entire efficiency = sum of all outputs (products)/ the sum of all inputs. The use of industrial networks has been tested in the Danish town Kalundborg, and the network (see Jørgensen 2006 for further details) is shown in Figure 16.4. The network makes it possible to utilize what is waste in one production as raw material in other industries, which makes the use of industrial networks a strong application of the cleaner technology idea.

It is of course possible to use the network idea also in agriculture and an example from Germany will be shown as illustration of the use of cleaner technology in agriculture. The example is "die Hermannsdorfer Landwerkstätte in Glonn" (abbreviated as LHL), which was founded in 1986. It is situated close to München. It is a farm based on sustainability principles. The characteristics of

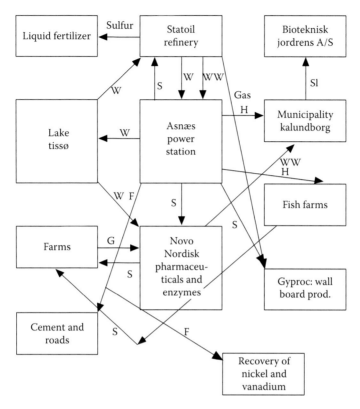

FIGURE 16.4
Industrial symbiosis at Kalundborg, Denmark. The flows of energy and matter make up an "ecological" network that implies that the overall efficiency of the utilization of the work energy = exergy input to the network is high. W, water; F, flyash; WW, waste water; S, steam; Sl, sludge; H, heat; G, grains.

the farm and calculation of the use of work energy (exergy; see Chapter 18 for more details, as exergy is used as an ecological indicator) of this agricultural system will be presented next according to Vestergaard (2005).

The characteristics of LHL can be summarized in the following points:

1. *The principles of organic farming are applied.*
2. *All distributions are locally ensuring that waste of energy for transportation and conservation is avoided* and all products are tasteful and fresh, when delivered to the customers.
3. *The local network of skilled workers, customers, and others is utilized.* Modern technology (provided nonlocally) is utilized but only if it has clear advantages. The various production units are linked in a symbiotic ecological network, which facilitates the possibilities to

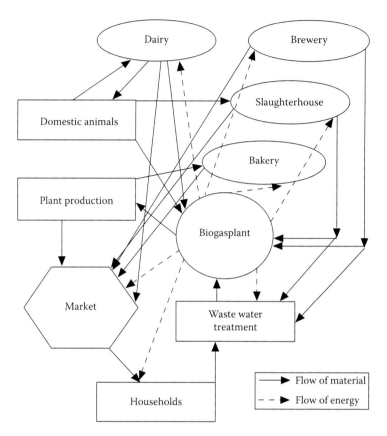

FIGURE 16.5
Material flows (full line arrows) and energy flows (dashed lines arrows) in the LHL. The figure illustrates that LHL forms a symbiotic ecological network that facilitates the possibilities to obtain high efficiencies for the use of material and direct exergy, that is, it yields a high overall exergy efficiency.

obtain a high efficiency of material and direct exergy (work energy) use; see Figure 16.5. The selection of products and their quality is a result of a dialogue between LHL and the consumers.

4. *A wide spectrum of animal and vegetable products is produced.*

5. *LHL encompasses a slaughterhouse, a bakery, a dairy, a brewery, and a restaurant.*

6. *LHL works with nature not against nature,* because it uses nitrogen fixation, hedgerows, and a natural mosaic of the landscape.

7. *The energy supply is based on biogas, wind energy, and sun energy. The water supply* is to a high extent covered by collection of rain water, and the wastewater is treated by constructed wetlands and reused to a high extent.

Vestergaard (2005) has shown how the eight below mentioned sustainability criteria are followed.

1. *The negative feedback principle. The use of fertilizers is limited to the available natural fertilizers* from the animals and nitrogen fixation. The animal production is entirely based on the feed available in the region.

2. *The production is reasonable constant.* No quantitative growth.

3. *The function is to deliver the desirable food items to the local population.* A dialogue with the local population ensures that the right products are produced.

4. *LHL works with nature not against nature, because it uses nitrogen fixation, hedgerows, and a natural mosaic in the landscape.*

5. *The multiple use is applied in the energy supply.* Waste heat from power stations and from refrigerating plants is used for room heating and collection of rain water is used for watering cattle and for irrigation.

6. *Recycling is applied to the highest possible extent.* Any waste product is considered a raw material and is wherever possible applied in the production. All the waste products from the slaughterhouse, the brewery, the dairy, and the restaurant are used as animal food or fertilizers. The constructed wetland used for wastewater treatment is harvested. Solid waste is sorted and used in the power plant as fuel, as animal food, as fertilizer, or reused directly or used for biogas production.

7. *Symbiosis. The animal production is of course in symbiosis with the plant production by use of manure as fertilizer and the plants as food for the animals.* The waste from the slaughterhouse, the brewery, and the dairy is used for biogas production that is applied for electricity production in the power plant. The residues from the biogas are used as fertilizers (compare with the Chinese agriculture presented and discussed in Chapter 7). The production units are linked in a symbiotic ecological network that yields a higher efficiency for the use of exergy and matter than it is possible for the individual production units. This is shown in Figure 16.5. It is attempted as it is shown to close all cycles of matter and energy.

8. *LHL works like nature and in harmony with nature, because it uses nitrogen fixation, hedgerows, and a natural mosaic of the landscape.* LHL has high diversity and the components are working symbiotically. A biological design is to a certain extent applied.

Vestergaard (2005) has calculated the exergy efficiency = exergy of products/ exergy of inputs and resources efficiency = exergy of products/emergy used. Emergy is the amount of solar energy used (see Chapter 18 for further information). The results are shown in Table 16.1. The table contains also information about the percentage of animal production, because the production of animal products is not surprisingly more energy expensive than plant

TABLE 16.1

Comparison LHL with IEA
(Industrialized European Agriculture)

Factor	LHL	LDK
Exergy efficiency	2.64	0.3
Resource efficiency	3.8×10^{-5}	1.6×10^{-6}
% Animal products	29.2	33.6

products. IEA indicates for comparison the same efficiencies for a normal western European (industrialized) agriculture. It has a slightly higher animal production than LHL, but far from sufficient to explain the enormous difference in efficiencies. A comparison of exergy stored/emergy input for different ecosystems shows that nature is much more effective to store exergy in its structure than man-controlled or man-influenced systems per unit of solar energy (Bastianoni and Marcehttini 1997).

The difference in exergy efficiency between LHL and LDK (almost a factor 9!) is explained by the use of artificial fertilizers, pesticides, import of animal food, use of ground water, and a significant transportation of the products for LDK.

Generally, it is possible to conclude from this short overview of cleaner technology that it offers many possibilities to reduce pollution problems of aquatic ecosystems by cost moderate methods.

References

Bastianoni, S. and Marchettini, N. 1997. Emergy/exergy ratio as a measure of the level of organization of systems. *Ecol. Model.* 99:33–40.

Hauschild, M. and Wenzel, H. 2000. Life cycle assessment—Environmental assessment of products. In: Jørgensen, S.E. (Ed.), *A System Approach to the Environmental Analysis of Pollution Minimization.* Lewis Publications, Boca Raton, FL, pp. 155–190, 256pp.

Jørgensen, S.E. 2000. *Principles of Pollution Abatement.* Elsevier, Amsterdam, the Netherlands, 520pp.

Jørgensen, S.E. 2006. *Eco-Exergy as Sustainability.* WIT, Southampton, U.K., 220pp.

Vestergaard, B. 2005. Sensitivitets modellen (The sensitivity model). Master thesis, Roskilde University Center, Roskilde, Denmark, 92pp.

17

Application of Ecotechnology in Ecosystem Management of Inland Waters

17.1 Ecotechnology: Definition and Classification

What is *ecotechnology or ecological engineering?* Mitsch and Jørgensen (2003) defined *ecological engineering* as *the design of sustainable ecosystems that integrate human society with its natural environment for the benefit of both.* It requires, on the one hand, that we understand nature and ensure a sustainable development of natural resources and ecosystems and, on the other hand, that we make use (but not abuse) of natural resources to the benefit of the human society. Thus, our inevitable interactions with nature must be made under the comprehensive consideration of the sustainability and balance of nature.

Ecotechnic is another often applied word, but which also encompasses the development of all types of "soft" technology applied in society, in addition to ecotechnology or ecological engineering. These types of technology are often based on ecological principles (e.g., all types of cleaner technology), particularly if they are applied to solve an environmental problem. The use of ecological principles in the development of technology is denoted as industrial ecology. Recently UNEP and UNESCO have introduced two other terms relevant to this discussion:

- *Phytoremediation*—the use of plants in ecological engineering (e.g., using wetlands to treat wastewater pollutants or for removing toxic substance from contaminated soil)
- *Ecohydrology*—the use of a combination of ecological and hydrological principles to obtain ecologically sound environmental management

Both phytoremediation and ecohydrology are subdisciplines within the discipline ecological engineering—or ecotechnology, which is often used synonymously with ecological engineering.

Ecological engineering may be based on one or more of the following four classes of ecotechnology:

1. Ecosystems are used to reduce or solve a pollution problem that otherwise would be (more) harmful to other ecosystems. A typical example is the use of wetlands for wastewater treatment.
2. Ecosystems are imitated or copied to reduce or solve a pollution problem, leading to constructed ecosystems. Examples are fish-ponds and constructed wetlands for treating wastewater or diffuse pollution sources.
3. The recovery of ecosystems after significant disturbances. Examples are coal mine reclamation and restoration of lakes and rivers.
4. The use of ecosystems for the benefit of humanity without destroying the ecological balance (i.e., the utilization of ecosystems on an ecologically sound basis). Typical examples are the use of integrated agriculture and development of organic agriculture; this type of ecotechnology finds wide application in the ecological management of renewable resources.

The rationale behind these four classes of ecotechnology is illustrated in Figure 17.1. It is noted that ecotechnology or ecological engineering operates in the environment and its ecosystems. It is in this domain that ecological engineering has its toolbox.

Section 17.2 is covering Classes 1 and 2, including the use of wetlands in ecotechnology (see, however, also Chapter 6), while Section 17.3 presents Class 3. Class 4 is at least partially covered under cleaner technology.

Illustrative examples of all four classes of ecological engineering may be found in situations where ecological engineering is applied to replace environmental engineering or environmental technology, mainly because the ecological engineering methods offer often an ecologically more acceptable

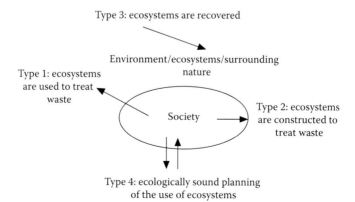

FIGURE 17.1
Illustration of the four types of ecological engineering.

solution than environmental technology. There are, however, also situations where ecological engineering is the only method that can offer a proper solution to a problem. This does not imply that ecological engineering should replace environmental engineering completely. On the contrary, the two technologies should work hand-in-hand to solve environmental management problems, better than they could do if applied individually.

Examples of ecotechnology/ecological engineering of Classes 1 and 2 are given to illustrate the concepts.

Class 1 may be illustrated by wetlands utilized to reduce diffuse nutrient loads to lakes. This problem cannot be solved by environmental technology.

The application of constructed wetlands to cope with diffuse pollution is a good example of ecological engineering Class 2. Again, this problem cannot be solved by environmental technology. The application of subsurface wetlands also denoted root zone plants for treating small quantities of wastewater is another example of Class 2 ecological engineering, in which the environmental technological solution is replaced by an ecotechnological solution.

Sound solutions in environmental management of lakes and reservoirs will often imply the use of a combination of environmental and ecological technology. The first tool is often used to solve the problem of point pollution and the second to solve the problem of nonpoint pollution and for restoration of lakes and reservoirs. Figure 17.2 gives an example of the use

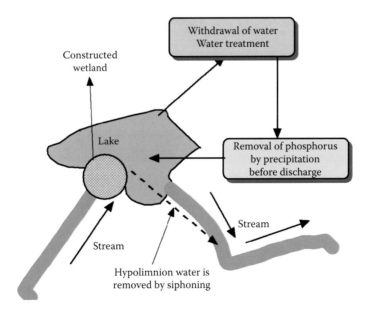

FIGURE 17.2
(See color insert.) Control of lake eutrophication, illustrating a combination of chemical precipitation for phosphorus removal from wastewater (environmental technology), a wetland to remove nutrients from the inflow (ecotechnology, Class 1 or 2), and siphoning of nutrient-rich hypolimnetic water downstream (ecotechnology Class 3).

of a combination of methods, namely, environmental technology for the treatment of wastewater and the use of a constructed wetland to reduce the nonpoint pollution from agriculture and brought to the lake by a tributary. Simultaneously, restoration of the lake is applied. The restoration methods will as mentioned previously be covered in Section 17.3.

17.2 Application of Class 1 and Class 2 Methods in Ecotechnology

Wetlands are probably the ecosystems that have received most attention in ecological engineering efforts, for the following reasons:

1. Wetlands have been drained on a very large scale in Europe, North America, and in many other countries, mostly because of the expansion of agricultural land areas. It is recognized today that wetlands are important for reducing the probability for flooding, and as buffer zones between nature and human activities, including agriculture.

2. Constructed wetlands can be used to treat wastewater. This is of particular interest in developing countries because (1) these treatment plants are cost moderate, at least where the cost of land is modest; (2) they have higher pollutant removal efficiencies than constructed wetlands in the temperate zone, due to the warm climate characterizing many developing countries, where they have also been applied to treat wastewater from small villages far away from the main sewage system; and (3) they require only minor maintenance, because they are as natural systems that are self-designing and self-regulating.

3. According to ecological engineering principles, the use of wetlands—natural or constructed—is the best means of removing nonpoint source pollutants, particularly those originating from the agricultural use of fertilizers and pesticides.

Nonpoint or diffuse environmental pollutants will inevitably flow toward lakes, rivers, and/or coastal aquatic ecosystems. However, a transition zone (denoted ecotone) is able to transform and/or adsorb such pollutants partially or entirely (see Figure 6.2). Figure 17.3 shows a riparian wetland along a river in Malaysia. It is easy to understand that the dense vegetation along the river is an excellent buffer zone for pollutants coming from land. Thus, ecotones will significantly reduce the overall irreversible effects on the aquatic ecosystems. The most important processes—which also explain the benefits of using constructed wetlands for treating wastewater

FIGURE 17.3
(See color insert.) Riparian wetland in Malaysia. The dense vegetation along the river is adsorbing much of the pollution coming from land.

or for recovering previously drained wetlands—may be summarized as follows (reference again to Figure 6.1):

1. Nitrate is denitrified by the anaerobic conditions in the wetlands. Accumulated organic matter in the wetland converts nitrate to free nitrogen (e.g., see Chapter 6).

2. Clay mineral is able to adsorb ammonium and metal ions.

3. Organic matter is able to adsorb metal ions, pesticides, and phosphorus compounds. Metal ions form complexes with humic acids and other polymer organic substances, which significantly reduces the toxicity of these ions.

4. Biodegradable organic matter is decomposed aerobically or anaerobically by the microorganisms in the transition zone.

5. Pathogens are outcompeted by the natural microorganisms in the transition zone.

6. Macrophytes can take up heavy metals with high efficiency. Although other toxic substances may also be removed by macrophytes, it is not possible to provide any general rule regarding their removal efficiency.

7. Toxic organic compounds will be decomposed, to a certain extent, by anaerobic processes in wetlands, dependent on the biodegradability of the compounds and the wetland water retention time.

8. Phosphorus sorption by soil with a high total metal content. Among the four major metal ions (magnesium, calcium, iron, and aluminum), calcium has the strongest correlation to the phosphorus sorption capacity. High pH values also imply an increasing sorption capacity to soil with high metal content. This relationship between high calcium content and high pH on the one hand, and high phosphorus sorption capacity on the other hand, may be utilized in constructing artificial wetlands. It is often beneficial to transport a soil with high phosphorus sorption capacity to the wetland under construction, which may increase the phosphorus removal capacity by more than one magnitude (e.g., as much as 1–3 g P/kg soil can be obtained with 200–600 g of total metal/kg soil).

The denitrification potential of wetlands is often surprisingly high. As much as 2000–3000 kg of nitrate–nitrogen can be denitrified per ha of wetland per year, dependent on the hydraulic conditions (see Mitsch and Jørgensen 2003). This is of great importance for protecting lakes, since significant quantities of nitrate are generated and released from agricultural activities. As much as 100 kg nitrate-N/ha may be found in the drainage water from intensive agricultural areas. Since denitrification is accompanied by a stoichiometric oxidation of organic matter, significant amounts of organic matter also are removed by this process, which may make the agricultural land less fertile. The phosphorus bound as organic matter, or adsorbed to the organic matter, however, may be released by these processes (Jørgensen and Bendoricchio 2001). Thus, these processes should be examined carefully and quantitatively in each case, including consideration of whether the released phosphorus will flow toward a lake or toward groundwater. These two possibilities should be considered in developing management strategies.

It should be underlined that the adsorption capacity of the transition zone offers significant protection against pollution from toxic substances, including metals and toxic organic substances (primarily pesticides originating from agricultural activities). The ratio of the concentration of heavy metals or pesticides in organic matter to the concentration in water at equilibrium is strongly dependent on the composition of the organic matter. It is, however, usually between 50 and 5000, indicating the transition zone has an enormous binding capacity for these pollutants.

Ecotones serve as a buffer zone not only for pollutants but also for species present in the adjacent ecosystems. Thus, preservation of wetlands at the lake and river shoreline may be crucial for maintenance of biodiversity in a lake ecosystem—a function that should not be overlooked by a manager in developing an appropriate management strategy.

The importance of wetlands adjacent to aquatic ecosystems has resulted in the cessation of wetland drainage in many countries, as well as the restoration of previously drained wetlands. The construction of artificial wetlands offers a solution for addressing diffuse pollution originating from agriculture, septic tanks, and other sources (examples are given in Mitsch and Jørgensen 2003). Legislation in the United States, for example, does not allow wetlands to be drained, unless another wetland of the same size is installed elsewhere.

Construction of artificial wetlands is an attractive, cost-effective solution to pollution from diffuse sources and even wastewater. First of all, wetlands are able to cope with nitrogen and heavy metal pollution from these sources. It is essential, however, to ensure proper planning on placement of an artificial wetland, since their effectiveness depends on their hydrology (i.e., they should be covered by water most of the year and should have sufficient water retention time to allow them to take care of key pollution problems) and the landscape pattern (i.e., they should protect the most vulnerable ecosystems, which often are lakes and reservoirs). Thus, as previously mentioned, it is important to ensure that the wetlands do not release other components (e.g., phosphorus).

Emergent types of plants proposed for use in constructed wetlands include cattails, bulrush, reeds, rushes, papyrus, and sedges. Submerged species can be applied in deep-water zones. Species previously used for such purposes include coontail or horn wart, redhead grass, widgeon grass, wild celery, and water milfoil.

It also should be kept in mind that, in most cases, a wetland will reduce the water budget because of evapotranspiration (Mitsch and Jørgensen 2003). However, wetlands also reduce the wind speed at the water surface, thereby potentially also reducing the evaporation. It is important to consider these factors in planning for artificial wetlands. Finally, it should not be forgotten that development of an artificial wetland takes time. In most cases, it will require 2–4 years for an artificial wetland to obtain sufficient plant coverage and biodiversity to be fully operational. However, it is clear from the experience gained in the relatively few constructed wetlands that application of models for a wetland, encompassing all the processes reviewed earlier, as well as for the lake, is mandatory, if positive results are to be expected.

Wetlands encompassing the so-called root zone plants also may be utilized as wastewater treatment facilities. This application of soft technology seems particularly advantageous for application in developing countries, due to its moderate cost.

The self-purification ability of wetlands has found wide application as a wastewater treatment method in several developing countries (China, Philippines, Burma, India, and Thailand). Constructed wetlands in the tropic regions generally exhibit a higher efficiency than wetlands in the temperate regions, due mainly to the higher photosynthetic and microbiological activity of the former. In addition, seasonal variations in the pollutant removal capacity of constructed wetlands in the tropical regions are generally much smaller than for the temperate regions. Constructed wetlands in the northern part of Europe in the winter period (December–February), for example, have a capacity for removing nitrogen or COD (expressed as kg N or kg COD removed/ha 24 h), which may be less than 30% of the summer capacity. This implies that constructed wetlands in the temperate zone must be constructed with an overcapacity, relative to the average situation, in order to be able to meet the required efficiency during the winter period. These differences between constructed wetlands in tropic regions and temperate regions explain why it usually is more expensive to use constructed wetlands in the temperate zone.

The use of floating species of aquatic plants for wastewater treatment also has been proposed. Different types of duckweed and water hyacinths (*Eichhornia crassipes*) have been applied as an alternative to waste stabilization ponds. Use of water hyacinths, however, requires strict control, since they can easily be spread widely as a nuisance aquatic weed whose growth can get completely out of control.

Water hyacinths are able to absorb the inorganic nitrogen and phosphorus brought in by sewage and decomposed from organic pollutants by microorganisms. Experiments in China, for example, showed that an average yield of up to approximately $10 \, kg/m^2$ may be obtained during the growing period (May–November). Such production is able to absorb 1500–3600 kg of nitrogen, 150–500 kg of phosphorus and 100–250 kg of sulfur/ha (see Mitsch and Jørgensen 2003). The water hyacinths, with microorganisms and organic pollutants attached or coagulated on the root surfaces, are harvested to serve as feed in fish culture ponds, duck farms, pig farms, and oxen farms. During May–November, the concentrations of COD, total nitrogen, ammonium, total phosphorus, and orthophosphate were less than half the concentrations in the inlet. On the other hand, when the water hyacinths were absent during December–April, the differences in the COD, nitrogen, and phosphorus concentrations between the inlet and outlet were very small. In tropical areas, it may be possible to obtain the removal efficiencies indicated earlier for a wetland with floating plants.

Models are widely used to design surface and subsurface wetlands, or to recover wetlands. It is recommended, however, that before a model is used, some "back of the envelop calculations" be made to get a first estimation of the wetland capacity for removing various pollutants, particularly nutrients and organic matter.

The following equation may be applied to surface wetlands to estimate the nitrogen removal (Jørgensen 2000):

$$\frac{N}{No} = \exp(-kt) \tag{17.1}$$

where
  N is the nitrogen concentration in the effluent
  No is the nitrogen concentration in the influent
  k is the rate constant (typically 0.2–0.6 1/24 h) dependent on the plant density (i.e., 0.2 at 3 t/ha, 0.5 at 10 t/ha, and 0.6 at 20 t/ha)
  t is the retention time (days)

Equation (17.1) assumes harvest of the plants on at least an annual basis. The removal of BOD_5 can be estimated at 50 kg–250 kg/ha, dependent on the plant density (from 3 to 20 t/ha). The phosphorus removal is about 5–10 times lower than the nitrogen removal, based on at least an annual plant harvest. The equations referred to have been developed for water hyacinth–based surface wetlands in tropic regions. However, they also can be used—as they only give a first estimation—for other types of surface wetlands in the tropical region. In the temperate zone, the estimations should be reduced on the basis of the climate (i.e., the number of months of the plant growing season).

Many heavy metals present in very low concentrations in water are concentrated and accumulated in water hyacinths. However, the heavy metal enrichment in this plant varies with its aquatic habitat. Water hyacinths with high residual quantities of heavy metals obviously cannot be used as fodder, limiting the application of this ecological engineering approach for treating water polluted by organic pollutants (mainly municipal wastewater).

The root zone plant or subsurface wetland has found its application in treating small volumes of municipal wastewater far from the main sewage system in industrialized countries, or for more general wastewater treatment in developing countries. The decomposition of organic matter and denitrification usually do not cause any problems, provided the plant is 2–10 m² per person equivalent, dependent on the climatic conditions. Phosphorus is only removed with an efficiency not exceeding 10%–20% by a subsurface wetland. However, with the addition of iron chloride, the removal efficiency may be increased to 80% or more, due to precipitation of iron phosphate. It is also possible to increase the phosphorus removal efficiency by using soil with a high phosphate adsorption capacity (e.g., calcium or iron rich soil).

Constructed subsurface wetlands or root zone plants are currently designed beneficially with the use of models that have been calibrated on the basis of observations from other constructed wetlands in the same climatic conditions. A rough, first estimate of BOD-removal can be obtained with the following equations, which should/could be applied to get some idea of the needed area before the model is applied (Jørgensen 2000):

$$\frac{C}{Co} = \exp\left(\frac{-K*V}{Q}\right) \tag{17.2}$$

$$Qw + p - b*d*HC*i \tag{17.3}$$

where
   C and Co are the BOD_5 or COD concentration of the inflowing water and
      treated water, respectively
   K is the rate constant (1/24 h)
   V is the volume (m³) of plant (depth 0.4–1.0 m)
   Q is the waste flow expressed in m³/24 h
   Qw + p is the flow rate of wastewater plus rain, often estimated to be three
      to four times Q

b is the width of the plant
d is the depth (0.5–1.0 m)
HC is the hydraulic conductivity of the soil
i is the slope in cm/m

K is estimated to be 0.2–0.5 at approximately 20°C for root zone plants and constructed wetlands. K at temperature t can be estimated from the following equation:

$$Kt - K20 * 1.06^{(t-20)} \tag{17.4}$$

17.3 Ecotechnology: Restoration Methods

The most important restoration methods are listed next with a brief description of their application, advantages, and disadvantages. Most restoration methods are applied on lakes and reservoirs, although they also may be applied on other freshwater ecosystems.

1. *Diversion* of wastewater has been extensively used, often to replace wastewater treatment. Discharge of effluents into an ecosystem which is less susceptible than the one used at present is, as such, a sound principle, which under all circumstances should be considered, but quantification of all the consequences has often been omitted. Diversion might reduce the number of steps in the treatment but cannot replace wastewater treatment totally, as discharge of effluents, even to the sea, always should require at least mechanical treatment to eliminate suspended matter. Diversion has often been used with a positive effect when eutrophication of a lake has been the dominant problem. Canalization, either to the sea or to the lake outlet, has been used as solution in many cases of eutrophication. However, effluents must be considered as a freshwater resource. If it is discharged into the sea, effluent cannot be recovered; if it is stored in a lake, after sufficient treatment of course, it is still a potential water resource. It is far cheaper to purify eutrophic lake water to an acceptable drinking water standard than to desalinate seawater.

   Diversion is often the only possibility when a *massive* discharge of effluents goes into a susceptible aquatic ecosystem (a lake, a river, a fjord, or a bay). The general trend has been toward the construction of larger and larger wastewater plants, but this is quite often an ecologically unsound solution. Even though the wastewater has received multistep treatment, it will often still have a large amount of pollutants relative to what the ecosystem can absorb, and the more massive the discharge is at one point, the greater the environmental impact will be. If it is considered that the canalization is often a significant part of the overall cost of handling wastewater, it might often turn out to be a both better and cheaper solution to have smaller treatment units with individual discharge points. Although diversion is not considered an ecotechnological method based on sound ecological principles, a number of successful applications of diversion has been reported in the limnological literature. The most frequently quoted case is probably the restoration of Lake Washington resulting from diversion of the wastewater.

2. *Removal of superficial sediment* can be used to support the recovery process of very eutrophic lakes and of areas contaminated by toxic substances (for instance, harbors). This method can only be applied with great care in small ecosystems due to the stirring up of suspended matter. Sediments have a high concentration of nutrients and many toxic substances, including trace metals. If a wastewater treatment scheme is initiated, the storage of nutrients and toxic substances in the sediment might prevent recovery of the ecosystem due to exchange processes between sediment and water. Anaerobic conditions might even accelerate these exchange processes; this is often observed for phosphorus, as iron(III) phosphate reacts with sulfide and forms iron(II)-sulfide by release of phosphate. The amount of pollutants stored in the sediment is often very significant, as it reflects the discharge of untreated wastewater for the period prior to the introduction of a treatment scheme. Thus, even though the retention time of the water is moderate, it might still take a very long time for the ecosystem to recover.

The removal of sediment can be made mechanically or by use of pneumatic methods. The method is, however, costly to implement and has therefore been limited to smaller ecosystems. Maybe the best-known case of removal of superficial sediment is Lake Trummen in Sweden—40 cm of the superficial sediment was removed. The transparency of the lake was improved considerably but decreased again due to the phosphorus in overflows from storm-water basins. Probably, treatment of the overflow after the removal of superficial sediment would have been needed.

3. *Uprooting and removal of macrophytes* have been widely used in streams and also to a certain extent in reservoirs, where macrophytes have caused problems in the turbines. The method can, in principle, be used wherever macrophytes are a significant result of eutrophication. A mass balance should always be set up to evaluate the significance of the method compared with the total nutrient input. Collection of the plant fragments should be considered under

all circumstances. A simultaneous removal of nutrients from effluents should also be considered.

4. *Coverage of sediment by an inert material* is an alternative to removal of superficial sediment. The idea is to prevent the exchange of nutrients (or maybe toxic substances) between sediment and water. Polyethylene, polypropylene, fiberglass screen, or clay is used to cover the sediment surface. The general applicability of the method is limited due to the high costs, even though it might be more moderate in cost than removal of superficial sediment. It has only been used in a few cases and a more general evaluation of the method is still lacking.

5. *Siphoning of hypolimnetic water* is more moderate in cost than Methods 2 and 4. It can be used over a longer period and thereby gives a pronounced overall effect. The effect is dependent on a significant difference between the nutrient concentrations in the epilimnion and the hypolimnion, which, however, is often the case if the lake or the reservoir has a pronounced thermocline. This implies, on the other hand, that the method will only have an effect during the period of the year when a thermocline is present (in many temperate lakes from May to October/November), but as the hypolimnetic water *might* have a concentration fivefold or higher than the epilimnetic water, it might have a significant influence on the nutrients budget to apply the method anyhow. The method is used in combination with environmental technology and wetlands in Figure 17.2.

   If there are lakes or reservoirs downstream, the method cannot be used, as it only removes, but does not solve the problem. A possibility in such cases would be to remove phosphorus from the hypolimnetic water before it is discharged downstream. The low concentration of phosphorus in hypolimnetic water (maybe 0.5–1.0 mg/L) compared with wastewater makes it almost impossible to apply chemical precipitation. However, it will be feasible to use ion exchange, because the capacity of an ion exchanger is more dependent on the total amount of phosphorus removed and the flow than on the total volume of water treated. Figure 17.4 illustrates the use of siphoning for restoration of Lake Bled in Slovenia. Figure 17.5 shows a schematic drawing of siphoning hypolimnic water followed by removal of phosphate from the hypolimnic water by ion exchange/adsorption.

   Several lakes have been restored by this method, mainly in Austria, Slovenia, and Switzerland, with significant decrease of the phosphorus concentration as a result. Generally, the decline in the total phosphorus concentration in epilimnion is proportional to the amount of total phosphorus removed by siphoning and to the time the process has been used. The method has relatively low costs and is relatively

FIGURE 17.4
(See color insert.) Siphoning of hypolimnic water in the Lake Bled, Slovenia.

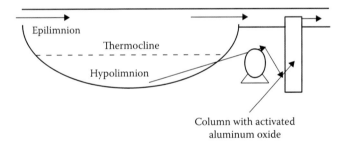

FIGURE 17.5
Application of siphoning and ion exchange of hypolimnetic water. The dotted line indicates the thermocline. The hypolimnetic water is treated by activated aluminum oxide to remove phosphorus.

effective, but the phosphorus must of course as already underlined be removed from hypolimnetic water before it is discharged, if there are other lakes downstream.

6. *Flocculation* of phosphorus in a lake or reservoir is another alternative. Either aluminum sulfate or iron(III)-chloride can be used. Calcium hydroxide cannot be used, even though it is an excellent precipitant for wastewater, as its effect is pH-dependent and a pH of 9.5 or higher is required. The method is not generally recommended as (1) it is not certain that all flocs will settle and thereby incorporate the phosphorus in the sediment and (2) the phosphorus might be released from the sediment again at a later stage.

7. *Circulation* of water can be used to break down the thermocline. This might prevent the formation of anaerobic zones, and thereby the release of phosphorus from sediment.

8. *Aeration* of the lake sediment is a more direct method to prevent anaerobic conditions from occurring. Aeration of highly polluted rivers and streams has also been used to avoid anaerobic conditions. Pure oxygen has been used in the Danish Lake Hald and in Lake Fure close to Copenhagen, instead of air. The water quality of the lake has been permanently improved since the oxygenation started.

9. *Regulation of hydrology* has been extensively used to prevent floods. Lately, it has also been considered as a workable method to change the ecology of lakes, reservoirs, and wetlands. If the retention time in a lake or a reservoir is reduced with the same annual input of nutrients, eutrophication will decrease due to decreased nutrient concentrations. Another possibility is to use a variable retention with the shortest retention time from shortly before to shortly after the spring and summer blooms. In most cases, the eutrophication can be reduced in the order of 20%–35% by this method, but the result is case dependent. The role of the depth, which can be regulated by use of a dam, is more complex. Increased depth has a positive effect on the reduction of eutrophication, but if the retention time is increased simultaneously, the overall effect cannot generally be quantified without the use of a model. The productivity of wetlands is highly dependent on the water level, which makes it highly feasible to control a wetland ecosystem by this method.

10. *Application of wetlands or impoundments as nutrient traps* in front of a lake could be considered as an applicable method, wherever the nonpoint sources are significant. It is known that wetlands effectively remove nitrogen by denitrification. Removal of phosphorus by adsorption is also a possibility. Both nitrogen and phosphorus can be removed more effectively if the wetland is harvested at the fall.

11. *Shading by use of trees* at the shoreline is a cost-effective method, which, however, only can give an acceptable result for small lakes due to their low area/circumference ratio.

12. *Biomanipulation* can only be used in the phosphorus concentration range from about 50 to 130 µg/L dependent on the lake. In this range, two ecological structures are possible. This is illustrated in Figure 17.5. When the phosphorus concentration initially is low and increases, zooplankton is able to maintain a relatively low phytoplankton concentration by grazing. Carnivorous fish is also able to maintain a low concentration of planktivorous fish which implies relatively low predation on zooplankton. At a certain phosphorus concentration (about 120–150 µg/L), zooplankton is not any longer

able to control the phytoplankton concentration by grazing and as the carnivorous fish (for instance, Nile perch or pike) is hunting by the sight, and the turbidity increases, the planktivorous fish become more abundant, which involves more pronounced predation on zooplankton. In other words, the structure is changed from control by zooplankton and carnivorous fish to control by phytoplankton and planktivorous fish. When the phosphorus concentration decreases from a high concentration the ecological structure is initially dominated by phytoplankton and planktivorous fish. This structure can, however, be maintained until the phosphorus concentration is reduced to about 50 μg/L. There are therefore two possible ecological structures in the phosphorus range of approximately 50–130 μg/L. Biomanipulation (de Bernardi and Giussani 1995) can be used in this range—and only in this range—to make a "short cut" by removal of planktivorous fish and release carnivorous fish. If biomanipulation is used above 130 μg P/L, some intermediate improvement of the water quality will usually be observed, but the lake will sooner or later get the ecological structure corresponding to the high phosphorus concentration, that is, a structure controlled by phytoplankton and planktivorous fish. Biomanipulation is a relatively cheap and effective method provided that it is applied in the phosphorus range where two ecological structures are possible. de Bernardi and Giussani (1995) give a comprehensive presentation of various aspects of biomanipulation. Simultaneously, biomanipulation makes it possible to maintain relatively high biodiversity which does not change the stability of the system, but a higher biodiversity gives the ecosystem a greater ability to meet future, unforeseen changes without changes in the ecosystem function.

There is a number of cases where biomanipulation has been successful, but only if the phosphorus loading was reduced simultaneously and the total phosphorus concentration is below 130 μg/L, as mentioned earlier. Biomanipulation is denoted a top-down approach, because a regulation of the top trophic level, the carnivorous fish, is assumed to an effect on the first trophic level, the phytoplankton. This is in contrast to the bottom-up approach, which is the regulation of the nutrient concentration by the use of wastewater treatment to reduce the discharge of nutrients. The two approaches, both bottom-up and top-down, have to been applied simultaneously in most cases of practical environmental management. This statement is in most cases completely in accordance with the results presented in Figure 17.6.

The results in this figure can be explained theoretically by use of eco-exergy as goal function in a eutrophication model (Jørgensen and de Bernardi 1998). Lake Annone, Italy, and Lake Søbygaard,

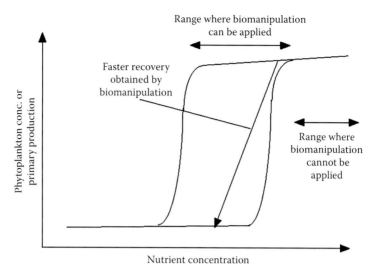

FIGURE 17.6
Hysteresis relation between nutrient level and eutrophication measured by the phytoplankton concentration is shown. The possible effect of biomanipulation is shown. An effect of biomanipulation can only be expected in the range approximately 50–150 μg P/L. Biomanipulation can hardly be applied successfully above 150 μg/L (see also de Bernardi and Giussani 1995, Jørgensen and de Bernardi 1998).

Denmark, represent two cases, where clear improvements in algal biomass and total phosphorus were observed. The zooplankton concentration increased significantly in both cases and simultaneously with a decrease in the phytoplankton concentration. Both cases were modeled by a structurally dynamic model (see Jørgensen and de Bernardi 1998, Jørgensen 2002).

For shallow lakes, a corresponding hysteresis takes place between submerged vegetation and phytoplankton (Scheffer et al. 2001). The two structures are between about 0.1 mg and 0.25 mg P/L both a possible solution to the prevailing conditions, provided that phosphorus is the limiting element for plant growth. Submerged vegetation is preferable, because it provides what is denoted as a clear water stage (see Figure 17.7). The selection of structure is dependent on the history (Zhang et al. 2003). A eutrophied lake with phytoplankton blooms and a concentration between 0.1 mg and 0.25 mg P/L can therefore be recovered by planting submerged vegetation.

13. *Biological control methods.* Water hyacinths and other macrophytes are pests in many tropic lakes, reservoirs, and lagoons. Many methods have been tested to abate this pollution problem. The best method tested up to now seems to be the use of beetles, that is, a biological control method. The method has given at least partial success in Lake Victoria.

FIGURE 17.7
(See color insert.) Shallow lake with dominance of submerged vegetation is named a clear water stage.

Biological control has also been used as removal process for heavy metals. Freshwater mussels can be applied for cadmium clearance. *Lemna trisulca* is able to accumulate as much as 3.8 mg cadmium/g dry weight.

Previously, algicides were applied to reduce the eutrophication of lakes. It is not recommended today due to the toxic effect of algicides. In this context, it should also be mentioned that addition of calcium hydroxide to lakes and rivers has been applied as restoration method for the adjustment of pH in acidified lakes, particularly in the regions, where acid rain has caused a significant decline of the pH.

17.3.1 Selection of Restoration Methods

It is not possible to give general recommendations to which restoration to apply in a specific case. Most restoration problems are associated with eutrophication, and it is necessary in each individual case to use a eutrophication model to assess the effect of the restoration method and compare the effects and the costs to decide which method gives most "pollution abatement" for the money. It is with other words necessary to set up a cost/benefit analysis.

The following modifications in the eutrophication model must be carried out to account for effect resulting from the application of the restoration method:

1. *Diversion*: the forcing functions (a) input of nutrients but also (b) the hydraulic retention time will be changed.
2. Removal of superficial sediment implies that the sediment contains less phosphorus and nitrogen, which will of course change the release rate of these nutrients from the sediment to the water phase.

3. Removal of macrophytes corresponds to a removal of the amount of phosphorus and nitrogen in the harvested plants.

4. Coverage of sediment by inert material will have the same effect as (2) but will in many cases be more cost moderate particularly for deeper lakes.

5. Siphoning of hypolimnic water corresponds in the model to removal of more nutrient (the concentration in hypolimnion to replace the concentration in epilimnion) with the out flowing water. It is of course necessary to examine what the effect of the higher nutrient concentration will be down streams. If there are other lakes downstream it is inevitably that the nutrient must be removed which is possible by a number of methods. For instance, for phosphorus can be used adsorption on activated aluminum oxide and/or precipitation of phosphate with aluminum sulfate, iron(III) chloride, or polyaluminates (see Figures 17.2 and 17.5). This method is obviously only applicable to lakes with a thermocline at least for part of the year.

6. Flocculation of phosphorus in the water phase implies that the phosphorus is once removed from the water phase to the sediment. Usually, it is necessary to apply this method several times.

7. and 8. Circulation and aeration of hypolimnion implies that the release rate of phosphorus and nitrogen from the sediment to the water phase is changed. Aerobic conditions usually imply that the release rate is lower particularly for phosphorus than under anaerobic conditions.

9. Changes of the hydrology mean that the forcing function hydraulic retention time in the model becomes shorter.

10. Construction of a wetland to cope with the nonpoint pollution of nutrients implies that the input of nutrients is reduced corresponding to the removal of nutrients by the wetland. The forcing functions in the model expressing the input of nutrients are changed correspondingly.

11. Shading by the use of trees change the photosynthetic activity in the lake. The forcing function, solar radiation, in the model is reduced corresponding to the shading effect.

12. Biomanipulation is often a cost-moderate method with a good effect, provided as mentioned previously that the phosphorus concentration is in the range about 50–130 µg P/L when removal of planktivorous fish is actual and in the range of about 100–250 µg P/L when plantation of submerged vegetation is actual.

Numerous eutrophication models have been published in the literature (see Jørgensen and Bendoricchio 2001, Jørgensen 2011, Jørgensen and Fath 2011). It is of course necessary to apply a model with a complexity that is balanced with the problem, the ecosystem and the data.

17.4 Selection of the Most Appropriate Methods or Combination of Methods in Environmental and Ecological Management: A Conclusion of Chapters 14 through 17

It is presumed, as previously pointed out, that the problems associated with the freshwater ecosystem have been defined not only qualitatively but also quantitatively and all sources of pollution have been determined and measured. Moreover, it is assumed that a model or at least mass balance considerations have been applied to assess how much the focal pollutants have to be reduced to achieve the desirable water quality preferably by application of ecological indicators. It is recommended to go through all the possible methods mentioned in this chapter and in the two previous chapters and see which methods we could use to solve the defined problem(s). The list should of course include all methods mentioned in Chapters 14–17 including the reestablishment of wetlands, ecotones, and littoral zones. The next step would be to select not only the applicable methods but also the possible combinations of these methods, as a combination of methods often may offer an environmentally better or more cost-moderate solution. In this context, it is extremely important to include all the sources of pollution in the discussion about suitable solution methods. It is furthermore important that no methods or combinations of methods are excluded from the list before it is clear that they are not useful to solve the given pollution problem.

The described considerations will lead to a list of possible solutions and the next step is of course to calculate the costs of the solutions. Having all the needed information about the possible solutions, it is not only a question about selecting the most cost-moderate method, but it is necessary to answer for all the possible methods and combination of methods the following questions:

1. What are differences for the possible methods in their short-term and long-term effects? Time is an important factor as different methods may require different amounts of time to obtain an acceptable result.
2. Are there additional benefits by the various methods, such as for instance local interests?
3. Have the solution methods other socioeconomic implications? Which?
4. If the economic resources are limited, would it be possible to obtain almost the same effect by a considerably more cost-moderate method?
5. Are there other additional advantages and disadvantages associated with the proposed solutions?

After the discussion of these questions, it will be possible to conclude on which method or combination of methods would be advantageous to use.

17.5 Example to Illustrate the Selection of Pollution Abatement Methods

The treatment of the wastewater discharged to Lake Glumsø, situated 70 km south of Copenhagen, Denmark, had of course to follow the general legislation for the treatment of wastewater in Denmark and the European Union. Therefore, removal of 95% of the BOD_5, 90% of the phosphorus, and 85% of the nitrogen is established. The question was, however, which methods should be used to cope with the problems of nonpoint pollution to supplement the wastewater treatment. The drainage area of Lake Glumsø consists of mainly agricultural areas. The volume of the lake is about 1 million m^3 (depth 2 m and a surface area of about 500,000 m^2).

Table 17.1 gives the result of a comparison of the effect and cost of five ecotechnological methods, using the model presented in Jørgensen and Bendoricchio (2001) and Jørgensen and Fath (2011). The lake was very eutrophied but the wastewater has since 1983 been diverted to the River Sus, which is downstream the lake. Before 1983 was the primary production about 1000–1100 g C/m^2 year, while it was reduced to about 500 g C/m^2 year during the period 1983–1988. The lake has a retention time of about 6 months, which implies that the period 1983–1988 corresponds to about 10 times exchange of the water. The transparency was in the same period increased from 18 cm at spring and summer blooms to about 60 cm. Maximum chlorophyll a concentration was reduced from about 850 to 360 µg/L as a result of the diversion. Encouraged by these results, the community considers various restoration methods.

TABLE 17.1

Comparison of Restoration Methods[a]

Method	Effects Found by a Eutrophication Model		Investment Million $	Running Costs 1000 $/year
	Primary Prod. (g C/m^2 year)	Max. Chlorophyll a (µg/L)		
Without restoration	500	360	0	0
Coverage of sediment	320	350	1	0
Removal of sediment	320	350	3.5	0
Precipitation of P in lake	460	360	0.6	0
Wetland	210	270	1.0	15
Twenty-five percent reduction of retention time	400	350	0.6	20

[a] See also Mitsch and Jørgensen (2003).

The effects were compared by the applied eutrophication model. The result of this investigation is summarized in Table 17.1, where the effects of the third year after the restoration that has been applied are shown. Biomanipulation was of course not considered because the phosphorus concentration was not sufficiently low. The results clearly indicate that erection of a wetland would give the best results. It would be possible to increase the minimum transparency from 60 cm to about 0.9 m and reduce the primary production 210 g C/m² year from about 500 g C/m² year at a cost that is relatively moderate in an industrialized country.

17.6 Conclusions

A wide spectrum of possible solutions that can be recommended to reduce lake pollution problems, particularly eutrophication, is available today. In most cases, it is, however, necessary to apply a combination of methods that may be based on environmental legislation, cleaner technology, environmental technology, and ecotechnology. The selection of the best combination of all the available methods should be made on the basis of a quantification of the problem preferably by development of a well-balanced ecological model of the freshwater system (see Chapters 14 and 19).

References

de Bernardi, R. and Giussani, M. 1995. *Guideline of Lake and Management: Biomanipulation.* ILEC and UNEP, New York, 220pp.

Jørgensen, S.E. 2000. *Pollution Abatement in the 21st Century.* Elsevier, Amsterdam, the Netherlands, 488pp.

Jørgensen, S.E. 2002. *Integration of Ecosystem Theories: A Pattern.* Kluwer, Dordrecht, the Netherlands, 432pp.

Jørgensen, S.E. (Ed.). 2011. *Handbook of Ecological Models Used in Ecosystem and Environmental Management.* CRC Press, Boca Raton, FL, 620pp.

Jørgensen, S.E. and Bendoricchio, G. 2001. *Fundamentals of Ecological Modelling,* 3rd edn. Elsevier, Amsterdam, the Netherlands, 528pp.

Jørgensen, S.E. and de Bernardi, D. 1998. The use of structural dynamic models to explain successes and failures of biomanipulation. *Hydrobiologia* 359:1–12.

Jørgensen, S.E. and Fath, B. 2011. *Fundamentals of Ecological Modelling,* 4th edn. Elsevier, Amsterdam, the Netherlands, 400pp.

Mitsch, W.J. and Jørgensen, S.E. 2003. *Ecological Engineering and Ecosystem Restoration*. John Wiley & Sons, New York, 412pp.

Scheffer, M.; Carpenter, S.; Foley, J.A.; Folke, C.; and Walker, B. 2001. Castrophic change of ecosystems. *Nature*. 413:591–596.

Zhang, J.; Jørgensen, S.E.; Tan C.O.; and Beklioglu, M. 2003. Hysteresis in vegetation shift-lake Mogan Prognoses. *Ecol. Model*. 164:227–238.

18

Application of Ecological Indicators in Environmental Management of Freshwater Ecosystems

18.1 Introduction

If there is a conflict between different interests in the selection of a management strategy for an aquatic ecosystem, it is necessary to solve the possible conflicts by a brainstorming meeting. An alternative would be to develop a mediated or institutionalized model (see Chapter 19), which as a step in the development of the model presumes such brainstorming meetings. Models are synthesizing tools. Without a model, it will hardly be possible to overview the problems and the roots of the problems in the society in relation to the possible solutions, the development of the drainage area, and the weighting of different interests.

For all the possible solutions to the environmental problems, it is recommended to check if the following points are fully considered:

1. Are the water withdrawals and diversion managed?
2. Is water pollution controlled and prevented now and in the future?
3. Is the fishery managed sustainable?
4. Is the ecosystem biodiversity conserved?
5. Are invasive species controlled?
6. Are health risks prevented?
7. Are garbage and litter controlled?

Provided that all the needed information would be available, it should be possible to develop a model that is able to tell the users what management plan is best able to solve the problems to the highest possible extent under the given financial constraints. The model will be able to make prognoses for different management plans and for different future development (see Chapter 19). It would of course require a very careful selection of the model, which is discussed in Chapter 19 and in Jørgensen (2011) and Jørgensen and Fath (2011).

It is furthermore recommended to select a number of ecological indicators. The current use of indicators should be integrated in the management mechanism for the sustainable use of freshwater ecosystems and their resources. The model will be able to give the user information about the future values for the ecological indicators when a given plan is followed. By selection of good indicators, it is therefore possible to follow to what extent the management plan is successful, which enable a current adjustment of the management plan.

18.2 Tools to Synthesize and Overview: Models and Indicators

All aquatic ecosystems are complex systems and their management is a complex issue. It is therefore necessary to apply a synthesizing and overviewing tool to be able to come up with a close to optimal management strategy. Fortunately, mathematical models have been developed during the last three to four decades to the level of today, where the models can be used to get a good overview of the management conditions and at the same time—provided that a good database is available—set up prognoses for the application of different management strategies. Ecological/environmental models of today are powerful management tools and it can therefore be recommended to apply models to support the management decisions (Jørgensen 2011).

In the last 20 years or so, it has been increasingly common to apply ecological indicators to assess the ecosystem health. The idea is basically that the indicators summarize quantitatively the information about possible problems, for instance, for freshwater ecosystems: eutrophication, acidification, declining water level, toxic substances, endangered species, biodiversity, etc., and the general conditions of the focal ecosystem.

It is crucial for the success of the management that the selected model and the selected indicators fit to the available database, the problem, and the characteristics of the ecosystem (area, depth, retention time, and hydrodynamics). The selection of models and indicators is therefore an important step that requires contemplations.

Based upon the answers to the seven aforementioned questions, it should be possible to select a good model and a good set of indicators. Notice that it is hardly possible to get a good picture of ecosystem conditions by two to four indicators. In most cases, up to about 10 indicators are probably needed to assess the ecosystem health and to follow how the environmental strategy will be able to reduce or eliminate the problems as a function of time. In most cases, a selection of very specific indicators, for instance, concentration of a specific toxic substance or the concentration of chlorophyll a, is needed, but they should in most cases be supplemented with holistic or semiholistic indicators, for instance, biodiversity, annual production/biomass, or even thermodynamic indicators. The discussion on the seven questions listed previously will probably in many cases lead to the selection of a mediated or institutionalized model; see Chapter 19. The answers to these seven questions form the first stage of the development of a mediated model. It is important that a team of modelers, managers, and local scientists participate in the selection also of a standard model or an expert model and discuss which one to select among the several possible including which complexity to select and in this context also to decide on the need for a supplement of the available database.

18.3 Ecological Indicators: Application and Classification

The ecological indicators applied today in different context, for different ecosystems, and for different problems can be classified on eight levels from the most reductionistic to the most holistic indicators. The book *Handbook of Ecological Indicators for the Assessment of Ecosystem Health* by Jørgensen et al. (2010) gives a very comprehensive overview of the applied ecological indicators on all levels. A short overview of the eight levels of indicators is presented.

Ecological indicators were introduced in the environmental management to take "the temperature and pulse" on ecosystems. The results of environmental management can of course be followed easily when we are dealing with a specific problem. If we are focusing on the solution of the eutrophication problem for instance, we will follow the primary production, the concentration of phytoplankton, the concentrations of phosphorus and nitrogen, and the transparency, but all five state variables can of course be considered as indicators. When several problems are of our concern and the problems are interactive, we need to use a wider spectrum of ecological indicators to follow the result of our management.

The use of ecological indicators may be compared with the use of indicators by our doctor when he wants to assess our health. He will measure the pulse, the temperature, the blood pressure etc. He will use the values of the indicators to set up a diagnosis and later he will use the indicators to be able to assess the progress I our health. The use of ecological indicators is based on the same idea when we are applying them to assess the ecosystem health. Similar to the use of indicators for human health, it is necessary to use a spectrum of indicators, which may be simple to select when one clearly defined problem is in focus. It is however more difficult to select a set of good indicators when the diagnosis is not assessed or, as mentioned earlier, several problems are interacting.

The ecological indicators applied today in different context, for different ecosystems, and for different problems can be classified on eight levels from the most reductionistic to the most holistic indicators. The book *Handbook of Ecological*

Indicators for the Assessment of Ecosystem Health (EHA) by Jørgensen et al. (2010) gives a very comprehensive overview of the applied ecological indicators on all levels. Ecological indicators for EHA do not include indicators of the climatic conditions, which in this context are considered entirely natural conditions.

Level 1 covers *the presence or absence of specific species.* The best known application of this type of indicators is the saprobic systems (Hynes 1970), which classify streams in four classes according to their pollution by organic matter causing oxygen depletion: oligosaprobic water (unpolluted or almost unpolluted), beta-mesosaprobic (slightly polluted), alpha-mesoprobic (polluted), and poly-saprobic (very polluted). This classification was originally based on observations of species that were either present or absent. The species that were applied to assess the class of pollution were divided into four groups: organisms characteristic of unpolluted water, species dominating in polluted water, pollution indicators, and indifferent species. Records of fish in European rivers have been used to find by ANN (artificial neural network) a relationship between water quality and presence (and absence) of fish species. The result of this examination has shown that presence or absence of fish species can be used as strong ecological indicators for the water quality.

Level 2 uses *the ratio between groups of organisms.* A characteristic example is Nygaard Algae index.

Level 3 is based on *concentrations of chemical compounds.* Examples are assessment of the level of eutrophication on basis of the total phosphorus concentration assuming that phosphorus is the limiting factor for eutrophication. When the ecosystem is unhealthy due to too high concentrations of specific toxic substances the concentration of one or more focal toxic compounds is of course a very relevant indicator For example, the PCB contamination of the Great North American Lakes has been followed by recording the concentrations of PCB in birds and in water. It is often important to find a concentration in a medium or in organisms where the concentration can be easily determined and has a sufficiently high value that is magnitudes higher than the detection limit, which facilitates a clear indication.

Level 4 applies *concentration of entire trophic levels as indicators,* for instance, the concentration of phytoplankton (as chlorophyll a or as biomass per m^3) is used as indicator for the eutrophication of lakes. A high fish concentration has also been applied as indicator for a good water quality or birds as indicator for a healthy wetland ecosystem.

Level 5 uses *process rates as indication, for instance, primary production determinations* are used as indicator for eutrophication either as maximum g C/(m^2 day) or g C/(m^3 day) or g C/(m^2 year) or g C/(m^3 year). A high annual growth of trees in a forested wetland is used as indicator for a healthy forested wetland ecosystem and a high annual growth of a selected population may be used as indicator for healthy environment. A high mortality in a population can on the other side be used as indication of an unhealthy environment. High respiration may indicate that an aquatic ecosystem has tendency to oxygen depletion.

Level 6 covers *composite indicators* as, for instance, represented by many of E.P. Odum's attributes; see Table 18.1 The early stage corresponds for lakes and reservoirs to a vulnerable, nonsustainable stage and the mature stage to a sustainable stage. Examples are biomass, respiration/biomass, respiration/production, production/biomass, and ratio primary producer/consumers. E.P. Odum uses these composite indicators to assess whether an ecosystem is at an early stage of development or a mature ecosystem.

TABLE 18.1

Differences between Initial Stage and Mature Stage Are Indicated[a]

Properties		Early Stages	Late or Mature Stage
A	*Energetic*		
	P/R	$\ll 1$ or $\gg 1$	Close to 1
	P/B	High	Low
	Yield	High	Low
	Specific entropy	High	Low
	Entropy production per unit of time	Low	High
	Exergy	Low	High
	Information	Low	High
B	*Structure*		
	Total biomass	Small	Large
	Inorganic nutrients	Extrabiotic	Intrabiotic
	Diversity, ecological	Low	High
	Diversity, biological	Low	High
	Patterns	Poorly organized	Well organized
	Niche specialization	Broad	Narrow
	Size of organisms	Small	Large
	Life cycles	Simple	Complex
	Mineral cycles	Open	Closed
	Nutrient exchange rate	Rapid	Slow
	Life span	Short	Long
C	*Selection and homeostatis*		
	Internal symbiosis	Undeveloped	Developed
	Stability (resistance to external perturbations)	Poor	Good
	Ecological buffer capacity	Low	High
	Feedback control	Poor	Good
	Growth form	Rapid growth	Feedback controlled
	Growth types	R-strategists	K-strategists

[a] A few attributes are added to those published by Odum (1969, 1971).

Level 7 encompasses *holistic indicators such as resistance, resilience, buffer capacity, biodiversity,* all forms of diversity and size, and connectivity of the ecological network, turnover rate of carbon, nitrogen, etc., and of energy. As it will be discussed in the next section, high resistance, high resilience, high buffer capacity, high diversity, big ecological network with a medium connectivity, and normal turnover rates are all indications of a healthy ecosystem.

Level 8 indicators are *thermodynamic variables,* which we may call superholistic indicators as they try to see the forest through the trees and capture the total image of the ecosystem without inclusion of details. Such indicators are exergy, eco-exergy or work capacity, emergy, entropy production, power, mass, and/or energy system retention time. The economic indicator cost/benefit (which includes all ecological benefits—not only the economic benefits of the society) belongs also to this level.

18.4 Emergy and Exergy

It is needed here to give the definition for emergy and exergy or eco-exergy, because these concepts have been used in previous chapters with reference to this chapter.

Emergy was introduced by Odum (1983) and attempts to account for the solar energy required for formation of an organism or a product by multiplying their actual energy content by their solar energy transformation ratios. The more transformation steps there are between the two kinds of energy, solar energy and energy content of the organism or the product, the greater the quality and the greater is the solar energy required to produce one unit of energy (J) of that type. Embodied energy or emergy flows are, as seen, from this definition, determined by the biogeochemical energy *flow* into an ecosystem component, measured in solar energy equivalents. The stored emergy, Em, per unit of area or volume to be distinguished from the emergy flows can be found from

$$Em = \sum_{i=1}^{i=n} \Omega i * ci \tag{18.1}$$

where
  Ωi is the quality factor = transformity (seJ/J), which is the conversion to solar equivalents, as illustrated in Table 18.2
  ci is the concentration expressed per unit of area or volume

Exergy is defined as the amount of work (=entropy-free energy) a system can perform when it is brought into thermodynamic equilibrium with its environment (Jørgensen 2011, 2012). For ecosystems, the environment is the next

TABLE 18.2

Embodied Energy Equivalents for Various Types of Energy

Type of Energy/Item	Embodied Energy Equivalents or Transformity (seJ/J)
Solar energy	1.0
Winds	315
Gross photosynthesis	920
Coal	6,800
Tide	11,560
Electricity	27,200
Kinetic energy of spring flow	7,170
Detritus	6,600
Gross plant production	1,620
Net plant production	4,660
Herbivores	1,27,000
Carnivores	4,090,000
Top carnivores	40,600,000

ecosystem and it is therefore necessary to assume another reference system when we are using exergy for natural systems. In that case, we apply a reference environment that represents the same system (ecosystem) but at thermodynamic equilibrium, which means that all the components are inorganic at the highest possible oxidation state and homogeneously distributed in the system (no gradients). It is illustrated in Figure 18.1.

As the chemical energy embodied in the organic components and the biological structure contribute far more to the exergy content of the system, there seems to be no reason to assume a (minor) temperature and pressure difference between the system and the reference environment. Under these circumstances, we can calculate the exergy, which we will name eco-exergy to distinguish from the exergy applied in technology for power plants, as coming entirely from the chemical energy.

According to this definition, eco-exergy density becomes

$$Ex = RT \sum_{i=0}^{i=n} c_i \ln \frac{c_i}{c_{io}} = [ML^{-1}T^{-2}] \tag{18.2}$$

where
   R is the gas constant
   T is the temperature of the environment, while c_i is the concentration of the ith component expressed in a suitable unit, for example, for phytoplankton in a lake c_i could be expressed as mg/L or as mg/L of a focal nutrient
   c_{io} is the concentration of the ith component at thermodynamic equilibrium
   n is the number of components

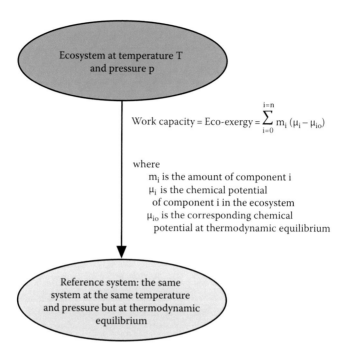

FIGURE 18.1
Exergy content of the system is calculated in the text for a system relative to a reference environment of the same system at the same temperature and pressure but as an inorganic soup with no life, biological structure, information, or organic molecules.

c_{io} is very low for living components because the probability that living components are formed at thermodynamic equilibrium is very low. It implies that living components get a high exergy. c_{io} is not zero for organisms but will correspond to a very low probability of forming complex organic compounds spontaneously in an inorganic soup at thermodynamic equilibrium. c_{io} on the other hand is high for inorganic components, and although c_{io} still is low for detritus, it is much higher than that for living components. It can be shown (Jørgensen et al. 2010) that $ß = RT \ln c_i/c_{io}$ in Equation 18.2 is very closely correlated to the free energy (energy that can work) of the amino acid sequences that are determining the composition of the enzymes. The enzymes control the life processes.

Table 18.3 gives an overview of the exergy of various organisms expressed by the weighting factor ß that is introduced to be able to cover the exergy for various organisms in the unit chemical exergy detritus equivalent per unit of volume or unit of area (we find the exergy density). The following equation can be applied:

$$Ex - total\text{-}density = \sum_{i=1}^{N} \beta_i c_i$$

(as detritus equivalents at the temperature $T = 300$ K)

TABLE 18.3

ß-Values = Exergy Content Relative to the Exergy of Detritus and
ß-Values = Eco-Exergy Content Relative to the Eco-Exergy of Detritus

Organisms	Plants		Animals
Detritus	1.00		
Viroids	1.0004		
Virus		1.01	
Minimal cell		5.0	
Bacteria	8.5		
Archaea	13.8		
Protists (algae)	20		
Yeast		17.8	
		33	Mesozoa, placozoa
		39	Protozoa, amoeba
		43	Phasmida (stick insects)
Fungi, moulds	61		
		76	Nemertina
	91		Cnidaria (corals, sea anemones, jelly fish)
Rhodophyta	92		
		97	Gastrotricha
Prolifera, sponges	98		
		109	Brachiopoda
		120	Plathyhelminthes (flatworms)
		133	Nematoda (round worms)
		133	Annelida (leeches)
		143	Gnathostomulida
Mustard weed	143		
		165	Kinorhyncha
Seedless vascular plants	158		
		163	Rotifera (wheel animals)
		164	Entoprocta
Moss	174		
		167	Insecta (beetles, flies, bees, wasps, bugs, ants)
		191	Coleoidea (Sea squirt)
		221	Lepidoptera (butter flies)
		232	Crustaceans
		246	Chordata
Rice	275		
Gymnosperms (incl. pinus)	314		
		310	Mollusca, bivalvia, gastropoda
		322	Mosquito

(continued)

TABLE 18.3 (continued)

ß-Values = Exergy Content Relative to the Exergy of Detritus and
ß-Values = Eco-Exergy Content Relative to the Eco-Exergy of Detritus

Organisms	Plants	Animals	
Flowering plants	393		
		499	Fish
		688	Amphibia
		833	Reptilia
		980	Aves (Birds)
		2127	Mammalia
		2138	Monkeys
		2145	Anthropoid apes
		2173	*Homo sapiens*

Source: Jørgensen, S.E. et al., *Ecol. Model.*, 185, 165, 2005a.

The ß-value embodied in the biological/genetic information is found on the basis of the genome that determines the amino acid sequences of the various organisms (see Jørgensen 2012). Detritus has, in accordance with the previously shown equation, the ß-value = 1.0. By multiplication of the result obtained by the previously shown Equation by 18.7, the exergy can be expressed in kJ per cubic meter or square meter assuming, because the chemical energy (free energy, work energy, exergy) of detritus is 18.7 kJ/g.

18.5 Selection of Ecological Indicator

The selection of the ecological indicators should be based on the following criteria:

1. Relevant, meaning that it would cover the problem, the use, and the characteristics of the lake or reservoir. It is, for instance, important that the indicators focus on nonpoint sources if the lake problems are dependent on nonpoint sources.

2. Simple and easily understood by laymen to allow the laymen to follow the water quality and other improvements. It is important that all stakeholders participate in the selection of the indicators.

3. Scientifically justifiable to ensure that the indicators are selected not only on basis of what is easily "sold" to the politicians and the population, but also on basis of what is important from an ecological evaluation of the problem.

4. Quantitative, because it will make it easier to express the possible progress in water quality and ecosystem health.

5. Sensitive to possible changes in the water quality and ecosystem health to be able to express actual improvements by the selected ecological indicators.
6. Acceptable in terms of costs.
7. Cover *all* relevant and actual problems of the entire drainage basin of the lake, river, estuary, or reservoir.

Before the ecological indicators are selected, it is recommended to answer the following questions:

1. What do we know about the problem of the aquatic ecosystem qualitatively as well as quantitatively? The selection of the indicators will inevitably be different when the problems are known and not known, because in the latter case a diagnosis is needed by the use of indicators.
2. Which aspects of environmental management must be included to cover the problems to the extent that they are known?
3. Which indicators will be able to cover the relevant and actual problems of the lake or reservoir?

The selection should at least in the first hand encompass rather too many than too little indicators, because it is more easy to see after the use the redundant indicators. All stakeholders (interested in a sustainable development of the ecosystem) should usually be involved in the selection of indicators. After the selection of a handful or more indicators each of them should be tested for the aforementioned seven criteria, which probably will imply that the number of selected indicators is reduced.

For the various environmental problems of freshwater ecosystems, the immediate selection of indicators is straight forward:

1. *Eutrophication*: The transparency, which is easily understood by laymen, should pr be supplemented by indication of phytoplankton concentration as biomass mg/L or as chlorophyll a in mg/m^3. Also the maximum primary production as mg C/((m^3 or m^2) 24 h) and/or as g C/((m^3 or m^2) year) would be very informative, although it is probably not the most understandable indicators for laymen.
2. *Toxic substances*: The concentration of the toxic substance in the water, sediment, and organisms late in the food chain. Due to biomagnification, it is recommended to use the concentration in carnivorous fish or in birds, when the concentration in the water is low and therefore very uncertain.
3. *Siltation*: Transparency and concentration of suspended matter in the water. It is recommended to determine the total amount of silt flowing to the lake per unit of time.

4. *Water level*: The water depth supplemented with calculations of the total volume of water in the lake as function of time. It may also be relevant to follow the oxygen concentration.

5. *Acidification of freshwater ecosystems*: pH, alkalinity, and pH-buffer capacity of the water.

6. *Introduction of exotic species*: The concentration of the introduced species and all the species that are influenced directly by the introduced species. For instance, when the Nile Perch was introduced in Lake Victoria it would have been relevant to follow also the concentration of Tilapia and Haplochromis.

For all environmental management problems, it may be relevant and recommendable to follow the biodiversity. Odum's attributes (see Table 18.1), including the thermodynamic indicators are very informative, and although they are not fully understandable by laymen, they may be important for environmental managers. At least some of this group of indicators should therefore be included in the selection of indicators.

18.6 Case Study

Lake Fure is close to Copenhagen and has therefore a high recreational value. Recently, a restoration project focusing particularly on the eutrophication was realized and the question in this context was: Which indicators should be selected to follow the restoration project? The questions relevant for the selection of indicators in this case are clear:

1. How can the reduction in the eutrophication, resulting from the restoration project, be followed closely?

2. How would it be possible to indicate and follow the recreational value of the lake, which is very important?

It was decided to select the following indicators to follow the reduction in the eutrophication: transparency, primary production, concentration of chlorophyll a, concentration of the phytoplankton determined by filtration as the suspended matter from 1 to 60 mm, total P, dissolved P, total N, dissolved ammonium-N, dissolved nitrate-N, and the concentration of zooplankton. All these variables were determined currently anyhow, which means that it was not an additional cost to use these indicators, but more a question of presenting the data as graphs and selecting the most representative data for the public. Many of these indicators are state variables in a eutrophication model, which was available as management tool.

FIGURE 18.2
(See color insert.) Lake Fure.

The recreational value was considered a question of maintaining the lake and the surrounding forests as close as possible to natural ecosystems. It was therefore decided to follow the biodiversity of (1) fish in the lake, (2) birds in the surrounding forests, and (3) zooplankton in the lake.

Lake Fure is situated 12–18 km from the center of Copenhagen and has a great recreational value; see Figure 18.2. The lake has an area of 9.9 km² and is the deepest lake of Denmark with a maximum depth of 39 m. The lake has therefore a thermocline from mid-May to the beginning of November. It is used for swimming, boating, sailing, and angling. Many people are jogging, biking, or walking in the forests and wetland areas surrounding the lake, while enjoying the beautiful lake sceneries. The lake water is not used for production of drinking water or for transport and the commercial fishery is minor and economically insignificant. The lake has almost only recreational value. The wastewater discharged during the 1950s and 1960s and early 1970s was mechanically biological treated but without nutrient removal. The lake was therefore gradually becoming more and more eutrophied after the Second World War and up to 1972. The transparency was early in the twentieth century measured to be 5 m when it was at the annual minimum (mid-May or early August), while it was determined to be 1.2 m in the beginning of the 1970s. In 1972, it was decided either to treat the wastewater by a very effective removal of nitrogen (about 80%–85%) and phosphorus (98%–99%) or to discharge to the sea by a 10 km pipeline. Two of the three municipalities, that were discharging treated wastewater to the lake, decided to apply the pipeline solution. The retention time of the water in the lake (volume 100 million m³) was in the 1960s 16 years and after the discharge of most of the wastewater to the sea in 1972 it was increased to 20–22 years. It implies that the full recovery of the lake after the measures taken in 1972 was estimated to take in the order of 60–65 years. It was mainly due to the significant amount of phosphorus accumulated in the sediment of the lake during the 1950s and 1960s when wastewater without removal of nutrients was discharged to the lake.

The annual external phosphorus loading in the 1960s was about 33 t, which was reduced to 3.0 t in 1972 and later to 2.0 t by construction of more storm water basin capacity. The internal loading from sediment, however, was in 1997 determined to be about 12.5 t P/year. The idea to restore the lake faster by one or more ecotechnological methods came up around year 2000. In this context, a model and ecological indicators were selected to follow the development of the eutrophication. The following eight questions (see also Jørgensen et al. 2005b) listed earlier will be used to understand the selection:

1. The lake problems are eutrophication and the slow restoration time after the treatment of wastewater commenced. The minimum transparency (usually in May and/or August) was since 1972 and up to year 2002—a period of 30 years—only increased from about 1.2 to 2.0 m. Could ecotechnological methods be used to accelerate the restoration and which methods should be selected in accordance with a cost/benefit criteria?

2. The lake is deep with one main tributary (river) and several minor streams. The hydrology is relatively simple although a part of the lake is shallow. In this part, submerged vegetation was dominant earlier before the Second World War.

3. The nutrient source was definitely the internal sources—the nutrient accumulated in the sediment. It was measured in 2004 still to be more than 7 t P/year.

4. The lake has an enormous recreational value due to its adjacency to Copenhagen (1.5 million inhabitants). There are no other socioeconomic implications and the relationship to the problem (eutrophication) is simple—swimming, boating, and sailing is more unpleasant when the transparency is low. Also the angling suffers from the eutrophication. In August, when occasionally there is blue–green blooms, swimming is not allowed. It would therefore be very beneficial to reduce the eutrophication, increase the transparency, and avoid the blue–green bloom.

5. The amount of data available for the development of a good model is almost overwhelming. The lake has been investigated limnologically since 1890. The observations are therefore not limiting the selection of a proper eutrophication model.

6. The following ecotechnological methods could be considered: removal of sediment, aeration of hypolimnion during the summer, biomanipulation (removal of planktivorous fish), and siphoning of the hypolimnic water.

7. The model should be able to assess the reduction in eutrophication and increase in water transparency as function of time for the four ecotechnological methods mentioned under (6) including combinations of these methods.

8. A reduction of blue–green algae bloom and an increased transparency would be an advantage for the use of the lake for all recreational activities.

Based upon these answers to the eight core questions, it was decided to apply the International Environmental Technology Center (IETC)-United Nations' Environmental Program (UNEP) structurally dynamic model to cover the development of the deep and shallow areas of the lake. The resources available for the development of models were limited and it was therefore decided to use a standard model. A tailored, not too complex, model focusing particularly on the sediment water exchange of phosphorus by the application of the different ecotechnological methods was, however, also developed to get more accurate assessment of the development of the internal phosphorus loading.

It was decided to use biomanipulation and aeration of the lake. Removal of sediment would be too expensive due to the depth of the lake. Siphoning of hypolimnic water had complex implications because two lakes downstream would be affected. The restoration project was launched in April 2004. The results of the model were published (Gurkan et al. 2006). The main results indicate that the internal phosphorus loading is reduced from about 7.2 t/year in 2003 to about 2 t in year 2006 and to less than 1 t in year 2009–2010. It was therefore according to the model results expected that the restoration project would have a major effect on the water quality of the lake.

As indicators to interpret the model results and follow the lake restoration were selected: transparency, chlorophyll a, number of days with blue–green bloom occurrence, zooplankton (to see if the biomanipulation was successful) and exergy/biomass to assess that the lake developed toward reduced concentration of primitive organisms (phytoplankton, blue–green algae), and a higher concentration of zooplankton and carnivorous fish (see also Jørgensen et al. 2005b).

References

Gurkan, Z.; Zhang, J.; and Jørgensen, S.E. 2006. Development of a structurally dynamic model for forecasting the effects of restoration of lakes. *Ecol. Model.* 197:89–103.

Hynes, H.B.N. 1970. *Ecology of Running Waters*. University of Liverpool Press, Liverpool, U.K., 555pp.

Jørgensen, S.E. (Ed.). 2011. *Handbook of Ecological Models Used in Ecosystem and Environmental Management*. CRC Press, Boca Raton, FL, 620pp.

Jørgensen, S.E. 2012. *Fundamentals of Systems Ecology*. CRC Press, Boca Raton, FL, 320pp.

Jørgensen, S.E. and Fath, B. 2011. *Fundamentals of Ecological Modelling*, 4th edn. Elsevier, Amsterdam, the Netherlands, 400pp.

Jørgensen, S.E.; Ladegaard, N.; Debeljak, M.; and Marques, J.C. 2005a. Calculations of exergy for organisms. *Ecol. Model.* 185:165–176.

Jørgensen, S.E.; Löffler, H.; Rast, W.; and Straškraba, M. 2005b. *Lake and Reservoir Management*. Developments in Water Sciences, Vol. 54. Elsevier, Amsterdam, the Netherlands, 502pp.

Jørgensen, S.E.; Ludovisi, A.; and Nielsen, S.N. 2010. The free energy and information embodied in the amino acid chains of organisms. *Ecol. Model.* 221:2388–2392.

Jørgensen, S.E.; Xu, F.-L.; and Costanza, R. 2010. *Handbook of Ecological Indicators for the Assessment of Ecosystem Health*. Elsevier, New York, 464pp.

Odum, H.T. 1983. *System Ecology*. Wiley Interscience Publishers, New York, 510pp.

19

Application of Ecological Models in Management of Aquatic Inland Ecosystems

19.1 Models as Tools in Ecological Management

As already mentioned in Chapter 18, models are powerful management tools that can synthesize and integrate our knowledge about the environmental problems of ecosystems. It should be emphasized that models not only integrate the knowledge based on observations (data) but also are able to include in the synthesis our scientific knowledge, knowledge about allometric principles, knowledge about food items and food chains, and so on. Models can be applied to answer relevant environmental management problems—how will the ecosystem be changed if these and these forcing functions (ecosystem impacts) are changed? It is therefore clear that aquatic ecosystem management without models would be almost impossible. Models are an indispensable tool in environmental management in general.

The next section will present institutionalized and mediated models, which are particularly strong management tools. The third section will discuss the model selection, but the bulk of this chapter presents to the reader models applicable to various management problems, with a focus on eutrophication models, toxic substance models, and then successively on models of acidification, wetlands, and fisheries. Special attention is given to one management option, biomanipulation. The complex water quality models most commonly used are discussed later in the chapter. Finally, recent models dealing with environmental risk assessment and considering structurally dynamic changes of ecosystems, denoted SDMs, are discussed.

How to go modeling, an applicable model procedure, the modeling elements, and an overview of available model types can be found in Jørgensen and Fath (2011) or in other textbooks of ecological modeling. Jørgensen (2011) gives an overview of available ecological models, of which many are developed for freshwater ecosystems. Jørgensen (2011) is therefore a useful supplement to this chapter.

19.2 Institutionalized or Mediated Modeling

There are hundreds of ecological–environmental models of aquatic ecosystems available in the literature. It is, however, better to develop a tailored model for a given environmental problem for a considered aquatic ecosystem. Every ecosystem is different and the environmental problems are different in the sense that they have different implications for different ecosystems and for the local population dependent on the natural resources of the freshwater ecosystems. The tailored model can be developed by an expert team, this model is denoted as an expert model, or by a brainstorming meeting of all local stakeholders together with a model team. This latter type of model is denoted mediated/ institutionalized models. The experience shows, however, that both standard (nontailored) models and expert models are often not applied at least not sufficiently to set up management plans, because they are not accepted by the local organization as an integrated part of their environmental management and they do not for most of the models consider the relevant socioeconomic aspects. The very best is therefore to develop a mediated or institutionalized model that is step by step developed and accepted by the local stakeholders together with a very few nonlocal modeling experts. The following procedure can be applied:

1. Arrange a 1 week brainstorming where the modelers, the NGOs, the local scientists, and managers formulate the problem and the characteristics of the ecosystem and how the problems associated with the ecosystem interfere with other local environmental-economic-social problems. A multidisciplinary and interdisciplinary team is needed. An overview of the knowledge about the ecosystem is also provided.

2. The result of the workshop is a conceptual diagram of the model that includes the other interfering local problems. It cannot be excluded that two or more models are needed to solve the complex of problems. It cannot be excluded that supplementary knowledge and observation are needed.

3. The modeling team develops an easy-to-go software for the model according to the conceptual diagram.

4. The model and its possibility as supporting tool in the lake management are demonstrated by the modelers for the entire brainstorming team. The entire team works with the models together for at least 3 days.

5. The managers and other personnel that should be able to run the models are trained (probably 4–10 days dependent on the model complexity) in the use of the model.

6. The entire team has an annual or biannual meeting to discuss the application of the model and possible minor modifications of the model.

The following questions arise generally as a result of model development:

- How is it possible to consider all the different aspects of the problem and its solution, when many aspects of the problem are considered?— it may be different natural science aspects, for instance, geology, zoology, botany, and chemistry, or it may even be a combination of environmental, economic, and social aspects of the problem.
- The answer is by implementing a very wide spectrum of expertise in the modeling team, but it gives rise to the next obvious question, how to ensure a good cooperation of the team members, when they represent many different disciplines and therefore many different opinions and "languages"?
- How is it possible to consider *all* relevant ecosystem properties at the same time?
- How is it possible to integrate insights from many different disciplines as for instance ecology, economy, and social science or even different disciplines of natural sciences?
- How can we ensure that all important stakeholders are included in the modeling process?
- How is it possible to integrate impacts and knowledge at different scales?
- How is it possible to understand the very root of the problems and their sources and have this understanding reflected in the modeling and the final model result?
- How is it possible to build the best of a consensus among the different opinions and disciplines?

Institutionalized or mediated modeling (abbreviated IMM) can give the answer to these questions. The institutionalized modeling process is described in detail in the following text. The main idea is to represent without exception *all* stakeholders, policymakers, managers, and scientists that have knowledge and ideas about the problem, the system, and the possible solutions in the modeling procedure. The model is developed as a result of an integrated brainstorming meeting, where all ideas, opinions, disciplines, and knowledge are represented. For most development of mediated models, but of course dependent on the complexity of the problem and the system, several days are required to reach a satisfactory model that can be used as a tool to solve the problem. The advantages of IMM are that (partly taken from Van den Belt 2004)

1. The level of shared understanding increases.
2. A consensus is built about the structure of a complex problem for a complex system, because all interests are represented in the step-wise model development.

3. The result of the modeling process, the model, serves as a tool to disseminate the insights gained by the modeling procedure.
4. The effectiveness of the decision making is increased, because the mediated model makes it possible for the policymakers and the stakeholders to see the consequences of the action plans over longer time scales.
5. Team building is developed parallel to the model development.
6. The process is emphasized over the product.
7. The state-of-the-art knowledge is captured, organized, and synthesized.

It is characteristic for a team developing a mediated model that what could be denoted "groupiness" is increased because

1. The individual members perceive clearly that they are a part of the group.
2. The members become oriented toward a common goal.
3. Interaction between group members take place.
4. The interdependence is realized and acknowledged.
5. A structure of roles/status and norms is built.

19.3 Model Selection

Standard models are available to solve the most basic aquatic ecosystem problems, for instance, eutrophication, water level problems, acidification, oxygen depletion, etc. Still it is better to tailor the model according to the combination of observations, data, problem, and ecosystem characteristic. Expert models are more expensive to apply than standard models because they are more time consuming. Development of mediated models based on brainstorming meetings will inevitably cost more than application of standard models or even more than expert models due to the cost of the brainstorming meetings. In most cases, however, institutionalized/mediated models are good investment because the models *are* applied and *accepted* by the local organization and they consider the link to other local environmental and socioeconomic problems and are therefore proposing more holistic and sustainable solutions to the environmental problems.

Which model to apply among the standard models is to a high extent a question about the database, the problem, and the characteristics of the lake. This question is discussed next on basis of a brief overview of the available models. A comprehensive overview of models can be found in Jørgensen et al. (2005), Jørgensen and Fath (2011), and Jørgensen (2011).

TABLE 19.1

Possible Model Strategies

Model	Database Required	Cost (10^3 Dollars)
Tailored expert model, high complexity	Comprehensive	400–1000
Tailored mediated model, high complexity	Comprehensive	500–1200
Tailored expert model, medium complexity	At least for the st. variables	300–600
Tailored mediated model, medium complexity	At least for the st. variables	350–700
Tailored expert simple model	Small	100–250
Tailored mediated simple model	Small	120–300
Standard model, complex	Comprehensive	100–250
Standard model, medium complexity	At least for the st. variable	50–150
Standard model, simple	Small	25–75

Table 19.1 gives an overview and synthesizing results of different environmental strategies and different model complexity.

The cost indicated in the table is a range, where the higher values should be valid for large ecosystems with a complex hydrodynamics and the lower values for relatively smaller ecosystems with a simple hydrodynamics. The cost indicated is only for the model development, but includes for the tailored models the cost of the meetings leading to the model. If the database is insufficient the cost of updating the database must of course be added to the overall costs. The costs for the standard models are mainly manpower for the calibration and validation of the model and organization of the database to be accessible for the model. For the expert models, the cost includes the cost of expert meeting and expert time and for the mediated models, the cost includes also the brainstorming meetings.

UNEP–IETC offers downloadable standard models of different complexity for lakes and reservoirs:

1. *A one layer model* that is based on four variables only (total nitrogen in the water phase, total phosphorus in the water phase, phosphorus in the sediment, and nitrogen in the sediment). The model is not able to offer a high accuracy of the prognoses but can be developed based on a limited database. The model is not considering two or more layers, but it is possible to account for the formation of a thermocline by a correction factor. Chlorophyll-a concentration, zooplankton concentration, productivity, fish concentration, and transparency can all be calculated based upon general correlations. It means that the calculations of course have a relatively high standard deviation.

2. *A two layer model* that gives better prognoses but also requires a much more comprehensive database.

3. *A structurally dynamic model (SDM) for deep lakes* that considers adaptations and shifts in species composition. It is easier to use than (2) because it contains facilities to automatically calibrate some of the model parameters. It is recommended to use this standard model instead of (2). It can capture the structural changes as a result of biomanipulation and other ecotechnological restoration methods.

4. *A structurally dynamic model for shallow lakes*, that is also easier to use than (2). It is able to cover the structural changes that may result from competition between submerged vegetation and phytoplankton.

If eutrophication or acidification is the problem, and it is not possible to provide sufficient economy for the development of a mediated or institutionalized model, it is recommendable to use a standard or expert model, but it is *very important* that the selection is made by a meeting between local scientists, local managers, and model experts. It is, however, the best long-term solution as indicated earlier to develop a mediated/institutionalized model.

Table 19.1 gives an overview of the model possibilities and gives some first guidance on which model to apply in which case. Follow, however, the discussion of the selection problem presented later to make the final decisions together with the information summarized in Table 19.1. Compare the cost of the model with the economic importance of the ecosystems, including the values of the ecosystem services. If the ecosystem is large and of great local importance, it would most probably be the best long-term solution to invest more money in a proper management model including the erection of a suitable database that is able to support the model. The cost of developing a good model for a given case is often 1% or less of the economic value of the lake. In other words, if we avoid a more than 1% mistake in our management decisions by using a good model, we have more than paid for the model. If the freshwater ecosystem is only important for a limited population, and it is a smaller ecosystem, it may be sufficient to use a standard model and in some cases where the needed observations are not available for a complex standard model, a simple model seems to be the best selection at least in the first hand. For instance, when eutrophication is the problem for a lake, the aforementioned one layer model will most probably be able to do the job in the first hand. It is always beneficial to apply a model, which is easily accessible. The results can often be used to make a better model selection afterward because the unanswered questions are revealed.

Models are being used increasingly in environmental management, primarily because they are the only tools able to quantitatively relate the impacts on an ecosystem with the consequences for the state of the ecosystem. Aquatic ecosystem models have been developed particularly during the last few decades, and it is not surprising they have found a wide application in aquatic ecosystem management.

The extensive use of mathematical models in water quality management can be seen from the survey by Alasaarela et al. (1993). They assembled questionnaires from 100 institutions. The results document the use of 105 models applied in 800 situations and requiring approximately 500 person work years. The average modeling team consisted of four persons.

The field of ecological and environmental modeling has developed rapidly during the last two to three decades due essentially to the following factors:

- The development of computer technology, enabling us to handle very complex mathematical systems.

- A general understanding of pollution problems, including the knowledge that a complete elimination of all pollution ("zero discharge") is not feasible, but rather that a proper pollution control with often limited economic resources requires serious considerations of the influence of various pollution impacts on ecosystems.

- Our knowledge of environmental and ecological problems has increased significantly. We have particularly gained more knowledge about quantitative relations in ecosystems, and between ecological properties and the environmental factors.

The goal of this chapter is to demonstrate to managers of aquatic ecosystems, limnologists, chemists, sanitary engineers, and other interested parties of the possibilities that mathematical modeling offers as a tool to support management decisions. More technical modeling issues are discussed by other researchers, including Jørgensen (1992), Orlob (1983, 1984), Straskraba and Gnauck (1985), Jørgensen (1994, 2011), Chapra (1997), and Jørgensen and Fath (2011).

19.4 Models as a Strong Management Tool: Problems and Possibilities

A management problem to be solved often can be formulated as follows: If certain forcing functions (management actions) are varied, what will be their influence on the state of an ecosystem? The model is used to answer this question or, in other words, to predict what will change in a system when the control functions that are managed by humans are varied over time and space.

Typical *control functions* are the consumption of fossil fuel, regulation of water level in a river by a dam, discharge of pollutants, or fisheries policy.

It is important that, to a certain extent, the manager should take part in the entire development of a management model, since he will ultimately define the modeling objectives and select the modeling scenarios. The success of the application of a management model, to a large degree, is dependent on an open dialogue between the modeler and the manager.

A further complexity is the construction of ecological–economic models. As we gain more experience in constructing ecological and economic models, more and more of them will be developed. It often is feasible to find a relation between a control function and economic parameters. If a lake is a major water resource, an improvement in its water quality will inevitably result in a reduction in the treatment costs of drinking water if the same water quality is to be provided. It is also possible to sometimes relate the value of a recreational area to the number of visitors, and to how much money they spend on average in the area. In many cases, however, it is difficult to assess a relationship between the economy and the state of an ecosystem. For example, how can we assess the economic advantages of an increased transparency in a water body? Ecological–economic models are useful in some cases, but should be used with caution, and the relations between the economy and environmental conditions critically evaluated, before the model results are applied.

Data collection is the most expensive component of model construction. For many lake models, it has been found that needed data collection comprises 80%–90% of the total model costs. Because complex models require much more data than simple ones, the selection of the complexity of environmental management models should be closely related to the costs involved in the environmental problem to be solved.

Thus, it is not surprising that development of the most complex environmental management models have generally been limited to large ecosystems, where the economic involvement is great.

The predictive capability of environmental models can always be improved in a specific case by expansion of the data collection program, and by a correspondingly increased model complexity, provided that the modelers are sufficiently skilled to know in which direction further expansion of the entire program must develop to order to improve the model's predictive capabilities.

The relation between the economy of the project and the accuracy of the model is presented in the form shown in Figure 19.1. The reduction in the discrepancy between model predictions and reality is lower for the next dollar invested in the project, because the log (cost) versus difference between model results and observations gives a straight line. But it is also clear from the shape of the curve that the associated errors can hardly be completely eliminated. All model predictions have a standard deviation associated with them. This fact is not surprising to scientists, but it often is not understood or appreciated by decision makers, to whom the modeler typically presents his or her results.

Engineers use safety factors to assure that a building or a bridge will last for a certain period of time, with a very low probability of breakdown, even under extreme conditions. No reputable engineer would propose using a smaller, or no, safety factor to save some concrete and reduce the costs. The reason is obvious: Nobody would want to take the responsibility for even the smallest probability of a building or bridge to collapse.

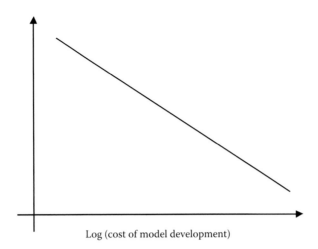

Log (cost of model development)

FIGURE 19.1
More a modeler invests in a model and in data collection, the closer he or she will come to realistic predictions. However, the modeler will always gain less for the next dollar invested and will never be able to give completely accurate predictions. With good approximations, log (costs) versus the difference between model results and observations is a straight line, corresponding to an exponentially decreasing difference with an increasing investment in the model development.

When decision makers are going to make decisions on environmental issues, the situation is strangely different. Decision makers in this situation want to use the standard deviation to save money, rather than assuring a high environmental quality under all circumstances. It is the modeler's duty, therefore, to carefully explain to the decision maker all the consequences of the various decision possibilities. A standard deviation of a prognosis for an environmental management model can, however, not always be translated into a probability, because we do not know the probability distribution. It might be none of the common distribution functions, but it is possible to use the standard deviation qualitatively or semiquantitatively, translating the meaning of the model results by the use of words. Civil engineers are more or less in the same situation and have been successful in the past convincing decision makers of appropriate steps to be taken in various situations. There is no reason that environmental modelers cannot do the same.

It is often advantageous to attack an environmental problem in the first place with the use of simple models. They require much fewer data, and can give the modeler and decision maker some preliminary results. If the modeling project is stopped at this stage for one or another reason, a simple model is still better than no model at all, because it will at least give a survey of the problem.

Simple models, therefore, are good starting points for the construction of more complex models. In many cases, the construction of a model is carried out as an iterative process, and a step-wise development of a complex structurally dynamic model may be the result. As previously mentioned, the first

TABLE 19.2

Examples of Sources of Water Pollutants

Source	Examples
Point sources	Wastewater (nitrogen, phosphorus, biochemical oxygen demand); sulfur dioxide from fossil fuels; discharges of toxic substances from industries
Nonpoint man-made sources	Agricultural use of fertilizers; deposition of lead from vehicles; contaminants in rain water
Nonpoint natural sources	Runoff from natural forests

step is development of a conceptual model. It is used to get a survey of the processes and state variables in the ecosystem of concern. The next step is development of a simple calibrated and validated model. It is used to establish a data collection program for a more comprehensive effort closer to the final selected version. However, the third model will often reveal specific model weaknesses, the elimination of which is the goal of the fourth version of the model. At first glance, this seems to be a very cumbersome procedure. However, because data collection is the most expensive part of modeling, constructing a preliminary model for optimization of the data collection program will ultimately require fewer financial resources.

A first, simple mass balance scheme is recommended for biogeochemical models. The mass balance will indicate what possibilities exist for reducing or increasing the concentration of a chemical or pollutant, which is a crucial issue for environmental management.

Point sources of pollution are usually easier to control than anthropogenic nonpoint sources, which, in turn, are more easily controlled than natural pollutant sources, as shown in Table 19.2. Distinction can be made between local, regional, and global pollutant sources. Because the mass balance indicates the relative quantities from each source, it is possible to identify which sources should receive the initial attention (e.g., if a nonpoint regional source of pollutants is dominant, it would be pointless to concentrate first on eliminating small, local point sources, unless the latter also might have some political influence on regional decisions). Table 19.3 gives an overview of the spectrum of eutrophication models that have been developed on basis of the Glumsø Model. The table shows how the various versions of the model includes different processes, which may be of importance in different case studies. The table illustrates therefore also the generality of eutrophication models.

It has already been recognized that the modeler and the decision maker should communicate with each other. It is recommended, in fact, that the decision maker be invited to follow the model construction process from its very first phases, in order to become acquainted with the model strength and limitations. It also is important that the modeler and the decision maker together formulate the model objectives and interpret the

TABLE 19.3

Survey of Eutrophication Studies Based upon the Application
of a Modified Glumsø Model

Ecosystem	Modification	Level[a]
Glumsø, version A	Basic version	7
Glumsø, version B	Nonexchangeable nitrogen	7
Ringkøbing Firth	Boxes, nitrogen fixation	5
Lake Victoria	Boxes, thermocline, other food chain	4
Lake Kyoga	Other food chain	4
Lake Mobuto Sese Seko	Boxes, thermocline, other food chain	4
Lake Fure	Boxes, nitrogen fixation, thermocline	7
Lake Esrom	Boxes, Si-cycle thermocline	4
Lake Gyrstinge	Level fluctuations, sediment exposed to air	4–5
Lake Lyngby	Basis version	6
Lake Bergunda	Nitrogen fixation	2
Broia Reservoir	Macrophytes, 2 boxes	2
Lake Great Kattinge	Resuspension	5
Lake Svogerslev	Resuspension	5
Lake Bue	Resuspension	5
Lake Kornerup	Resuspension	5
Lake Søbygaard	SDM	7
Lake Balaton	Adsorption to suspended matter	2
Roskilde Fjord	Complex hydrodynamics	4
Lagoon of Venice	Ulva/Zostera competition	6
Lake Annone	SDM	6
Lake Balaton	SDM	6
Lake Mogan, Ankara	Only P cycle, competition submerged vegetation/ phytoplankton + SDM	6
Stadsgraven, Copenhagen	Four to six interconnected basins (level 6: 93)	5
Internal lakes of Copenhagen	Five to six interconnected basins	5

[a] Level 1: Conceptual diagram selected. Level 2: Verification carried out. Level 3: Calibration using intensive measurements. Level 4: Calibration of entire model. Level 5: Validation. Object function and regression coefficient are found. Level 6: Validation of a prognosis for significant changed loading or development of SDM (structurally dynamic model; see Section 19.10). Level 7: Validation of a prognosis and development of SDM.

model results. Holling (1978) has demonstrated how such teamwork can be developed and used phase by phase. Although his recommendations are not reiterated here, it is recommended that a modeling team should acquaint themselves well with the procedures outlined by him. The conclusions are clear: The modeler and decision maker should work together in all phases of the modeling exercise. Having the modeler first build a model and then transfer it to a decision maker accompanied by a small report on the model is not recommended.

Communication between the decision maker and the modeler can be facilitated in many ways, and it often is the primary responsibility of the modeler to do so. If a model is built as a menu system, it might be possible to teach the decision maker how to use the model in only a few hours, thereby also increasing his or her understanding of the model and its results. If an interactive approach is applied, it is possible for the decision maker to visualize a wide range of possible decisions. The effect of this approach is increased by the use of various graphic methods to illustrate the best possible decision in regard to what happens with the use of various management strategies. Under all circumstances it is recommended that time be invested in developing a good graphic presentation of the model results to a decision maker. Even if he or she has been currently informed about a model project through all its phases, the decision maker will not necessarily understand the background and assumptions of all the model components. Thus, it is important that the model results, including the main assumptions, shortcomings, and standard deviations underlying them, are carefully presented with the use of an illustrative method.

It is clear that we are not yet sufficiently advanced in environmental modeling efforts to solely use model results to define management programs, even utilizing expert systems and decision-support systems. A model should never be used in this way by a decision maker, but rather should be considered one useful tool in the management decision-making process. This implies that modeling results should be clearly and illustratively presented, and be considered a significant component in discussions about selecting specific courses of action. Other elements to be considered in such discussions would include potential side effects, interpretation of model predictions, and the implications of the accuracy of the prognosis.

A good environmental model can be a powerful tool in the decision-making process for management actions. A wide range of environmental problems has been modeled to varying degrees over the last 10–15 years. They have generally been of important assistance to decision makers. With the continuing rapid growth in the use of environmental models, the situation will only improve in the future. However, we have not achieved the same level of experience for all environmental problems.

The use of models in environmental management is definitely growing. They have been widely used in several European Countries, in North America, and in Japan. Further, environmental agencies in more and more countries are making use of model applications. Through such journals as *Ecological Modelling* and the *International Society for Ecological Modelling* (ISEM), it is possible to follow the progress in the field. This "infrastructure" of the modeling field facilitates communication and accelerates the exchange of experiences, thereby enhancing the growth of the entire field of ecological modeling.

It appears there will soon be a need for a model "bank" or database, where users can obtain information about existing models, their uses, characteristics, etc. At the same time, it is not possible to readily transfer a model from one case study to another, as is reiterated in this book. It is furthermore even difficult to transfer a model from one computer to another unless it is exactly the same type of computer. However, it is often very helpful to learn from experiences of others involved in similar modeling situations in other parts of the world. Journals such as those mentioned earlier facilitate such exchanges.

The issue of the generality of models also deserves discussion. Few models have been used on multiple case studies, in order to gain wide experience on this important matter. The eutrophication model presented by Jørgensen and Fath (2011) has been used on a wide range of lakes in the temperate and the tropical zone, as well as for shallow and deep lakes, and even fjords. The experience gained in these case studies is illustrative, but does not necessarily represent general properties of models. A further discussion on the generality of models is presented later in this chapter.

It should not be forgotten in this context that models have always been applied in science. The difference between present and previous models is that today, with modern computer technology, we are able to work with very complex models. At the same time, however, it is a continuing temptation to construct models that are too complex. It is easy to add more and more equations and state variables to a computer program, but much harder to get the data needed for calibration and validation of the model. Even if very detailed knowledge about a problem is available, it is not possible to develop a model capable of accounting for the complete input–output behavior of a real ecosystem and being valid for all frames (Ziegler 1976). This type of model is called 'the base model' by Ziegler, and it would be very complex and require such a large number of computational resources that it would be almost impossible to simulate. The base model of a problem in ecology will never be fully known because of the complexity of natural ecosystems and our inability to observe all its states. Thus, a model may be made more realistic, up to a point, by adding more connections. However, addition of new parameters after that point is reached does not contribute further to an improved simulation capacity. Indeed, more parameters imply more uncertainty because of the possible lack of information about the flows the parameters quantify.

For a given quantity of data, the addition of new state variables or parameters beyond a certain model complexity does not increase our ability to model an ecosystem but only adds to the unaccountable uncertainty. The question that can be formulated in relation to this problem is *how can we select the complexity and the structure of the model to assure the optimum for knowledge gained or the best answer to the question posed by the model?*

Costanza and Sklar (1985) have examined 88 different models and were able to show that the theoretical discussion behind these considerations

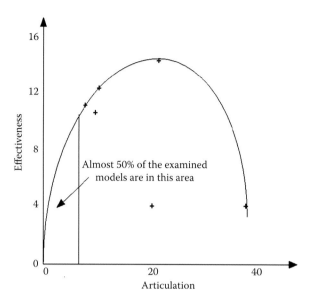

FIGURE 19.2

Plot of articulation index versus effectiveness = articulation * certainty for the models reviewed by Costanza and Sklar (1985). Articulation is expressing complexity as number of state variables + 0.1 * number of boxes + 0.01 * time steps. Certainty = 1/uncertainty = 1/standard deviation found by the validation. As almost 50% of the models were not validated, they had an effectiveness of 0. These models are not included in the figure but are represented by the line effectiveness = 0. Notice that almost another 50% of the models have a relatively low effectiveness due to too little articulation and that only one model had too high articulation, which implies that the uncertainty by drawing the effectiveness frontier as shown in the figure is high at articulations above 25.

actually is valid in practice; see Figure 19.2. Selection of the right degree of complexity is of great importance for environmental and ecological models. Jørgensen and Fath (2011) present methods that could be used to select the complexity in a specific case. However, the selection will always require that the application of these methods is combined with a good knowledge of the system being modeled. The methods must work hand-in-hand with an intelligent answer to the question.

Which components and processes are most important for the problem of concern? Know the ecosystem and the problem being addressed before a model is selected, including the model complexity. Thus, the conclusion: A proper overview and a holistic picture of a given ecosystem are crucial for the right selection of the model complexity. Some details are of course needed to understand how the ecosystem works at the system level. Thus, the additional conclusion is therefore that, although we can never know everything about an ecosystem needed to make a complete model (i.e., with inclusion of all details). However, good, workable models that expand our knowledge of ecosystems can be produced, particularly because of their properties as systems. Ulanowicz (1979) points out that a correct, very precise predictive model cannot be constructed. Thus, it would often be most fruitful to build a model that illustrates general trends, taking into account the probabilistic nature of the environment.

Further, models are, and can be applied, as management tools (e.g., see Straskraba and Gnauck 1985, Jørgensen 1986, Chapra and Canale 1991). All in all, models should be considered as tools—to provide an overview of complex ecosystems. For a complex system, a few interactive state variables already make it impossible to get an intuitive or logical overview on how the system reacts to perturbations or other changes. In such cases, models are good tools to help obtain an overview or picture of the properties of an ecosystem on the systems level.

There are only two possibilities for getting around the dilemma rooted in the model complexity issue: either limit the number of state variables in the model or describe the system by the use of holistic methods and models, preferably by using higher-level scientific laws. *The tradeoff for the modeler is between knowing much about little, or little about much!*

More complex models require more data and more knowledge about the ecosystem. They imply higher costs, but they should be justified by the importance of the pollution study. Thus, it is recommended to proceed in a step-wise manner toward a more complex learning from simpler models before the complex model is constructed. A procedure for this step-wise development is presented and discussed in Jørgensen and Fath (2011).

Another crucial problem is associated with model generality. Can a model, used in, for instance, one lake, be used unchanged for other case studies? The answer is not a simple yes or no. However, experience shows that simpler models can be used more generally than more complex models. They contain a description of the basic processes characterizing aquatic ecosystems (e.g., nutrient uptake by phytoplankton, which is dependent on the nutrient concentration, the concentration of phytoplankton, and mineralization of detritus). The more complex models inevitably will contain more site-specific process descriptions, which may not be important for all lake and reservoir case studies. Thus, more complex models will generally have to be modified from case to case. This is illustrated in Tables 19.2 and 19.3, in which the experience gained with the general use of an eutrophication model containing 17–20 state variables is presented.

Problems of interest for management of aquatic inland ecosystems include (1) eutrophication, (2) ecotoxicological effects, (3) acidification, (4) fishery management, (5) the oxygen concentration and possibly oxygen depletion, (6) management of wetland processes, (7) the bacteriological quality of water, and (8) hydrodynamic problems related to water discharge or to the use of water for cooling or production of drinking water.

All eight of these problems have been modeled extensively. Table 19.4 provides an overview of the modeling activity for each of these eight problems and a couple of other problems associated with aquatic ecosystems, on a scale from 0 to 5, where 5 means very intense modeling efforts,

TABLE 19.4

Models of Environmental Problems of Freshwater Ecosystems

Problem	Modeling Effort
Oxygen balance	5
Eutrophication	5
Heavy metal pollution, all types of ecosystems	4
Pesticide pollution of freshwater ecosystems	4
Other toxic compounds include. ERA	5
Endangered species (includes population dynamic models)	3
Ground water pollution	5
Acid rain	5
Consequences of climate changes	4
Fishery models	5
Bacteriological water quality	3
Hydrodynamic problems	5

TABLE 19.5

Models of Freshwater Ecosystems

Ecosystem	Modeling Effort
Rivers	5
Lakes, reservoirs, ponds	5
Wetlands	5
Lagoons	5
Arctic ecosystems	1–2
Wastewater systems	5
Estuaries	5

4—intense modeling effort, 3—some modeling effort, 2—a few models have been studied, 1—one or a few studies with not well calibrated or validated models, and 0—no modeling efforts have been done. The table indicates that all eight problems have been modeled to relatively high extent. The same scale is applied in Table 19.5, which gives an overview of the modeling effort for different aquatic inland water ecosystems: lakes (and reservoirs), rivers and streams, wetlands, and lagoons.

Thus, a spectrum of models is available for all the mentioned problems and for the seven aquatic inland ecosystems listed in Table 19.5. The selection of the appropriate complexity dictated by the problem, the characteristics of the ecosystems, and the quality and quantity of the available data, as discussed is crucial for the modeling results.

The first step in selecting a model for a specific case is clear specification of the goals for which the model will be used. The selected model must be one intended to answer the questions of interest, as defined by the lake manager. Another consideration in selecting a model is the specific data available for use with the model. Using most advanced models is impossible, for example, if the quantity and quality of the water inflows have not been measured. In some cases, it may be possible to obtain data required for the model, but this inevitably involves expenditures of money and time, since most lake inflows vary considerably over time. The availability of personnel skilled in the use of models also can be a significant limitation. Learning the modeling procedure, and understanding the applied basis for model development, can be both difficult and time consuming. Thus, appropriate model selection requires balancing the importance of the problem, the money and the time needed to be invested in the model development, the available personnel, and the existence of appropriate models.

As emphasized in previous sections, models only provide a gross simplification of reality. Thus, caution is always necessary when considering model results. Moreover, three levels of uncertainties influence the use of even the best model (Hilborn 1987), including

- Noise—the natural variability that occurs sufficiently frequently to be routine (various sampling schemes and statistical analyses are required to accommodate this uncertainty)
- The state of nature that is not well known
- Surprises—unanticipated events (flexible management strategies cope with unanticipated events more effectively than rigid, dogmatic strategies)

These uncertainties are inherent in any complex system, and may occur in any specific case. By proper use of validation techniques, it is possible in most cases to quantify the uncertainty, which is very valuable for the interpretations of the model predictions.

Models useful for water quality management may be classified as follows:

- *Simple static calculation models* consisting of algebraic equations or graphs. These models are based predominantly on the statistical elaboration of data sets, sometimes large ones. Thus, they are limited by the extent of the materials covered in the analysis. For example, many models based on empirical relationships between phosphorus and chlorophyll presented in the literature. Examples are the models provided by Thornton et al. (1999) for estimating nonpoint source pollution loads.
- *Complex dynamic models* providing analysis of the timing of water quality conditions. Several of the models presented in the next sections belong to this class, and represent a wide spectrum of complexity.

- *Geographical information systems (GIS)* used for problems requiring spatial resolution. This is typical for estimating pollutant loads, particularly for nonpoint source pollution. The basis of GIS is computerized maps and procedures for entry and treatment of spatial data. For example, a specific watershed can be included in the database and corresponding pollution sources indicated. Through the use of models (usually more complex dynamic models), the expected pollution input can be calculated. Several attempts to apply GIS in watershed management have emerged in recent years.

- *Prescriptive models* can be used to calculate water quality conditions without indicating appropriate management options for a given situation. By means of scenario analysis, it is possible to test management alternatives and to predict the potential consequences of water quality. This can be useful in selecting the most appropriate management possibilities. These models are based either on simple static calculations or dynamic simulations, depending on whether a time-independent or time-dependent solution is required.

- *Optimization models* incorporate selection procedures to choose the most suitable option based on a set of criteria. Such models, which are often based on complex dynamic models, can allow simultaneous analysis of several management alternatives or goals. The major component of an optimization model is called a goal function, which is a function the user seeks to minimize or maximize critical water quality variables (e.g., oxygen concentration in a lake) or the money spent on attaining a specified level of water quality improvement. Optimization with constraints means that some, or all, of the management parameters are limited (i.e., they are forced to remain within specified limits due to natural conditions, management limitations, etc.). The following examples illustrate these kinds of constraints:
  a. Inflow phosphorus levels cannot be reduced more than the capabilities of treatment plants or other available reduction methods.
  b. It is impossible to mix a water body beyond its greatest depth; thus, there is a natural depth of mixing.
  c. There are no feasible reasons to reduce the chlorophyll concentrations, or increase the oxygen concentrations, above a certain limit.

The *standard optimization* procedure is used when a management decision involves only one variable (e.g., reservoir outflow) and is determined by one characteristic of this variable (e.g., outflow rate by turbines). The goal in this case, for example, can be to keep the reservoir water level within certain limits. *Multiparameter* optimization also concerns one variable, although the optimal performance is searched for in a multidimensional parameter space. It can be asked, for example, to what extent should the bottom outflow gate of the reservoir be opened, and a turbine run, to keep the water level in the reservoir within certain limits. The constraints are given by the maximum capacity of the bottom outlets and the turbines and by the minimum and maximum flow rates prescribed for the river downstream of the reservoir. The model formulation with respect to water quality, for example, may be retaining some water quality variables below a certain limit, while also modifying several parameters characterizing the use of various management options. As an example, the model GIRL OLGA is intended for making dynamic, time-dependent estimates of the best combination of the time sequence of using five different management options, each characterized by one parameter that can be manipulated within certain limits. *Multivariable* optimization simultaneously takes into account several variables. Examples would include keeping the oxygen concentration, quantity of algae, and content of organic matter in a lake within certain limits. *Multigoal* optimization is the most complex, with several goals having to be simultaneously achieved. In this latter case, a compromise set is sought, rather than a unique optimal solution, with the model user driven to make a proper selection. A combination of multiparameter, multivariable, and multigoal formulation also is possible, although such a solution has not yet been attempted for water quality modeling. Up to the present time, more complex formulations have generally been used to address water quantity problems (e.g., optimal operation of a reservoir cascade). Nevertheless, progress in developing optimization techniques is moving forward rapidly, and useful water quality formulations will doubtless be available in the future. Table 19.6 provides a listing of optimization water quality management models.

TABLE 19.6

Optimization of Water Quality Management Models

Dynamic optimization of eutrophication by *phosphorus removal*. Used for a Japanese lake (Matsumura and Yoshiuki 1981)

Optimal control by *selective withdrawal* (Fontaine 1981)

Optimizing reservoir operation for *downstream aquatic resources*. Applied on Lake Shelbyville, Illinois, United States (Sale et al. 1982)

GIRL OLGA for cost minimization of eutrophication abatement, using time-dependent selection from five management options (Kalceva et al. 1982, Schindler and Straskraba 1982). Applied on several reservoirs in the Czech Republic

Stochastic optimization of *water quality* (Ellis 1987)

COMMAS for prediction of environmental multiagent system (Bouron 1991)

DELWAG-BLOOM-SWITCH for management of eutrophication control of shallow lakes (Van der Molen et al. 1994)

GFMOLP, a fuzzy multiobjective program for the optimal planning of reservoir watersheds (Chang et al. 1996)

It is emphasized that optimization procedures only allow to select among the possibilities included in the model and are limited by the validity of the model, including its assumptions and formulations and its imposed constraints.

Thus, the model conclusions should be used with caution, with the user considering the model limitations, possible inadequacies, and possible insufficiency of input data. The model can be run before any decision is taken and several alternatives investigated. Alternatively, the model can be connected to automatic devices that activate water quality management options. As examples, chlorophyll concentrations and meteorological variables can be automatically recorded and put into a computer model. Short-term predictions by the model can be used to switch water mixing devices on or off, or to specify the intensity of phosphorus purification.

- *Expert systems* use qualitative and quantitative expressions to guide the model user toward relevant answers to complex water quality questions. The major advantage of expert systems is their ability to consider qualitative characteristics, in addition to quantitative characteristics, and to handle complex decision rules. The name of this model group originated from the basis in which these answers are obtained, namely, the judgment of experts in a given field. An interactive mode is available in which the user interacts with the computer software, selecting questions offered by the computer, obtaining answers, and answering the questions. Each expert system is devoted to a particular problem. They are called *empty expert systems* or *expert shells* directed to handling the *knowledge base*, which is the specific decision tree on what to do, how to decide in the given instance, etc. A general review of expert systems applicable to different environmental problems is provided by Hushon (1990), Davis (1993), and Davis and Guariso (1994).

- *Decision support systems (DSS)* represent a further extension of expert systems. They incorporate other computer software products relevant for a specific water quality decision problem for which the system was constructed (e.g., simple models, simulation models, optimization models, and GIS systems). A graphics package that generates explanatory drawings and texts can be an integral part of *DSS*, and all the model types mentioned earlier can be incorporated in the system. The entire decision support system is driven by questions offered by the computer and answers provided by the model user.

DSS were named for their ability to *support* decisions. They are not intended to *make* decisions. Wise, experienced people must always make the final decisions. In order to make acceptable decisions, however, people need varying amounts of information that often is not easily obtainable. For complex systems such as water quality systems, it is difficult to preview the consequences of different options because there are many nonlinear relationships and complicated interactions among their components (Loucks and da Costa 1991, Simonovic 1996).

DSS is a tool that provides managers with the necessary information regarding the potential consequences of various management decisions. DSS uses both the experiences of numerous experts and the capabilities of the computer to rapidly calculate many complex relationships. The interactive function allows the user to assess the various versions of decisions under different possible situations.

19.5 Eutrophication Models

As eutrophication is one of the most important problems of freshwater ecosystems, it is not surprising that particularly many eutrophication models have been developed. They represent a particularly wide spectrum of complexity. Table 19.7 reviews the eutrophication models representing the spectrum of complexity and the characteristics of the models (i.e., the number of state variables, the nutrients considered, the number of lake segments, or number of water layers, whether constant stoichiometric or independent nutrient cycles were applied, whether the model has been calibrated and validated, the number of case studies to which the model has been applied, etc.). It is assumed, particularly for the most complex models, that some modifications from case to case are needed to reflect specific lake conditions or properties (also see discussion in Section 19.4 on model generality).

As seen in Table 19.7, the model complexity is wide and the models include a different number of submodels. Three possible core submodels that should be considered for inclusion in the model are presented here. The characteristics of each case study, as discussed previously, should determine whether these three submodels should be included or not. They represent typical considerations for selecting the complexity for eutrophication models.

TABLE 19.7

Spectrum of Eutrophication Models

From 1 to more than 40 state variables
P, N, C, and Si considered as nutrients
From 1 to 7 layers—with or without thermocline
From 1 to many boxes or segments
Some have been calibrated and validated and even prognoses has been validated
Stoichiometric and nonstoichiometric description of phytoplankton growth, that is, dependent or independent cycling of nutrients
From 1 to many case studies

The application of independent nutrient cycles inevitably increases the complexity of the model, as this situation requires that nitrogen, phosphorus, carbon, and perhaps silica should be included as state variables in each trophic level. In most models that include independent nutrient cycles, however, only one state variable is considered for zooplankton and fish. The application of independent nutrient cycles implies that the growth of phytoplankton is described as a two-step process as follows:

- Phytoplankton nutrient uptake, in accordance with Monod's kinetics
- Phytoplankton growth determined by the internal substrate concentration

This complication obviously requires that the underlying data are of sufficient quality and quantity. Di Toro (1980) has shown that the application of independent nutrient cycles is particularly important when the model is used for shallow, very eutrophic lakes. In contrast, this complication can be omitted for deep, mesotrophic or oligotrophic lakes.

Because lake bottom sediments accumulate nutrients, it is important to describe quantitatively the processes determining the mass flows from sediment to water. This is of particular importance, when the mass flows sediment $\rightarrow$ water are significant compared with the other mass flows. As the relative amount of nutrients stored in the sediment is most significant for shallow, eutrophic lakes, the more detailed description presented later should almost always be included in models for such types of lakes.

The sediment–water submodel attempts to answer the following crucial question: To what extent will accumulated compounds in the sediment be redissolved into the lake water? Because these processes are important for lake eutrophication, the phosphorus and nitrogen exchange processes between mud and water have been extensively studied. Chen and Orlob (1975) ignored the exchange of nutrients between mud and water; as pointed out by Jørgensen et al. (1975), this will inevitably give a false prognosis. Ahlgren (1973) applied a constant flow of nutrients between sediment and water, and Dahl-Madsen and Strange-Nielsen (1974) used a simple first-order kinetic to describe the nutrient exchange rate.

A more comprehensive submodel (Figure 19.3) for the exchange of phosphorus has been developed by Jørgensen et al. (1975). The settled material, S, is divided into $S_{detritus}$ and S_{net}; the first being mineralized by microbiological activity in the lake, and the latter being the material actually transported to the sediment. Snet can also be divided into two flows as follows:

$$S_{net} = S_{net,s} + S_{net,e} \tag{19.1}$$

where
  $S_{net,s}$ is the flow to the stable nonexchangeable sediment
  $S_{net,e}$ is the mass flow to the exchangeable unstable sediment

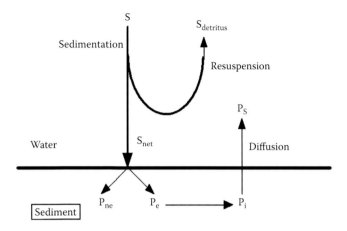

FIGURE 19.3
Submodel of the phosphorus exchange between sediment and water. The model distinguish between P_{ne}—nonexchangeable phosphorus—and P_e—exchangeable phosphorus. The latter is decomposed microbiologically and yields the interstitial phosphorus P_i, which by diffusion is transferred from sediment to water.

Correspondingly, the nonexchangeable, P_{ne}, and exchangeable, P_e, phosphorus concentrations, both based on the total dry matter in the sediment, also can be distinguished. Increases in the stabilized sediment can be found by numerous methods. The analysis of lead isotopes, for example, is a fast and reliable method.

The exchangeable phosphorus is similarly mineralized to detritus in a lake, and a first-order reaction gives a reasonably good description of the conversion of P_e into interstitial phosphorus, P_i.

Finally, the interstitial phosphorus, P_i, will be transported by diffusion from the pore water in the sediment to the lake water. This process, which has been studied by Kamp-Nielsen (1975), can be described by means of the following empirical equation (valid at 7°C):

$$\text{Phosphorus release} = 1.21\ (P_i - P_s) - 1.7\ (\text{mg P/m}^2\text{day}) \qquad (19.2)$$

where P_s is the dissolved phosphorus in the lake water.

This submodel of water–sediment exchange was validated in three case studies (Jørgensen et al. 1975), based on examining sediment cores in the laboratory. Kamp-Nielsen (1975) has added an adsorption term to these equations.

A similar submodel for sediment nitrogen release has been developed by Jacobsen and Jørgensen (1975). The nitrogen release from sediment is expressed as a function of the nitrogen concentration in the sediment and the temperature under both aerobic and anaerobic conditions.

Figure 19.4 shows a sediment profile from these examinations, illustrating the interpretation of the profile that can be used in the model to distinguish between exchangeable and nonexchangeable sediment phosphorus.

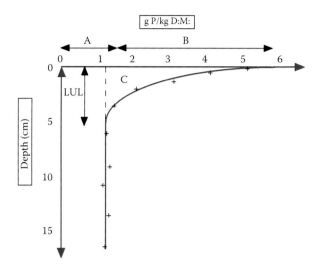

FIGURE 19.4
Typical profile of phosphorus in sediment. A corresponds to the nonexchangeable phosphorus and B to the exchangeable phosphorus. The phosphorus amount indicated as C can be decomposed microbiologically and be transferred as phosphorus in the pore water of the sediment to the water phase.

Grazing of phytoplankton by zooplankton (Z) and the predation of zooplankton by fish (F) are both expressed by a modified Monod expression, which considers a threshold concentration, KT, below which grazing or the predation does not occur. The expression for phytoplankton grazing by zooplankton (see Jørgensen and Fath 2011) is expressed as follows:

$$\mu Z = \mu Z_{max} \frac{(Phyt - KT)}{(Phyt + KM)} \tag{19.3}$$

where KM is the Michaelis–Menten constant.

A zooplankton carrying capacity often must be introduced to give a better simulation of zooplankton and phytoplankton. Although carrying capacities are often observed in ecosystems, the need to introduce them in this case may be due to a too-simple representation of the grazing process. Phytoplankton might not be grazed, for example, by all the zooplankton species present, and some zooplankton species might use detritus as a food source. The zooplankton growth rate, mZ, is computed in accordance with these modifications as

$$\mu Z = \mu Z_{max} \cdot FPH \; FT2 \; F2CK \tag{19.4}$$

where
FPH is the expression in Equation 19.3
FT2 is a temperature regulation expression
F2CK is the accounts for the carrying capacity

Now,

$$F2CK = \frac{CK - ZOO}{CK} \tag{19.5}$$

where CK is the carrying capacity.

   If the data are not sufficient to include the three important submodels presented earlier or other relevant submodels that an individual case study might require, it is recommended that the needed data be obtained with an intensive measuring effort that can provide high-quality data for the specific time period when the eutrophication processes are most dynamic.

   An intensive measuring period, with several sets of measurements each week during the spring and summer bloom period, can first of all be applied to improve parameter estimation, which is often a focal problem in developing ecological models. The experience from conducting intensive measuring periods has identified the following advantages:

- Different optional expressions of simultaneously limiting factors were tested, and only two gave an acceptable maximum growth rate for phytoplankton and an acceptable low standard deviation. (1) Selection of the most limiting factor by using a min-expression (for further details see Jørgensen and Fath (2011) and (2) averaging the limiting factors).

- The previously applied expression for the influence of temperature on phytoplankton growth gave unacceptable parameters, with standard deviations that were too high. A better expression was introduced as a result of the intensive measuring period.

- It was possible to improve the parameter estimation, which gives more realistic values for some parameters. Whether this would give an improved validation when observations from a period with drastic changes in the nutrients loading are available could not be determined.

- The other expressions applied for process descriptions were confirmed.

It is important to validate models against another set of measurements. A typical validation for a model with a medium to high complexity, developed on basis of good data, would yield a standard deviation of about 15%–35%. It means that the difference between one observation and the corresponding model result would be 15%–35%. As the overall image of the annual variation of eutrophication in a freshwater ecosystem would be based on several hundred comparisons—let us say 400 sets of observations of state variables and the corresponding model results—the standard deviation for the annual image would be the square root of 400 or 20 times less than the standard deviation for one comparison—it means it would be about 1%–1.8%, which is an acceptable validation.

   A prognosis for the development of eutrophication in Lake Glumsø (the validation results referred to in the previous text are based on this case study)

TABLE 19.8

Model Predictions in Two Cases for Concentrations of Treated
Wastewater: Case A: 0.4 mg P/L; Case B: 0.1 mg P/L

	Third Year		Ninth Year	
	Case A	Case B	Case A	Case B
g C/m² Year	650	500[a]	500	320[a]
Minimum transparency (cm)	50	60	60	75

[a] An error of 3% on this value could be expected if the validation
results hold.

by different removal efficiencies for phosphorus, nitrogen, or phosphorus
and nitrogen simultaneously have been made. The validation of this prognosis is presented here to illustrate the reliability of prognoses made on the
basis of well-developed eutrophication models.

It was previously stated that nitrogen removal had little or no effect on the
lake, while phosphorus removal would give substantial reductions in the
phytoplankton concentration. The results of two cases are summarized in
Table 19.8 as follows:

Case A: The treated waste water has a concentration of 0.4 mg P/L, corresponding to about 92% removal efficiency, which should be achieved with
proper chemical precipitation.

Case B: The treated wastewater has a concentration of 0.1 mg P/L, corresponding to about 98% removal efficiency, which will require chemical precipitation, for example, in combination with ion exchange.

As seen in Table 19.8, the water quality will improve significantly in accordance with the predictions. Case B, with a 98% removal of phosphorus, is
obviously preferred. In the third year, Case B will give a reduction in production from 1100 to 500 g C/m² year, with the water transparency increasing from a minimum value of 20–60 cm. The ninth year would even result
in reduction of the production to 320 g C/m² year, corresponding to almost
a mesotrophic lake, which is an acceptable improvement for a shallow lake
situated in an agricultural area.

The prognosis gives a pronounced effect of 98% phosphorus removal,
which could therefore be recommended to the appropriate environmental
authorities. Further improvements after nine years should not be expected
with this case study.

Conveyance of the wastewater also was considered but has the following
disadvantages:

- It is slightly more expensive than the Case B solution, taking interests, depreciation, and running costs into consideration.
- The phosphorus is *not* removed but only transported to the downstream Susaa River where its effects have not been considered.

- The sludge produced at the biological treatment plant will be less valuable as a soil conditioner, since the phosphorus concentration will be lower than when phosphorus removal is included.
- The freshwater is not retained in the lake from which it could have been reclaimed, if needed, after storage for some time. Freshwater is not presently a problem in this area, but it is foreseen that it might be in 20–40 years.

In spite of these observations, the community in this case study chose to convey its wastewater to the Susaa River, due to a preference for traditional methods. The pipeline was constructed in 1980 and began operations in April 1981, which has enabled a validation of the presented prognosis.

Lake Glumsø was ideal for these studies of model application due to its limited depth and size, but also because a reduced nutrient input to the lake could be foreseen. The limited retention time (about 6 months) makes it realistic to obtain a validation of a prognosis within a relatively short time interval (a few years). On April 1, 1981, the direct input of wastewater to the lake was stopped. Because the capacity of the sewage system is still too small, however, a minor input of mixed rain water and wastewater is discharged into the lake from time to time through an upstream tributary. Thus, the phosphorus loading is not reduced by 98% but rather only by 88% (determined by a phosphorus balance). The prognosis in Case A, therefore, can be used for comparison.

During the third year after the reduction in phosphorus load occurred, a pronounced effect on the lake was observed. Table 19.9 compares some of the most important data of the prognosis. It also includes data obtained during the first 2 months of the third year. The table identifies errors as ± for g C/m^2 day and chlorophyll maximum (mg/m^3) based on the validation. For the prognosis values, the results from the validation (8% for production and 15% for phytoplankton concentration) are used to determine standard deviations. For the measured values, an error of 10% was estimated.

The lake was previously dominated by *Scenedesmus*. After the conveyance of the wastewater, it was dominated by diatoms, which have a lower optimum temperature and, therefore, typically bloom earlier in the spring than *Scenedesmus*. This seems to explain the discrepancy between the predicted and measured values regarding this parameter. The model, therefore, might improve its predictions if it was possible to account for shifts in species composition. Results published by Jørgensen (1981), Jørgensen and Mejer (1981a,b), Jørgensen (1992) indicate this would be possible by application of models named "structurally dynamic models"(SDMs); see Section 19.10 and Jørgensen and Fath (2011). They have given very promising results and the application of SDM will therefore be presented later in this chapter. In the Lake Glumsø case, however, since diatoms take up silica, it could probably also be necessary to introduce a silica cycle into the model.

TABLE 19.9

Comparison of Model Prognosis and Measured
Data

	Prognosis	Measurement
	(Case A, 92% P Reduction)	Approximately 88% Reduction
Minimum transparency		
First year	20 cm	20 cm
Second year	30	25
Third year	45	50
g C/m²·day, maximum		
First year	9.5 ± 0.8	5.5 ± 0.5
Second year		
(Spring)	6.0 ± 0 5	11 ± 1.1
(Summer)	4.5 ± 0.4	3.5 ± 0.4
(Autumn)	2.0 ± 0.2	1.5 ± 0.2
Third year		
(Spring)	5.0 ± 0.4	6.2 ± 0.6
Chlorophyll in spring, mg/m³, maximum		
First year	750 ± 112	800 ± 80
Second year	520 ± 78	550 ± 55
Third year	320 ± 48	380 ± 38

The other production and chlorophyll values are well predicted except the spring production in the second year (Table 19.9). The predictions of minimum transparency are acceptable, as they show a difference of 5 cm or less. The general trends in the nutrient concentrations exhibit a good correlation between predicted and measured values.

This case study shows that it is possible to make reliable predictions using eutrophication models, provided that data of sufficient quality and quantities are available. On the other hand, an attempt should always be made to develop an eutrophication model, even if the available data are not sufficient to construct a complex model, because the model results can be used to compare different management strategies with an indication of the uncertainty resulting from the validation. If the uncertainties are not acceptable, the model can be used to assess the weak points and help identify modifications that should be introduced to improve the model results.

Thus, the model can be considered a tool that is able to synthesize the knowledge available on a lake ecosystem of concern and, by following the recommendations in Section 19.4, should be able to obtain the best possible model predictions under given circumstances.

19.6 Toxic Substance Models

Over approximately the last 25 years, models of toxic substances have emerged as a result of the increasing interest in the environmental management of toxic substances that can cause water pollution. Toxic substance models attempt to model the fate and/or effects of toxic substances in ecosystems. Toxic substance models are most often biogeochemical models because they attempt to describe the mass flows of the toxic substances being considered. There also are models of population dynamics, which include the influence of toxic substances on the birth rate, growth rate, and/or mortality and, therefore, should also be considered toxic substance models.

Toxic substance models have some characteristic properties:

- The need for parameters to cover all possible toxic substance models is great, because the modern society is using about 100,000 different chemicals. Thus, general estimation methods are widely used. Methods based on the chemical structure of chemical compounds are developed: the so-called QSAR and SAR methods.
- The safety margin should be high, for example, when expressed as the ratio between the actual concentration and the concentration that gives undesired effects.
- The possible inclusion of an effect component, which relates the output concentration to its effect. It is easy to include an effect component in the model. However, it is often a problem to find a well-examined relationship upon which to base it. Rather, it will require the use of good knowledge of the properties, particularly the toxicological properties for the component of concern.
- The possibility and need of simple models due to points 1 and 2, and to our limited knowledge of the processes, parameters, sublethal effects, and antagonistic and synergistic effects.

The decision regarding which model class to apply is based on the ecotoxicological problem that one wishes to solve with the model. However, it often is only the toxic substance concentration in one trophic level that is of concern. This includes the zero trophic level, which is understood to be the medium—for aquatic ecosystem it means the water. Figure 19.5 shows a model of copper contamination in a freshwater ecosystem as an example. The main concern is the ionic copper concentration in the water, since it might reach a toxic level for the plants including phytoplankton. Zooplanktons, fishes, and mammals are much less sensitive to copper contamination, so the initial concern focuses on the concentration level harmful to phytoplankton. However, only the ionic form of copper is toxic, and, therefore, it is necessary to model the partitioning of copper in ionic form,

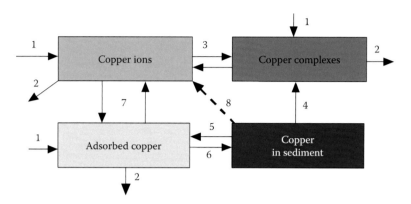

FIGURE 19.5
Model of aquatic copper contamination. Copper ions are the most crucial state variables, because copper ions are toxic to plants including phytoplankton, while ionic copper is much less toxic to zooplanktons, fishes, and mammals. The model gives the interactions between four forms of copper that are the state variables of the model: copper ions, copper complexes (with chloride, humic acid, and amino acids), copper adsorbed to suspended matter, and copper in the sediment. The transfer processes between the four state variables determine the copper ion concentration that determines the toxicity of the water.

complex-bound form, and adsorbed form (which easily can be done by the use of aquatic chemistry; see Chapters 10 and 11). The exchange between copper in the water phase and in the sediment also is included because the sediment is able to accumulate relatively large amounts of heavy metals. The amount released from the sediment may be significant under certain circumstances (e.g., under low pH).

The acknowledgement of the uncertainty is of great importance for all models, but it is particularly important for ecotoxicological models. It may be taken into consideration either qualitatively or quantitatively. Another problem is where does one take the uncertainty into account? Should the economy or the environment benefit from the uncertainty? Unfortunately, most decision makers up to now have used the uncertainty to the benefit of the economy. This is a completely unacceptable approach, since the same decision makers would, for example, never consider whether uncertainty should be used for the benefit of the economy or the strength of a bridge in a civil engineering project. Table 19.10 gives an overview of a number of ecotoxicological models applied in aquatic ecosystem management.

Ecotoxicological models are widely used to perform environmental risk assessments (*ERA*). This is of particular interest for lakes, reservoirs, and rivers that are used for drinking water supplies or for which the fisheries are significant. *ERA* typically answers questions such as what is the risk of contaminating drinking water or fish for a given application of pesticides in an agricultural project adjacent to a lake or reservoir?

TABLE 19.10

Examples of Ecotoxicological Lake Models

Toxic Substance	Model Characteristics	References
Cadmium	Food chain as in an eutrophication model	Thomann et al. (1974)
Mercury	Six state variables: water, sediment, suspended matter, invertebrates, plants, and fishes	Miller (1979)
Vinyl chloride	Chemical processes in water	Gillet (1974)
Heavy metals	Concentration factor, excretion, bioaccumulation	Ayoma et al. (1978)
Lead	Hydrodynamics, precipitation, toxicity of ionic lead on algae, invertebrates, and fishes	Lam and Simons (1976)
Radionuclides	Hydrodynamics, decay, uptake and release by various aquatic surfaces	Gromiec and Gloyna (1973)
Polycyclic aromatic	Transport, degradation, bioaccumulation hydrocarbons	Bartell et al. (1984)
Cadmium, PCB	Hydraulic overflow rate (settling), sediment interactions, steady state food chain submodel	Thomann (1984)
Hydrophobic organic	Gas exchange, sorption/desorption, hydrolysis compounds, photolysis, hydrodynamics	Schwarzenbach and Imboden (1984)
Mirex	Water–sediment exchange processes, adsorption, volatilization, bioaccumulation	Halfon (1984)
Toxins (aromatic hydrocarbons, Cd)	Hydrodynamics, deposition, resuspension, volatilization, photooxidation, decomposition, adsorption, complex formation, (humic acid)	Harris et al. (1984)
Persistent organic chemicals	Fate, exposure, and human uptake	Paterson and Mackay (1989)
pH, calcium and aluminum	Survival of fish populations	Breck et al. (1988)
Pesticides and surfactants	Fate in rice fields	Jørgensen et al. (1997)
Toxicants	Migration of dissolved toxicants	Monte (1998)
Growth promoters	Fate, agriculture	Jørgensen et al. (1998)
Toxicity	Effect on eutrophication	Legovic (1997)
Pesticides	Mineralization	Fomsgaard (1997)

19.7 Acidification Models

Lake acidification is caused by acid rain, which originates from emissions of sulfur and nitrogen oxides into the atmosphere. Thus, assessment of acidification for management purposes requires a chain of models, linking the emissions to the atmospheric pollution, to the effects of acid rain on soil chemistry in the catchment area, and further to the pH changes in the freshwater ecosystems. In this context, only the last submodels of the chain (i.e., freshwater ecosystem models and soil models) will be mentioned. The other models are mentioned in Jørgensen (2011) *Handbook of Ecological Modes Used in Ecosystem and Environmental Management*. See also more details of the other models in the model chain in Alcamo et al. (1990) and Jørgensen et al. (1995).

Several models have been developed to translate emissions of sulfur and nitrogen compounds into changes in soil chemistry (in the first hand to changes in the pH of soil water). Kauppi et al. (1984) used knowledge about buffer capacity and velocity to relate the emissions to soil water pH. A critical pH value of 4.2 is usually applied to interpret the results. From the results of the atmospheric deposition, it is possible to estimate the acid load in equivalents of hydrogen ions per square meter per year. In the European model, RAINS, the acid load is computed from the deposition after accounting for forest filtering and atmospheric deposition of cations (for details, see Alcamo et al. 1990).

The soil map of the world classifies European soils into 80 soil types. The fraction of each soil type within each grid element is computerized to an accuracy of 5%. The resolution of the European RAINS model is such that each grid element includes one to seven soil types, with a mean number of 2.2. The model goal is to keep track of the development of soil pH and buffer capacity, which is possible by aquatic chemical calculations; see Chapters 10 and 11. Further development of soil models is still needed, as the soil model must be considered the weakest model in the chain of models.

It can be concluded that models of soil processes and chemical compositions of soil water, including the concentrations of various ions and pH of the drainage water, are complex. It is not the intention here to present the models in detail—they are far too complex to do so—but rather to mention the difficulties and a few of the basic ideas behind them. Further details can be found in the references given in this section.

Henriksen (1980) and Henriksen and Seip (1980) developed an empirical model for the relationship between lake and river water pH and sulfur load. They plotted related values of pH and sulfur loads for a given catchment area with specific sensitivities for acidification. The resulting curves look like a titration curve. Sulfur load of less than 0.7 or even $0.5\,g/m^2$ year is needed to insure a pH of 5.3 or higher in sensitive areas, which means areas with a low buffer capacity of the water.

These simple empirical approaches are used without consideration of the drainage water. Weathering processes, however, can be incorporated in the empirical approaches, although the results have little generality since they are based on regression analysis of local measurements. The presented modeling approach has been used on Scandinavian lakes with good results.

Most biological models of lake acidification focus on fish. On the basis of data from 719 lakes, Brown and Salder (1981) developed an empirical model relating the fish population to the pH of the water. From their study lakes in southern Norway, they found that a 50% reduction of the sulfate emissions will give an average increase in pH of 0.2, which will only improve the fish populations in 9% of the lakes. However, a criticism of the model is that it underestimates the relationship between the reduction of sulfate emissions and pH.

Muniz and Seip (1982) developed another empirical model in which they distinguished between lakes of different conductivity. Chen et al. (1982) developed a very comprehensive pH-effect model, which considers the effects on all levels in the food chain and the total effects on the ecosystem.

In Sections 11.1 through 11.3, it is shown how it is possible relatively easy by a double logarithmic plot to find pH and buffer capacity for natural waters even in equilibrium with carbondioxide in the atmosphere.

19.8 Wetland Models

Nitrogen and phosphorus balances have shown that agriculture and other nonpoint sources contribute significantly to the eutrophication problem. These mass balance results imply that environmental technology alone is not sufficient to cope with nonpoint pollutant sources but must be supplemented with other methods. The results of comparative studies have shown that the use of wetlands is often a very effective method for reducing pollution (see also Chapters 6 and 17).

Mitsch (1976, 1983) provided a more comprehensive review of wetland models than we can present here. He distinguished between energy/nutrient models, hydrological models, models of spatial ecosystem, models of tree growth, process models, causal models, and regional energy models. Mitsch and Jørgensen (1988) also reviewed several types of wetland models.

A nitrogen balance for agricultural regions revealed that nitrogen from nonpoint sources plays a major role in lake pollution and that a solution to the eutrophication problem of freshwater and marine ecosystems cannot be realized without solving the problems associated with nonpoint sources of pollution. The entire spectrum of ecological engineering methods identified in Chapter 17 has been implemented in various cases to solve the problems. As indicated earlier, the methods may have different effects in different situations, depending on the actual mass balance and general properties of the lake or reservoir in question. In this context, there obviously is a need for a model that can be used to make predictions on the nutrient removal capacity of a wetland on the basis of certain information about an existing or a planned wetland. The goal is to develop as general a model as possible. However, because ecological models only have a certain generality, it is necessary to distinguish between the general relations and the more site-specific parameters and forcing functions. Thus, it is not possible to achieve a complete generality for wetland models.

Jørgensen and Fath (2011) present a relatively general wetland model that is able to predict nitrogen removal by wetlands. A wide spectrum of wetland models has been proposed lately (e.g., see Jørgensen et al. 1995, Jørgensen 2011). The various models that only focus on nitrogen removal can easily be expanded to also cover heavy metal and phosphorus removal by adsorption. A simple adsorption isotherm can be added, and consider the removal of phosphorus and nitrogen by harvest at day m of the wetland. The climatic information and some site-specific properties of the focal wetland(s) (e.g., content of organic matter, plant species, and hydraulic conductivity of the soil) must be assessed before a calibration, a validation, and a prediction with the use of the model can occur. The model software SubWet that can be applied to design constructed wetland has already been mentioned in Chapter 6. The details of this model are uncovered in Jørgensen and Fath (2011).

19.9　Fisheries Models

Augmented exploitation of freshwater fish resources by angling and commercial fishery and the deterioration of water quality have stimulated the concern about the depletion of fish stocks in freshwater ecosystems, particularly lakes, rivers, and reservoirs.

This reality has intensified the development of models which take into account the effects of fishery and water quality on the fish population. Models with a wide spectrum of complexity attempt to provide a management tool to assess an optimum fishery strategy.

For lakes with important commercial fisheries, landing records are often available for periods of multiple decades and may open the possibilities to develop a statistical fisheries model. However, a statistical model generally does not take into account a number of important factors, such as interactions among species, water quality, and changes in the concentrations of fish food. A statistical model would build on the assumption that present and past properties of the environment and the fish populations will be maintained. Statistical models will not be discussed further here, primarily because they do not consider the influences of water quality. It should be mentioned, however, that the International Lake Environment Committee (ILEC) and UNEP have developed software for a simple eutrophication model with four state variables, which also computes the fish population with a simple regression equation relating primary production and the fish population. Further details see also Section 19.3.

The simplest approach assumes that the entire fish population is homogeneous and does not consider the population dynamics and related age structure, which is essential for fishing policy. The more complex approaches consider the influences of water quality on the population dynamics and the age structure of the fish populations. A couple of simple fishery models with harvest are presented in Jørgensen and Fath (2011).

19.10 Structurally Dynamic Models

Our present models are built on generally rigid structures and a fixed set of parameters. No change or replacement of components is usually possible. However, we also need to introduce parameters (properties) which can vary according to changing general conditions for the state variables (components). The idea currently is to test whether a change of the most crucial parameters produces a higher so-called goal function of the system and, if so, to use that set of parameters.

As indicated in the introduction to this section, models that can account for change in species composition, and for the ability of the species (i.e., the biological components of the models) to change their properties (i.e., to adapt to the prevailing conditions imposed on them), are called *structural dynamic models*. They may also be called the *next* or *fifth generation* of ecological models, in order to underline that they are radically different from previous modeling approaches and can do more, including describing changes in species composition.

It could be argued that the ability of ecosystems to replace current species with others that are better adapted to the ecosystem can be addressed by constructing models that incorporate all possible species for the entire study period. However, there are two essential disadvantages. First, such models become very complex, since they will contain many state variables for each trophic level. Thus, many more of them need to be calibrated and validated, which will introduce a high uncertainty and render application of the model very case specific (Nielsen 1992a,b). In addition, the model will still be rigid and not replicate the ecosystem property of parameters that are continuously changing without changing species composition (Fontaine 1981).

Several goal functions have been proposed, but only a few models have been developed that can account for change in species composition or for the ability of the species to change their properties within some limits. Straskraba (1979) used maximization of biomass as the governing principle (the goal function). The model adjusts one or more selected parameters, in order to achieve maximum biomass at every instance. A modeling software routine is included which computes biomass for all possible combinations of parameters within a given realistic range. The combination that gives the maximum biomass is selected for the next time step, etc. However, the biomass cannot be used in models with more trophic levels. Adding the biomass of fish and phytoplankton together, for example, will lead to biased results.

The thermodynamic variable, exergy, or to emphasize its application in ecology, eco-exergy, has been used most widely as a goal function in ecological models. This thermodynamic concept is defined in Section 18.4. There are two pronounced advantages of eco-exergy as a goal function compared to entropy and maximum power (Odum 1983). These are that (1) it is defined far from thermodynamic equilibrium and (2) it is related to state variables that are easily determined or measured. One of lake case studies based on the SDM approach is discussed in the following.

Eco-exergy expresses work capacity. Eco-exergy assumes a reference environment representing the considered ecosystem but at thermodynamic equilibrium, see Section 18.4.

Eco-exergy measures the difference between free energy (given the same temperature and pressure) of an ecosystem compared with the same system as a dead homogenous system without gradients and without life. Survival implies the maintenance of biomass, while growth means its increase. Eco-exergy is needed to construct biomass, which then possesses eco-exergy (work capacity) transferable to support other processes. Thus, survival and growth can be measured by the use of the thermodynamic concept eco-exergy, which may be understood as *the free energy relative to the ecosystem at thermodynamic equilibrium.*

Darwin's theory of natural selection (1859), therefore, may be reformulated in thermodynamic terms and expanded to an ecosystem level as follows: *The prevailing conditions of an ecosystem steadily change. The system will continuously select those species (organisms) that contribute most to the maintenance or even to the growth of the eco-exergy (work capacity) of that system.*

Notice that the thermodynamic translation of Darwin's theory requires *populations* to possess properties of reproduction, inheritance, and variation. The selection of species that contribute most to the eco-exergy of the system under the prevailing conditions requires that there are sufficient individuals with different properties for selection to take place. Reproduction and variation must be high and, once a change has taken place, it must be conveyed to the next generation via better adaptation. It is also noted that a change in exergy is not necessarily ≥ 0 but rather depends on the resources of the ecosystem. However, the previous proposition claims that an ecosystem attempts to reach the highest possible eco-exergy level under the given circumstances with the available genetic pool (Jørgensen and Mejer 1977, 1979). It is not possible to measure eco-exergy directly. However, it is possible to compute it if the composition of the ecosystem is known.

The total eco-exergy of an ecosystem *cannot* be calculated exactly, as we cannot measure the concentrations of *all* the components or determine all possible contributions to the eco-exergy of an ecosystem. If we calculate the eco-exergy of a fox, for instance, the previously shown calculations will only give the contributions coming from the biomass and the information embodied in the genes, but what is the contribution from the blood pressure, the sexual hormones, and so on? These properties are at least partially covered by the genes, but is that the entire story? We can calculate the contributions from the dominant components, for instance, by the use of a model or measurements, that cover the most essential components for a focal problem. The *difference* in eco-exergy by *comparison* of two different possible structures (species composition) is here decisive in the development of SDMs. Moreover, eco-exergy computations give always only relative values, as the eco-exergy is calculated relatively to the reference system. Notice that the definition of eco-exergy is very close to free energy. Eco-exergy is, however, a *difference* in free energy between the system and the same system at thermodynamic equilibrium. The reference system used is different for every ecosystem according to the definition of eco-exergy. In addition, free energy is not a state function far from thermodynamic equilibrium. Consider for instance the immediate loss of free energy (or let us use the term eco-exergy as already proposed to make the use of the concepts more clear) when an organism dies. A microsecond before the death the information can be used and after the death the information is worthless and should therefore not be included in the calculation of eco-exergy. Therefore, eco-exergy cannot be differentiated.

It is possible to distinguish between the contribution to the exergy of information and of biomass (Svirezhev, 1998). p_i defined as c_i/A, where

$$A = \sum_{i=1}^{n} c_i \tag{19.6}$$

is the total amount of matter in the system is introduced as a new variable in Equation 18.2

$$Ex = ART \sum_{i=1}^{n} p_i \ln \frac{p_i}{p_{io}} + A \ln \frac{A}{A_o} \tag{19.7}$$

As $A \approx A_o$, exergy becomes a product of the total biomass A (multiplied by RT) and Kullback measure

$$K = \sum_{i=1}^{n} p_i \ln \left(\frac{p_i}{p_{io}} \right) \tag{19.8}$$

where p_i and p_{io} are probability distributions, a posteriori and a priori to an observation of the molecular detail of the system. It means that K expresses the amount of information that is gained as a result of the observations.

Table 19.4 gives an overview of the eco-exergy of various organisms expressed by the weighting factor ß = RTK (see Equations 19.7 and 19.8) that is introduced to be able to cover the exergy for various organisms in the unit detritus equivalent or chemical exergy equivalent per unit of volume or unit of area.

We find the exergy density as follows:

$$Ex-total-density = \sum_{i=1}^{N} ß i^*ci \quad \text{(detritus equivalents per unit of volume}$$
$$\text{expressed as g/L at the temperature 300 K)} \quad (19.9)$$

See also the introduction of this equation in Section 18.4. The ß-value embodied in the biological/genetic information is found on basis of (19.7) and (19.8) by adding the chemical and biological contributions to calculate the eco-exergy. Detritus has, in accordance with the equation, the ß-value = 1.0. By multiplication of the result obtained by (19.9) by 18.7, the exergy can be expressed in kJ per cubic meter or square meter assuming that (19.9) is giving g detritus equivalent.

ß = RTK is found from (1) the knowledge of the entire genome for 11 organisms and (2) for a number of other organisms by use of a correlation between various complexity measures and the information content of the genome (see Jørgensen et al. 2005). The values of ß for various organisms have been discussed in *"Ecological Modelling"* by several papers, but the latest published values in Jørgensen et al. (2005) are probably coming closest to eventually true ß-values. The previous ß-values were generally lower but the relative values between two organisms have not been changed very much by the recently published ß-values. As the ß-values have been used consequently as relative measures, the previous results obtained by exergy calculations are therefore still valid. Weighting factors defined as the exergy content relatively to detritus may be considered quality factors reflecting how developed the various groups are and to which extent they contribute to the exergy due to their content of information which is reflected in the computation. This is completely according to Jørgensen (2012). The ecosystem theory presented in this reference uses Boltzmann's equation to express the amount of work, W, that is embodied in the thermodynamic information

$$W = RT \ln M \ (ML^2T^{-2}) \quad (19.10)$$

where M is the number of possible states among which the information has been selected. M is as seen for species the inverse of the probability to obtain spontaneously the amino acid sequence valid for the enzymes controlling the life processes of the considered organism.

Notice that the ß-values are based on eco-exergy expressed in detritus equivalents per g of biomass, because by multiplication by the biomass concentration per liter you will obtain the eco-exergy in detritus equivalents per liter. The ß-values express therefore the specific eco-exergy, which is equal to RTK, where K is Kullbach's measure of information. R is 8.34 J/mol K or 8.34 kJ 10^{-8}/g, presuming an average molecular weight of 10^5 for the enzymes controlling the life processes. It can be shown that ß er is proportional to the free energy of the amino acids in the enzymes of various organisms (see Jørgensen et al. 2010). Three amino bases in the genome determine the selection of one amino acid among 20 possible. The number of amino acids in the right sequence can therefore also be used to find the information content.

It means that the ß-values or the specific eco-exergy in detritus equivalents

$$= RTK = 8.34 * 300 * 10^{-8} \ln20 * AMS/18.7 = 4.00 * 10^{-6} * AMS \quad (19.11)$$

where AMS is the number of amino acids in a coded sequence and ln 20 = 3.00.

Virus has coded about 2500 amino acids. The ß-value is therefore only 1.01. The smallest known agents of infectious disease are short strands of RNA. They can cause several plant diseases and are possibly implicated in enigmatic diseases of man and other animals. Viroid cannot encode enzymes. Their replication relies therefore entirely on enzyme systems of the host. Viroid has typically a nucleotide sequence of 360, which means that they would be able to code for 90 amino acids, although they are not translated. The ß-value can therefore be calculated to be 1.0004. It can be discussed of course whether viroid should be considered as a living material.

Eco-exergy calculated by use of the previously shown equations for ecosystems has some shortcomings:

1. We have made some although minor approximations in the equations presented earlier.
2. We do not know the non-nonsense gene and the details of the entire genome for all organisms.
3. We calculate only in principle the exergy embodied in the proteins (enzymes), while there may be other components of importance for the life processes that maybe or should be included. These components are, however, contributing less to the exergy than the enzymes and the information embodied in the enzymes control the formation of these other components, for instance, the hormones. It cannot be excluded that these components will contribute to the total exergy of the system.
4. We do not include the eco-exergy of the ecological network when we calculate the eco-exergy for an ecosystem. If we calculate the exergy of models, the network will always be relatively simple and the contribution coming from the information content of the network is therefore minor, but the information content of real ecological network may be significant due to their high complexity.

5. We will always use a simplification of the ecosystem, for instance, by a model or a diagram or similar. This implies that we only calculate the exergy contributions of the components included in the simplified image of the ecosystem. The real ecosystem will inevitably contain more components which are not included in our calculations.

It is therefore proposed that the exergy found by these calculations always should be considered a *relative minimum eco-exergy index* to indicate that there are other contributions to the total exergy of an ecosystem, although they may be of minor importance. In most cases, however, a relative index is sufficient to understand and compare the reactions of ecosystems, because the absolute exergy content is irrelevant for the reactions and cannot be determined due to the extremely high complexity. It is in most cases the change in eco-exergy that is of importance to understand the ecological reactions.

The weighting factors presented in Table 19.5 have been applied successfully to develop several SDMs. The relatively good results in application of the weighting factors, in spite of the uncertainty of their more exact values, seems only to be explicable by the robustness of the application of the factors in modeling and other quantifications. The differences between the factors of the microorganisms, the vertebrates, and invertebrates are so clear that it seems not to be important whether the uncertainty of the factors is very high—the results are robust.

The last pages have presented the theoretical background for the application and development of SDMs; for more details see Jørgensen and Fath (2011) and Jørgensen (2012). SDMs are important tools in environmental management, as they account for current changes of species composition and the properties of the organisms in the focal ecosystem. The idea of the new generation of models, SDMs, is to find a new set of parameters (limited for practical reasons to the most crucial ones, i.e., the most sensitive ones) better suited to the prevailing conditions of the ecosystem, as defined in the Darwinian sense by the ability of the species to survive and grow. As indicated earlier, this may be measured by the use of eco-exergy (Jørgensen and Mejer 1977, 1979, Jørgensen 1986, 1992, 2002, 2012). Figure 19.6 illustrates the proposed modeling procedure, which has been applied in the cases presented hereafter.

The use of eco-exergy calculations continuously to vary parameters has been employed in 25 case studies of biogeochemical modeling. Two of these cases (Søbygaard Lake and Fure Lake) are described next as an illustration of what can be achieved with this approach in modeling freshwater ecosystems.

Søbygaard is a shallow lake (depth of 1 m) with a short retention time (15–20 days). The phosphorus nutrient load was significantly lowered in 1982, from 30 g to 5 g P/m² year. However, the decreased load did not result in reduced nutrient and chlorophyll concentrations during the period 1982–1985, because of internal loading of stored nutrients in the sediment (Jeppesen et al. 1989, 1990).

Radical changes were then observed during the period 1985–1988. Recruitment of planktivorous fish was significantly reduced during the interval 1984–1988 because of a very high pH. The zooplankton increased

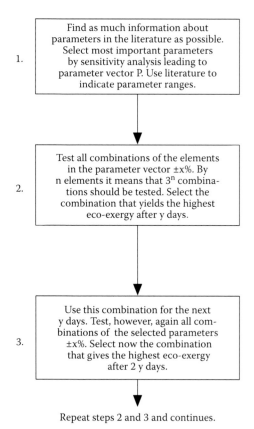

1. Find as much information about parameters in the literature as possible. Select most important parameters by sensitivity analysis leading to parameter vector P. Use literature to indicate parameter ranges.

2. Test all combinations of the elements in the parameter vector ±x%. By n elements it means that 3^n combinations should be tested. Select the combination that yields the highest eco-exergy after y days.

3. Use this combination for the next y days. Test, however, again all combinations of the selected parameters ±x%. Select now the combination that gives the highest eco-exergy after 2 y days.

Repeat steps 2 and 3 and continues.

FIGURE 19.6
Procedure used for the development of SDMs is shown. SDMs have been developed successfully in 23 cases. The observed structural changes have been predicted by the model with an acceptable standard deviation that is general for ecological modeling. The state variables are in addition generally predicted with a smaller standard deviation than for the same ecological models not considering structural changes.

and the phytoplankton decreased in concentration, while the average summer chlorophyll-a concentration reduced from 700 μg/L in 1985 to 150 μg/L in 1988. The phytoplankton population even collapsed during shorter periods because of extremely high zooplankton concentrations.

Simultaneously, phytoplankton species increased in size. Their growth rates declined and higher settling rates were observed (Jeppesen et al. 1990). In other words, this case study illustrates that pronounced ecosystem structural changes were caused by biomanipulation-like events. However, the primary production was not higher in 1985 than in 1988 because of pronounced self-shading by smaller algae. Thus, it was very important to include a self-shading effect in the model. Simultaneously, sloppier feeding of zooplankton was observed, with a shift from *Bosmina* to *Daphnia* taking place.

The model contains six state variables, all of which represent forms of nitrogen, including fish, zooplankton, phytoplankton, detritus nitrogen, soluble nitrogen, and sedimentary nitrogen. The model equations are given in Table 19.11. Because nitrogen is the limiting nutrient for eutrophication in this particular case, it may be sufficient to only include this element in the model.

The aim of the study is to describe, by use of a structural dynamic model, the continuous changes in the most essential parameters, using the

TABLE 19.11

Model Equations for Søbygaard Lake

```
fish = fish + dt * (-mort + predation)
INIT (fish) = 6
na = na + dt * (uptake - graz - outa - mortfa - settl
 - setnon)
INIT (na) = 2
nd = nd + dt * (-decom - outd + zoomo + mortfa)
INIT (nd) = 0.30
ns = ns + dt * (inflow - uptake + decom - outs + diff)
INIT (ns) = 2
nsed = nsed + dt * (settl - diff)
INIT (nsed) = 55
nz = nz + dt * (graz - zoomo - predation)
INIT (nz) = 0.07
decom = nd * (0.3)
diff = (0.015) * nsed
exergy = total_n * (Structural-exergy)
graz = (0.55) * na * nz/(0.4 + na)
inflow = 6.8 * qv
mort = IF fish > 6 THEN 0.08 * fish ELSE 0.0001 * fish
mortfa = (0.625) * na * nz/(0.4 + na)
outa = na * qv
outd = qv * nd
outs = qv * ns
pmax = uptake * 7/9
Predation = nz * fish * 0.08/(1 + nz)
qv = 0.05
setnon = na * 0.15 * (0.12)
settl = (0.15) * 0.88 * na
Structural-exergy = (nd + nsed/total_n) * (LOGN(nd + nsed/
 total_n) + 59) + (ns/total_n) * (LOGN(ns/total_n) -
 LOGN(total_n)) + (na/total_n) * (LOGN(na/total_n) + 60) +
 (nz/total_n) * (LOGN(nz/total_n) + 62) + (fish/total_n) *
 (LOGN(fish/total_n) + 64)
total_n = nd + ns + na + nz + fish + nsed
uptake = (2.0 - 2.0 * (na/9)) * ns * na/(0.4 + ns)
zoomo= 0.1 * nz
```

procedure shown in Figure 19.6. The data from 1984–1985 were used to calibrate the model. The two parameters that were intended to change for the
period 1985–1988 received the following values:

Maximum phytoplankton growth rate	2.2/day
Phytoplankton settling rate	0.15/day

The state variable, fish nitrogen, was kept constant at 6.0 mg N/L during the calibration period. During the period 1985–1988, however, an
increased fish mortality was introduced to reflect the increased pH. Thus,
fish stock was reduced to 0.6 mg N/L—notice the equation "mort = 0.08 if
fish > 6 (may be changed to 0.6) else almost 0." A time step of t = 5 days
and x% = 10% was applied. This means that nine runs were needed for
each time step in order to select the parameter combination that gives the
highest eco-exergy. Changes in parameters from 1985 to 1988 (summer)
are summarized in Table 19.12. It may be concluded that the proposed
procedure (Figure 19.6) can approximately simulate the observed change
in ecosystem structure.

The maximum phytoplankton growth rate is reduced by 50% from
2.2/day to 1.1/day, approximately in accordance to the increase in size. It was
observed that the average size was increased from a few 100 to 500–1000 μm^3,
a factor of 2–3 (Jeppesen et al. 1989). This would correspond to a specific
growth reduction by a factor $f = 2^{2/3} - 3^{2/3}$ according to the allometric principles (Peters 1983).

Thus,

$$\text{The growth rate in 1988} = \frac{\text{The growth rate in 1985}}{f} \tag{19.12}$$

where f is between 1.58 and 2.08. In the previous table, the value of 2.0 is found
with the use of the structurally dynamic modeling approach. Jeppesen et al.
(1989, 1990) observed that settling was 0.2 m/day (range 0.02–0.4) during 1985,
but 0.6 m/day (range 0.1–1.0) in 1988. Using the structural dynamic modeling
approach, the increase was found to be 0.15 m to 0.45 m/day, a slightly lower

TABLE 19.12

Parameter Combinations Giving the
Highest Exergy

	Maximum Growth Rate (Day^{-1})	Settling Rate (m/Day)
1985	2.2	0.15
1988	1.1	0.45

set of values. However, the phytoplankton concentration as chlorophyll a was simultaneously reduced from 600 to 200 µg/L, approximately in accordance with observations.

In this case, it may be concluded that structurally dynamic modeling gave an acceptable result. Validation of the model, and the procedure in relation to structural changes, was positive. Of course, the approach is never better than the model applied, and the model presented here may be criticized for being too simple and not accounting for changes in zooplankton.

For further elucidation of the importance of introducing parameter shifts, an attempt was made to run data for 1985 with parameter combinations for 1988 and vice versa. These results (Table 19.13) show that it is of great importance to apply the appropriate parameter set to given conditions. If those for 1985 are used for 1988, significantly less exergy is obtained, and the model behaves chaotically. The parameters for 1988 used under 1985 conditions give significantly less eco-exergy. Experience mentioned previously in this chapter shows that models can be applied to explain why biomanipulation may work under some circumstances and not others. Qualitatively, the results can be used to explain that hysteresis exists over an intermediate range of nutrient loadings, so that biomanipulation has worked properly over this range, but not above or below it. See also Chapter 17.

Another hysteresis behavior obtained with the use of SDMs for lakes have recently been published (Zhang et al. 2003a,b). It focuses on the structural change between a dominance of submerged vegetation and phytoplankton in shallow lakes. The model results show that between about 100 and 250 µg P/L both structures can exist—they show hysteresis in this range. This result is in accordance with observations from many shallow lakes (further details see Jørgensen and Fath 2011, Jørgensen 2012).

Ecosystems are very different from physical systems due mainly to their enormous adaptability. Thus, it is crucial to develop models that are able to account for this property in order to derive reliable model results. The use of eco-exergy as goal functions to cover the concept of fitness seems to offer a good possibility for developing a new generation of models, which is able to consider the adaptability of ecosystems and to describe shifts in species composition. The latter advantage is probably the most important, because a

TABLE 19.13

Eco-Exergy and Stability by Different Combinations of Parameters and Conditions

Year	Parameter 1985	Conditions 1988
1985	75.0—Stable	39.8 (average)—Violent fluctuations, chaos
1988	38.7—Stable	61.4 (average)—Only minor fluctuations

description of the dominant species in an ecosystem is often more essential than assessing the level of the focal state variables.

The structurally dynamic approach has also recently been used to calibrate eutrophication models. It is known that the different phytoplankton and zooplankton species are dominant in different periods of the year. Thus, a calibration based upon one parameter set for the entire year will not capture the succession that did take place over the year. By using eco-exergy optimization to capture the succession (i.e., the parameter giving the best survival for phytoplankton and zooplankton over the year), it has been possible to improve the calibration results (see Jørgensen 2002).

The structurally dynamic modeling approach generally has been most widely applied in eutrophication models. A recent software package named Pamolare launched by the International Lake Environment Committee (ILEC) and the United Nations Environment Programme's International Environment Technology Centre (UNEP-IETC) also contains a structurally dynamic model, in addition to a conventional two-layer model; see also Section 19.3. A test of this model has shown that it is calibrated and validated faster than the conventional two-layer model included in the package and gives better results, understood as a smaller standard deviation.

In other words, models are an appropriate tool in our efforts to understand the results of structural changes to ecosystems. In addition to the use of a goal function, it also is possible to base the structural changes on knowledge, for example, of what conditions under which specific classes of phytoplankton are dominant. This knowledge can be used to select the correct combination of parameters, as well illustrated by Reynolds (1992, 1998). That the combined application of expert knowledge and the use of eco-exergy as a goal function will offer the best solution to the problem of making models work more in accordance with the properties of real ecosystems cannot be ruled out. Such combinations would draw upon the widest possible knowledge at this stage.

Lake Fure (see Figure 19.7) is the deepest lakes in Denmark. It has a surface area of 941 ha and an average depth of 13.5 m. It consists of two ecologically different parts that are connected: the main basin, which is the large and deep part with an average depth of 16.5 m and the maximum depth of 37.7 m, and the shallow small part, which is called Store Kalv and has a mean and a maximum depth of 2.5 and 4.5 m, respectively. The lake is situated 12–17 km from the center of Copenhagen and has therefore great recreational value.

The eutrophication of the lake increased significantly in the 1950s and 1960s due to growing populations in the suburbs north of Copenhagen. The wastewater was treated mechanically and biologically but there was no removal of nutrients. It was therefore decided in the late 1960s to take measures to reduce the eutrophication. In 1970, it was decided that the three municipalities discharging wastewater to the lake could either choose to treat the wastewater effectively by removal of nutrients or pump the wastewater to the sea. A maximum phosphorus concentration of 0.2 mg/L and a

FIGURE 19.7
Lake Fure with high recreational value situated 12–17 km from Copenhagen in Denmark.

maximum nitrogen concentration of 8 mg/L were required for the treated wastewater if the municipality would select the treatment solution. One of the three municipalities selected the treatment solution, while the two other municipalities preferred to pump the mechanically, biologically treated water to the sea. A model developed at that time showed that the treatment solution was the best solution for the lake due to a faster recovery of the lake. The lake had a water retention of approximately 16 years, and the pumping of round 2 million m³ to the sea prolonged the retention to about 21 years, which implied of course a slower reduction of eutrophication.

The nutrient balances of the lake in 1972 just before the implementation of the aforementioned solution are shown in Figure 19.8. The measures taken reduced the phosphorus and nitrogen discharge to the lake by wastewater to less than 1 t and about 3 t, respectively. The three other sources of nutrients, drainage water, storm water overflow, and precipitation were unchanged. During the period from 1972–2000, the municipalities were able to reduce the other sources slightly as it can be seen in Figure 19.9 showing the nutrient balances for year 2000. The nutrient input from storm water was reduced by enlarging the storm water capacity. Notice, however, that the internal loading was not reduced, which is at least partly due to long retention time. It belongs to the story that the two municipalities that preferred to pump the wastewater to the sea in the early 1970s and did not accept the model result that the treatment solution would give a faster recovery were forced in 1986 to include nitrogen and phosphorus removal in the wastewater treatment due to introduction of maximum standard for discharge of all wastewater, even waste water discharge directly to the sea.

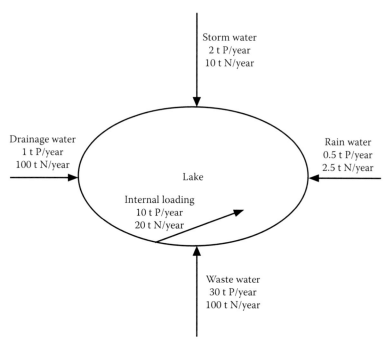

FIGURE 19.8
Nutrient balances in year 1972 before the loading of nutrients from wastewater was reduced.

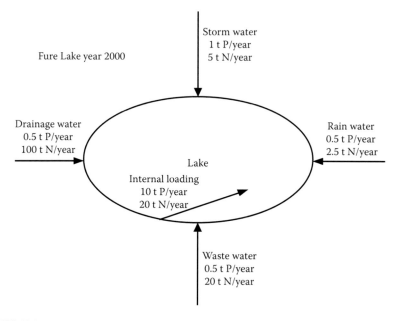

FIGURE 19.9
Nutrient balances in year 2000.

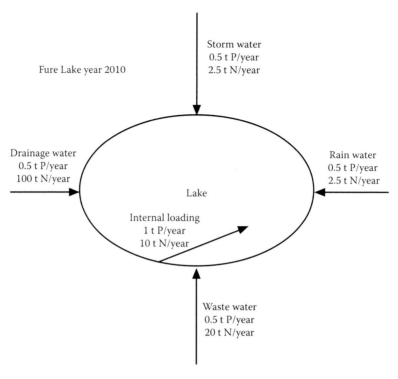

FIGURE 19.10
Nutrient balances according to the result of the SDM. The internal loading is reduced due mainly to the aeration of the hypolimnion. The phosphorus and nitrogen loading from storm water is reduced due to increase of the storm water capacity.

TABLE 19.14

Changes in Important State Variables from Year 2000 to 2010 according to the Results of the SDM Applied due to Restoration Project (Aeration of Hypolimnion, Biomanipulation, and Increased Storm Water Capacity)

State Variable	Year 2000	Year 2010
Total P mg/L	0.18–0.25	0.02–0.07
Chl. a maximum (early Aug.) mg/L	0.044	0.022
Transparency minimum (early Aug.) m	2.2	3.6
Zooplankton mg d.w./L	0.8	1.8

Shortly after year 2000 it was decided to use restoration methods to recover the lake faster. Two restoration methods were proposed:

1. Aeration of the hypolimnion by oxygen from late April to late October when the lake had a thermocline. The release of phosphorus from the sediment would thereby be reduced significantly.

2. Biomanipulation by massive removal of planktivorous fish.

The restoration of the lake started in 2003. An SDM was developed in year 2005; see the paper by Gurken et al. (2006). An SDM was naturally to apply in this case as structural changes were expected due to reduced internal loading, the removal of a significant part of the planktivorous fish, and further increase of the storm water capacity. The model was used for prognosis about the water quality in year 2010. Figure 19.10 shows the application of the model results on the nutrient balance and Table 19.14 shows the changes in the important state variables from 2004 to 2010 according to the prognosis made by the SDN model of Pamolare.

References

Ahlgren, I. 1973. *Limnologiska studier av Sjvn Norrviken. III. Avlastingster effecter.* Scipta Limnologica Upsaliensia.

Alasaarela, E.M.; Virtanen, M.; and Kopponen, J. 1993. The Bothnian bay project— Past, present and future. *Aqua Fenn.* 23:47.

Alcamo, J.; Shaw, R.; and Hordijk, L. (Eds.). 1990. *The Rains Model of Acidification.* Kluwer Academic Publishers, Dordrecht, the Netherlands.

Aoyama, I.; Yos. Inoue; and Yor. Inoue. 1978. Simulation analysis of the concentration process of trace heavy metals by aquatic organisms from the view point of nutrition ecology. *Water Res.* 12:837–842.

Bartell, S.M.; Gardner, R.H.; and O'Neill, R.V. 1984. The fates of aromatics model. *Ecol. Model.* 22:109–123.

Bouron, T. 1991. COMMAS: A communication and environment model for multi-agent systems. In: Mosekilde, E. (Ed.), *Modelling and Simulation 1991, European Simulation Multiconference*, June 17–19, 1991. The Society for Computer Simulations International, Copenhagen, Denmark, pp. 220–225.

Breck, J.E.; DeAngelis, D.L.; Van Winke, W.; and Christiansen, S.W. 1988. Potential importance of spatial and temporal heterogeneity in pH, Al and Ca in allowing survival of a fish population: A model demonstration. *Ecol. Model.* 41(1–2):1–16.

Brown, D.J.A. and Salder, K. 1981. The chemistry and fishery status of acid lakes in Norway and their relationship to European sulfur emission. *J. Appl. Ecol.* 18:434–441.

Chang, N.-B.; Wen, C.G.; and Chen, Y.L. 1996. A grey fuzzy multiobjective programming approach for the optimal planning of a reservoir watershed. Part B: Application. *Water Res.* 30(10):2335–2340.

Chapra, S.C. 1997. *Surface Water Quality Modelling*. McGraw Hill, New York.

Chapra, S.C. and Canale, R.P. 1991. Long-term phenological model of phosphorus and oxygen in stratified lakes. *Water Res.* 25(6):707–715.

Chen, C.W.; Dean, J.D.; Gherini, S.A.; and Goldstein, R.A. 1982. Acid rain model: Hydrologic module. *J. Environ. Eng Div. ASCE* 108:455–472.

Chen, C.W. and Orlob, G.T. 1975. Ecologic simulation of aquatic environments. In: Patten, B.C. (Ed.), *Systems Analysis and Simulation in Ecology*, Vol. 3. Academic Press, New York, pp. 476–588.

Costanza, R. and Sklar, F.H. 1985. Articulation accuracy and effectiveness of mathematical models: A review of freshwater wetland applications. *Ecol. Model.* 27:45–69.

Dahl-Madsen, K.I. and Strange-Nielsen, K. 1974. Eutrophication models for ponds. *VAND* 5:24–31.

Davis, J.R. 1993. Expert systems and environmental modelling. In: Jakeman, A.J.; Beck, M.B.; and McAleer, M.J. (Eds.), *Modelling Change in Environmental Systems*. John Wiley & Sons Ltd., Chichester, U.K., pp. 505–517.

Davis, J.R. and Guariso, J.R.D. 1994. Expert system support for environmental decisions. In: Zannetii, P. (Ed.), *Environmental Modeling*, Vol. 2. Computational Mechanics Publications, Southampton, U.K.

Di Toro, D.M. 1980. Applicability of cellular equilibrium and Monod theory to phytoplankton growth kinetics. *Ecol. Model.* 8:201–218.

Ellis, J.H. 1987. Stochastic water quality optimization using imbedded chance constraints. *Water Resour. Res.* 23:2227–2238.

Fomsgaard, I. 1997. Modelling the mineralisation kinetics for low concentrations of pesticides in surface and subsurface soil. *Ecol. Model.* 102:175–208.

Fontaine, T.D. 1981. A self-designing model for testing hypotheses of ecosystem development. In: Dubois, D. (Ed.), *Progress in Ecological Modelling*. Cebedoc, Liege, Belgium.

Gillet, J.W. 1974. A conceptual model for the movement of pesticides through the environment. National Environmental Research Center U.S. Environmental Protection Agency, Corvallis OR. Report EPA 660/3-74-024.

Gromiec, M.J. and Gloyna, E.F. 1973. Radioactivity transport in water. Final Report No 22 to U.S. Atomic Energy Commission, Contract AT (11-1)-490.

Gurkan, Z.; Zhang, J; and Jørgensen, S.E. 2006. Development of a structurally dynamic model for forecasting the effects of restoration of lakes. *Ecol. Model.* 197:89–103.

Halfon, E. 1984. Error analysis and simulation of Mirex behaviour in Lake Ontario. *Ecol. Model.* 22:213–253.

Harris, J.R.W.; Bale, A.J.; Bayne, B.L.; Mantoura, R.C.F.; Morris, A.W.; Nelson, L.A.; Radford, P.J.; Uncles, R.J.; Weston, S.A.; and Widdows, J. 1984. A preliminary model of the dispersal and biological effect of toxins in the Tamar estuary, England. *Ecol. Model.* 22:253–285.

Henriksen, A. 1980. Acidification of freshwaters—A large scale titration. In: Drablřs, D. and Tollan, A. (Eds.), *Ecological Impact of Acid Precipitation*. SNFS-project, Sandefjord, Norway. pp. 68–74.

Henriksen, A. and Seip, H.M. 1980. Strong and weak acids in surface waters of Southern Norway and Southwestern Scotland. *Water Res.* 14:809–813.

Hilborn, R. 1987. Living with uncertainty in resource management. *N. Am. J. Fish. Manage.* 7:1–5.

Holling, C.S. 1978. *Adaptive Environmental Assessment and Management*. John Wiley & Sons, New York.

Hushon, J.M. 1990. *Overview of Environmental Expert Systems*. American Chemical Society, Washington, DC.

Jacobsen, O.S. and Jørgensen, S.E. 1975. A submodel for nitrogen release from sediments. *Ecol. Model.* 31:147–151.

Jeppesen, E. et al. 1989. Restaurering af sřer ved indgreb i fiskebestanden. Status for igangvćrende undersřgelser. Del 2: Unsdersřgelser i Frederiksborg slotsř, Vćng sř og Sřbygĺrd sř. Danmarks Miljřundersřgelser, Silkeborg, Denmark, 114pp.

Jeppesen, E.J. et al. 1990. Fish manipulation as a lake restoration tool in shallow, eutrophic temperate lakes. Cross-analysis of three Danish Case Studies. *Hydrobiologia* 200/201:205–218.

Jørgensen, S.E. 1981. Application of exergy in ecological models. In: Dubois, D. (Ed.), *Progress in Ecological Modelling*. Cebedoc, Liege, Belgium, pp. 39–47.

Jørgensen, S.E. 1986. Structural dynamic model. *Ecol. Model.* 31:1–9.

Jørgensen, S.E. 1992. Development of models able to account for changes in species composition. *Ecol. Model.* 62:195–208.

Jørgensen, S.E. 1994. A general model of nitrogen removal by wetlands. In: Mitsch, W.J. (Ed.), *Global Wetlands: Old World and New*. Elsevier, Amsterdam, the Netherlands, pp. 575–583.

Jørgensen, S.E. 2002. *Integration of Ecosystem Theories: A Pattern*, 3rd revised edn., 1st edn. (1992), 2nd edn. (1997). Kluwer Academic Publishers, Dordrecht, the Netherlands, 428pp.

Jørgensen, S.E. (Ed.). 2011. *Handbook of Ecological Models Used in Ecosystem and Environmental Management*. CRC Press, Boca Raton, FL, 620pp.

Jørgensen, S.E. 2012. *Fundamentals of Systems Ecology*. CRC Press, Boca Raton, FL, 320pp.

Jørgensen, S.E. and Fath, B. 2011. *Fundamentals of Ecological Modelling*, 4th edn. Elsevier, Amsterdam, the Netherlands, 400pp.

Jørgensen, S.E.; Halling-Sorensen, B.; and Nielsen, S.N. (Eds.). 1995. *Handbook of Environmental and Ecological Modelling*. CRC Press, Boca Raton, FL.

Jørgensen, S.E.; Kamp-Nielsen, L.; and Jacobsen, O.S. 1975. A submodel for anaerobic mudwater exchange of phosphate. *Ecol. Model.* 1:133–146.

Jørgensen, S.E.; Ladegaard, N.; Debeljak, M.; and Marques, J.C. 2005. Calculations of exergy for organisms. *Ecol. Model.* 185:165–176.

Jørgensen, S.E.; Ludovisi, A.; and Nielsen, S.N. 2010. The free energy and information embodied in the amino acid chains of organisms. *Ecol. Model.* 221:2388–2392.

Jørgensen, S.E.; Lützhøft, H.C.; and Sřrensen, B.H. 1998. Development of a model for environmental risk assessment of growth promoters. *Ecol. Model.* 107:63–72.

Jørgensen, S.E.; Marques, J.C.; and Anastatcio, P.M. 1997. Modelling the fate of surfactants and pesticides in a rice field. *Ecol. Model.* 104:205–214.

Jørgensen, S.E. and Mejer, J.F. 1977. Ecological buffer capacity. *Ecol. Model.* 3:39–61.

Jørgensen, S.E. and Mejer, H.F. 1979. A holistic approach to ecological modelling. *Ecol. Model.* 7:169–189.

Jørgensen, S.E. and Mejer, H.F. 1981a. Application of exergy in ecological models. In: Dubois, D. (Ed.), *Progress in Ecological Modelling*. Cebedec, Liege, Belgium, pp. 311–347.

Jørgensen, S.E. and Mejer, H.F. 1981b. Exergy as key function in ecological models. In: Mitsch, W. et al. (Ed.), *Energy and Ecological Modelling*. Elsevier, Amsterdam, the Netherlands, pp. 587–590.

Kalceva, R.; Outrata, J.; Schindler, Z.; and Strakraba, M. 1982. An optimization model for the economic control of reservoir eutrophication. *Ecol. Model.* 17:121–128.

Kamp-Nielsen, L. 1975. A kinetic approach to the aerobic sediment-water exchange of phosphorus in Lake Esrom. *Ecol. Model.* 1:153–160.

Kauppi, P.; Posch, M.; Matzner, E.; Kauppi, L.; and Kdmdri, J. 1984. A model for predicting the acidification of forest soils: Application to deposition in Europe. IIASA Research Report.

Lam, D.C.L. and Simons, T.J. 1976. Computer model for toxicant spills in Lake Ontario. In: Nriago, J.O. (Ed.), *Metals Transfer and Ecological Mass Balances. Environmental Biochemistry*, Vol. 2. Ann Arbor Science, Ann Arbor, MI, pp. 537–549.

Legovic, T. 1997. Toxicity may affect predictability of eutrophication models in coastal sea. *Ecol. Model.* 99:1–6.

Loucks, D.P. and da Costa, J.R. (Eds.). 1991. *Decision Support Systems: Water Resources Planning*. Springer-Verlag, New York.

Matsumura, T. and Yoshiuki, S. 1981. An optimization problem related to the regulation of influent nutrient in aquatic ecosystems. *Int. J. Syst. Sci.* 12:565–585.

Miller, D.R. 1979. Models for total transport. In: Butler, G.C. (Ed.), *Principles of Ecotoxicology Scope*, Vol. 12. Wiley, New York, pp. 71–90.

Mitsch, W.J. 1976. Ecosystem modeling of water hyacinth management in Lake Alice, Florida. *Ecol. Model.* 2:69–89.

Mitsch, W.J. 1983. Ecological models for management freshwater wetlands. In: Jørgensen, S.E. and Mitsch, W.J. (Eds.), *Application of Ecological Modeling in Environmental Management, Part B*. Elsevier, Amsterdam, the Netherlands.

Mitsch, W.J. and Jørgensen, S.E. (Eds.). 1988. *Ecological Engineering. An Introduction to Ecotechnology*. John Wiley & Sons, New York.

Monte, L. 1998. Prediction the migration of dissolved toxic substances from catchments by a collective model. *Ecol. Model.* 110:269–280.

Muniz, I.P. and Seip, H.M. 1982. Possible effects of reduced Norwegian sulphur emissions on the fish populations in lakes in Southern Norway. SI-report 8103, 13-2.

Nielsen, S.N. 1992a. Application of maximum energy in structural dynamic models. PhD thesis. National Environmental Research Institute, Denmark.

Nielsen, S.N. 1992b. Strategies for structural-dynamic modelling. *Ecol. Model.* 63:91–100.

Odum, H.T. 1983. *System Ecology*. Wiley Interscience, New York, 510pp.

Orlob, G.T. 1983. *Mathematical Modeling of Water Quality: Streams, Lakes, and Reservoirs*. John Wiley & Sons, Chichester, U.K.

Orlob, G.T. 1984. Mathematical models of lakes and reservoirs. In: Taub, F. (Ed.), *Lakes and Reservoirs*. Ecosystems of the World 23. Elsevier, Amsterdam, the Netherlands, pp. 43–62.

Paterson, S. and Mackay, D. 1989. A model illustrating the environmental fate, exposure and human uptake of persistent organic-chemicals. *Ecol. Model.* 47(1–2):85–114.

Peters, R.H. 1983. *The Ecological Implications of Body Size*. Cambridge University Press, Cambridge, U.K.

Reynolds, C.S. 1992. Dynamics, selection and composition of phytoplankton in relation to vertical structure in lakes. *Arch. Hydrobiol. Beih. Ergebn. Limnol.* 35:13–31.

Reynolds, C.S. 1998. What factors influence the species composition of phytoplankton in lakes of different trophic status? *Hydrobiologia* 369/370:11–26

Sale, M.J.; Brill, E.D. Jr.; and Herricks, E.E. 1982. An approach to optimizing reservoir operation for downstream aquatic resources. *Water. Resour. Res.* 18:705–712.

Schindler, Z. and Straskraba, M. 1982. Optimalní Žizení eutrofizace údolních nadrří. *VodohospodáŽsk_ Ëasopis SAV* 30:536–548.

Schwarzenbach, R.P. and Imboden, D.M. 1984. Modeling concepts for hydrophobic pollutants in lakes. *Ecol. Model.* 22:145–170.

Simonovic, S.P. 1996. Decision support systems for sustainable management for water resources. *Water Int.* 24(4):223–244.

Straskraba, M. 1979. Natural control mechanisms in models of aquatic ecosystems. *Ecol. Model.* 6:305–322.

Straskraba, M. and Gnauck, A.H. 1985. *Freshwater Ecosystems: Modelling and Simulation.* Developments in Environmental Modelling 8. Elsevier, Amsterdam, the Netherlands.

Svirezhev, Yu. M., 1998. Thermodynamic orientors: How to use thermodynamic concepts in ecology? In F. Müller and M. Leupelt (Eds.), *Eco Targets, Goal Functions and Orientors.* Springer, Berlin, Heidelberg, pp. 102–122.

Thomann, R.V. 1984. Physico-chemical and ecological modeling the fate toxic substances in natural water system. *Ecol. Model.* 22:145–170.

Thornton, J.A.; Rast, W.; Holland, M.M.; Jolankai, G.; and Ryding, S.O. 1999. *Assessment and Control of Nonpoint Source Pollution of Aquatic Ecosystems.* UNESCO and The Parhenon Publishing Group, Paris, France.

Thomann, R.V.; Szumski, D.; DiToro, D.M.; and O'Connor, D.J. 1974. A food chain model of cadmium in western Lake Erie. *Water Res.* 8:841–851.

Ulanowicz, R.E. 1979. Prediction chaos and ecological perspective. In: Halfon E.A. (Ed.), *Theoretical Systems Ecology.* Academic Press, New York.

Van den Belt, M. 2004. *Mediated Modeling.* Island Press, Washington, DC, 340pp.

Van der Molen, D.T.; Los, F.J.; Van Ballegooijen, L.; and Van der Vat, M.P. 1994. Mathematical modelling as a tool for management in eutrophication control of shallow lakes. *Hydrobiologia* 275/276:479–492.

Zhang, J.; Jørgensen, S.E.; Tan, C.O.; and Beklioglu, M. 2003a. A structurally dynamic modelling—Lake Mogan, Turkey as a case study. *Ecol. Model.* 164:103–120.

Zhang, J.; Jørgensen, S.E.; Tan C.O.; and Beklioglu, M. 2003b. Hysteresis in vegetation shift—Lake Mogan prognoses. *Ecol. Model.* 164:227–238.

Ziegler, B.P. 1976. *Theory of Modeling and Simulation.* Wiley, New York.

Index

Printed and bound by CPI Group (UK) Ltd, Croydon, CR0 4YY

01/11/2024

01782637-0006